Mechanisms
and
Control
of
Cell Division

Mechanisms and Control of Cell Division

Edited by

Thomas L. Rost
and
Ernest M. Gifford, Jr.
University of California, Davis

Dowden, Hutchinson & Ross, Inc.
Stroudsburg, Pennsylvania

Copyright © 1977 by **Dowden, Hutchinson & Ross, Inc.**
Library of Congress Catalog Card Number: 76-48081
ISBN: 0-87933-267-0

79 78 77   1 2 3 4 5
Manufactured in the United States of America.

Library of Congress Cataloging in Publication Data
Main entry under title:
Mechanisms and control of cell division.
    Includes index.
    1. Cell division. 2. Cell cycle. 3. Cellular control mechanisms. I. Rost, Thomas
L. II. Gifford, Ernest M.
QH605.M425   574.8'762   76-48081
ISBN 0-87933-267-0

Exclusive distributor: **Halsted Press,**
A Division of John Wiley & Sons, Inc.
ISBN: 0-470-99082-1

Preface

Investigations of the cell cycle in both plants and animals is a blossoming field which is now beginning to occupy a central position in cytological research. Hundreds of papers are found in the literature, and new books surveying the field are appearing. The present volume is not a monograph, which perhaps is no longer possible to achieve, but instead is a compendium of several areas of cell cycle research, especially those dealing with plants. Some chapters are reviews while others present new ideas and the results of current research. The audiences we intend to reach are researchers and advanced students interested in cell behavior.

The volume is divided into three parts: I. Cell Cycle Regulation; II. Nuclear Structure and Chromosome Movements; III. Mechanisms of Cell Division. The first two chapters of Part I, by Gurley et al., and by Douvas and Bonner, the only ones dealing with animal cells, concern the possible roles of histone and acidic protein in cycle regulation. The chapters by Fosket and by Loy describe the regulatory roles of the plant hormones, cytokinin and gibberellic acid, in controlling cell cycle expression. Chapter 5 by Rost concerns induced cell cycle control from external environmental stress. Part II contains an excellent review of nuclear changes during cell cycle phases by Nagl, and a chapter on the structure of nucleoli, which persist during mitosis, by Braselton. Hepler's contribution concerns the role of membranes in spindle assembly, and Bajer proposes his "Zipper hypothesis" to explain chromosome movement. Part III contains three chapters dealing with mechanisms of cell division in higher plants (Giménez-Martín et al.), algae (Pickett-Heaps and Weik) and fungi (Wells).

Contents

Mechanisms
and
Control
of
Cell Division

Part I. Cell Cycle Regulation

Sequential Biochemical Events Related To Cell Proliferation

L. R. Gurley
R. A. Walters
C. E. Hildebrand
R. L. Ratliff
P. G. Hohmann   Los Alamos Scientific Laboratory
R. A. Tobey   University of California

INTRODUCTION

If we are to understand such physiological phenomena as tissue growth, cell differentiation, and neoplasia, we must first understand how cell proliferation is regulated. We know that this regulation is a complex operation composed of many interacting biochemical processes. The elucidation of these processes would be greatly simplified if investigations could be made using an *in vitro* biological system composed of a homogeneous population of cells which could be easily manipulated under controlled laboratory conditions. At the Los Alamos Scientific Laboratory, we have implemented the use of the Chinese hamster cell culture (line CHO) for this purpose (Petersen et al. 1969). This cell line offers the advantage of being especially amenable to a number of different synchronization techniques, thus facilitating the detection of cell-cycle regulatory events (Petersen and Anderson 1964; Tobey and Ley 1971; Tobey and Crissman 1972; Tobey 1973; Tobey et al. 1967, 1972; Kraemer and Tobey 1972).

CELL-CULTURE SYNCHRONIZATION METHODS

One synchronization method we have used to study the biochemistry of the cell cycle is mitotic selection (Tobey et al. 1967). This technique involves shaking the growth medium overlying a monolayer culture of cells attached to glass. This action suspends only the mitotic cells in the population, since the cells loosen their hold on the glass surface only after they have entered mitosis. By decanting this medium, a culture containing 99% mitotic cells can be obtained. The advantage of this method is that cell synchrony can be obtained without the use of drugs which might interfere with the normal metabolism of the cells. The disadvantage is that only a small number of cells can be obtained, thus greatly restricting the amount of material available for use in studies of cell-cycle regulation. Another disadvantage, frequently overlooked, is the effect of synchrony decay on such cultures. For example, while mitotic selection produces cultures highly synchronized through

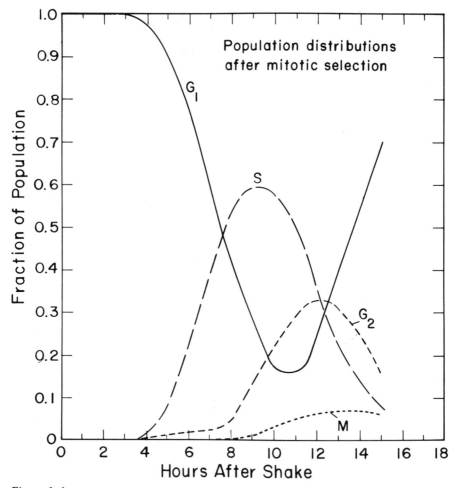

Figure 1-1.
Population distributions of CHO cells in culture after synchronization by mitotic selection. The fraction of cells in each phase of the cell cycle was determined from a combination of measurements of ^{3}H-thymidine autoradiography, cell concentration, and mitotic index.

G_1, the decay of this synchrony is significant by mid-S phase, and a maximum of only 33% G_2 cells is obtainable in "G_2-rich" cultures (Figure 1-1). However, we have developed methods which overcome these disadvantages and still produce minimum perturbation in the metabolism of the cell. Two such methods which we have used extensively are the isoleucine-deprivation method (Tobey and Ley 1971; Tobey 1973) and hydroxyurea treatment (Tobey and Crissman 1972; Tobey et al. 1972). In the isoleucine-deprivation method, cells are grown in medium containing suboptimal quantities of isoleucine. Under these conditions, CHO cells are arrested synchronously in early G_1. The properties of these cells superficially mimic the Properties of G_0 cells found *in vivo* to the extent that they (a) remain viable for prolonged periods of time in a G_1-like state; (b) do not enter a state of gross biochemical imbalance during arrest; (c) maintain high levels of biosynthetic activity for various macromolecules except DNA; and (d) can be stimulated to reenter the proliferative G_1 state (simply by restoring isoleucine to the culture) (Enger and Tobey 1972). The quantity of synchronized cells obtained by this method is limited only by the size of the culture vessel used, and the cell-cycle kinetics (Figure 1-2A) following reversal of isoleucine-deprivation are essentially identical to those of cultures synchronized by mitotic selection (Tobey et al. 1975).

Such highly synchronized cultures have allowed us to study both the biochemical events involved in the conversion of nonproliferating cells into proliferating cells as well as the events involved in the transition of cells from G_1 to S phase. As with mitotic selection, however, this techniques is unsuitable for studying late interphase events due to synchrony decay. Therefore, a second method has been employed to resynchronize these cells at the G_1/S boundary. Following release from isoleucine-deprivation, hydroxyurea can be added with the result that the synchronized cells traverse G_1 but do not enter S phase and thus accumulate at the end of the G_1 phase (Tobey and Crissman 1972; Tobey 1973; Gurley et al. 1973a,b, 1975). Because these cells fail to initiate DNA synthesis (Tobey and Crissman, 1972), they do not exhibit the cytotoxic effects observed when S-phase cells are treated with hydroxyurea (Sinclair 1967). Upon removal of hydroxyurea, the resynchronized cells enter the S phase immediately and commence dividing 6 hours later (Figure 1-2B). With this method, one can obtain G_2-rich CHO cultures containing a maximum of 70% G_2 cells.

CELL-CYCLE CHANGES IN DEOXYRIBONUCLEOSIDE TRIPHOSPHATE POOLS

The process by which hydroxyurea inhibits entry of CHO cells into S phase was discovered during studies on deoxyribonucleoside triphosphate pools (Walters et al. 1973). Suspension cultures (synchronized by mitotic selection) were allowed to traverse the cell cycle. At intervals during G_1 and S, aliquots were removed for measurements of the four deoxyribonucleoside triphosphates in the acid-soluble pools (Walters et al. 1973). Samples were also taken and pulse-labeled with ^{3}H-thymidine to determine the fraction of cells synthesizing DNA. The levels of dGTP, dATP, and dCTP were extremely low in G_1 (Figure 1-3). The level of dTTP

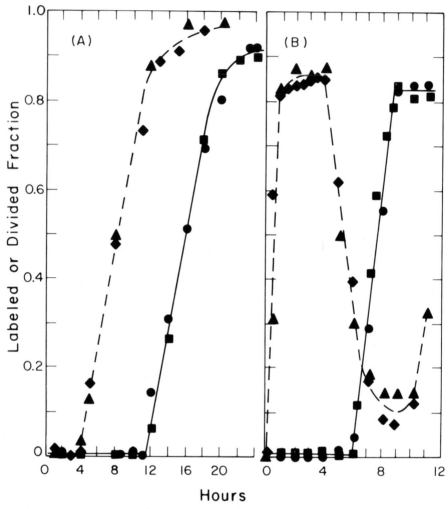

Figure 1-2.
Comparison of the techniques used to synchronize CHO cells. The fraction of cells synthesizing DNA (♦,▲) and the fraction of cells dividing (■,●). (A) Cell-cycle traverse from early G_1 to M of cultures synchronized by mitotic selection (♦,■) or by isoleucine-deficiency (▲,●). (B) Cell-cycle traverse from the G_1/S boundary to M of cultures synchronized by either mitotic selection/hydroxyurea (♦,■) or isoleucine-deficiency/hydroxyurea (▲,●). (Reprinted with permission from Tobey et al. 1975.)

was also quite small but detectable during this time (resulting from the fact that the F-10 culture medium is supplemented with thymidine). Approximately 30 minutes before the initiation of DNA synthesis, the level of all four deoxyribonucleoside triphosphates began to increase as determined by extrapolation of these pool levels back to the G_1 baseline values (Figure 1-3). In separate experiments, it was also

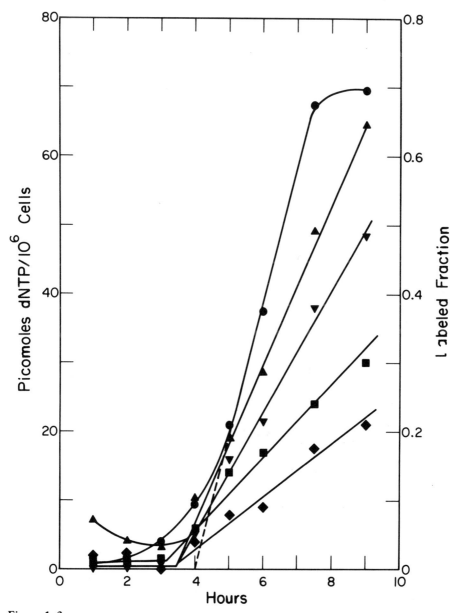

Figure 1-3.
Deoxyribonucleoside triphosphates in the acid-soluble pools of CHO cells traversing early interphase following synchronization by mitotic selection. The fraction of cells pulse-labeled with ^{3}H-thymidine (●). The levels of dTTP (▲), dCTP (▼), dATP (■), and dGTP (♦). (Reprinted with permission from Tobey et al. 1974.)

found that the levels of these nucleotides were similarly low in cells arrested in G_1 by isoleucine-deprivation (Walters et al. 1973).

When cells synchronized by mitotic selection were resynchronized at the G_1/S boundary by hydroxyurea treatment, it was found that the levels of dTTP, dCTP, and dGTP were elevated during hydroxyurea arrest, although DNA synthesis had not commenced (0.5 hour before zero time in Figure 1-4). However, the dATP level remained at a very low value. After removal of hydroxyurea, the level of dATP increased ninefold within 3 minutes, and DNA synthesis began simultaneously (Figure 1-4B). This coincided with a similar large increase in dGTP (Figure 1-4B) and a drop in dTTP (Figure 1-4A) early in S phase. Then the levels of dTTP, dGTP, and dATP increased through the latter part of S phase and G_2. The dCTP content also increased across S phase but began to decrease as cells entered G_2 (Figure 1-4A).

These trends were found to continue into mitosis (Figure 1-5A). Mitotic cells contained elevated levels of dTTP, dATP, and dGTP. However, the dCTP content continued to decrease in G_2 and was low in mitotic cells (zero time in Figure 1-5A). As the cells left mitosis and entered G_1, the pools of all four deoxyribonucleoside triphosphates rapidly dropped to the characteristic low G_1 levels. Thus, it appears that only the size of the dCTP pool correlates roughly with the DNA synthesis period in CHO cells. The function of elevated levels of the other three deoxyribonucleotides in G_2 and M is not understood at this time.

Since the level of dATP was observed to be low in cells synchronized at the G_1/S boundary by hydroxyurea (Figure 1-4B), the pools of deoxyribonucleoside triphosphates were measured in cells traversing G_1 in the presence of hydroxyurea (Figure 1-5B). As expected, the dATP levels remained at the G_1 levels during accumulation of cells at the G_1/S boundary, while the other three nucleotides increased at approximately the same time as cells not exposed to hydroxyurea. We conclude from these results that hydroxyurea prevents CHO cells from initiating DNA synthesis by specifically inhibiting the scheduled accumulation of dATP.

CELL-CYCLE CHANGES IN HISTONE SYNTHESIS

Both the isoleucine-deprivation synchronization method and the hydroxyurea resynchronization method have been used to advantage to study histone metabolism in CHO cells (Gurley et al. 1973a,b, 1974a,b, 1975). When cells synchronized in early G_1 by isoleucine-deprivation were released to traverse their cell cycle, histone f1 synthesis was observed to begin increasing simultaneously with entry of cells into S phase (Figure 1-6). Similar cell-cycle synthesis patterns were observed for the other four histone fractions (Gurley et al. 1974b). However, when hydroxyurea was present during G_1 traverse, it was found that scheduled S-phase synthesis of histone f1 was completely inhibited, as was the synthesis of DNA (Figure 1-6). No significant synthesis was observed for the other four histone fractions under these conditions as well (Gurley et al. 1974b). These results demonstrate that the hydroxyurea blockade occurs on the G_1 side of the G_1/S boundary, since two different S-phase operations (DNA and histone synthesis) are inhibited by this treat-

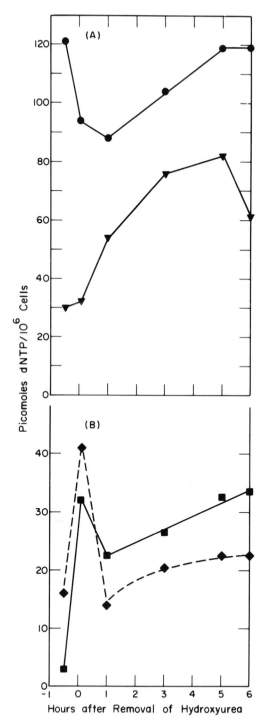

Figure 1-4.
Deoxyribonucleoside triphosphates in the acid-soluble pools of CHO cells traversing late interphase following synchronization at the G_1/S boundary by the mitotic selection/hydroxyurea technique. (A) The levels of dTTP (●) and dCTP (▼). (B) The levels of dGTP (♦) and dATP (■). (Reprinted with permission from Tobey et al. 1974.)

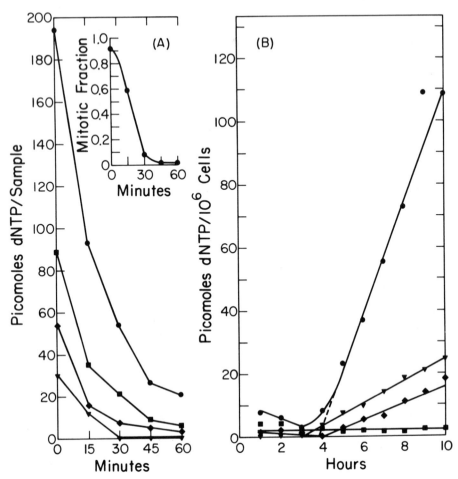

Figure 1-5.
Deoxyribonucleoside triphosphates in the acid-soluble pools of CHO cells undergoing division and traversing early interphase in the presence of hydroxyurea. (A) The levels of deoxyribonucleotides in cells progressing out of mitosis into G_1 following mitotic selection. The insert shows the rate of cells dividing. (B) The levels of deoxyribonucleotides in cells traversing early interphase in the presence of hydroxyurea; dTTP (●), dATP (■), dGTP (♦), and dCTP (▼). (Reprinted with permission from Tobey et al. 1974.)

ment. As shown in Figure 1-3, deoxyribonucleoside triphosphate synthesis does not begin in CHO cells until approximately 30 minutes before S phase (Walters et al. 1973). Since dCTP, dGTP, and dTTP synthesis begin in the presence of hydroxyurea (Figure 1-5B) but histone synthesis does not (Figure 1-6), it is concluded that the hydroxyurea block is in late G_1 less than 0.5 hour from the G_1/S boundary. This is supported by the fact that essentially all the cells in the population initiate DNA synthesis immediately following removal of hydroxyurea. Thus,

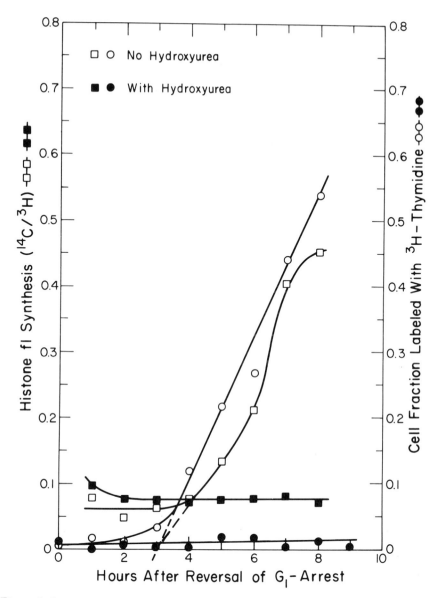

Figure 1-6.
Cell-cycle-dependent histone f1 synthesis. CHO cells prelabeled with ^{3}H-lysine were synchronized in early G_1 by isoleucine-deprivation and then released to traverse early interphase by adding isoleucine. Histone synthesis was measured hourly by a 1-hour incorporation of ^{14}C-lysine. Control cells received no hydroxyurea (open symbols); treated cells were prevented from entering S phase by adding 1 mM hydroxyurea (closed symbols). The fraction of S-phase cells in the culture was determined by a 1-hour incorporation of ^{3}H-thymidine (○ or ●). The relative specific activity (^{14}C/^{3}H) indicates the rate of f1 synthesis measured in f1 recovered following its purification by preparative polyacrylamide gel electrophoresis (□ or ■). (Reprinted with permission from Gurley et al. 1974b.)

hydroxyurea-treated cells are ideal for studying late G_1 biochemical events which precede DNA synthesis, as will be demonstrated in the experiments which follow.

CELL-CYCLE CHANGES IN HISTONE PHOSPHORYLATION

Data have accumulated recently which have led to the suggestion that chemical modifications of histones are involved in modulating the structure of chromatin, thereby exerting control on biological activity (Allfrey 1971; Bradbury and Crane-Robinson 1971). From this hypothesis, one would predict that changes in histone modification might occur when cells undergo transitions in their metabolic state, such as when nonproliferating cells are converted to proliferating cells or when changes in other cell-cycle phases occur. Support for this hypothesis was obtained in our laboratory (Gurley and Walters 1971) from the observation that doses of x-irradiation large enough to temporarily stop cell division also briefly inhibited histone fl phosphorylation but not the phosphorylation of other histones (Figure 1-7). Further support was provided by Balhorn et al. (1972), who observed a linear correlation between the growth rate of tumors and the extent of their histone f1 phosphorylation. In both of these cases, histone fl phosphorylation specifically was related to the proliferative capacity of the cell.

These experiments encouraged us to study the cell-cycle kinetics of histone phosphorylation to determine the details of a possible involvement between this type of histone modification and cell proliferation. Since histones extracted from chromatin contain significant amounts of phosphorus contaminants (Shepherd et al. 1970; Gurley and Walters 1971), it was necessary to purify the histones after their extraction in order to accurately measure their phosphorylation. By first extracting the arginine-rich and lysine-rich classes of histones from CHO cells (Gurley and Hardin 1968) and then subjecting each class to polyacrylamide gel electrophoresis in a Canalco preparative electrophoresis apparatus (Gurley and Walters 1971), we accomplished both the purification and fractionation of the histones into the five common component proteins (f1, f2a1, f2a2, f2b, and f3).

The phosphorylation of histones was first examined in cells arrested in early G_1 by isoleucine-deprivation (Gurley et al. 1973b). In these nonproliferating cells, only histine f2a2 was observed to be significantly labeled after a 1-hour pulse of $^{32}PO_4$ (Figure 1-8). When these synchronized cells were allowed to resume proliferation by adding isoleucine, f2a2 was still the only histone phosphorylated during early G_1 traverse (Figure 1-9). However, 10 hours after release, when these cells were in the S phase, histones f1 and f2a2 were both phosphorylated (Figure 1-10), suggesting that some relationship might exist between f1 phosphorylation and DNA synthesis.

To determine the exact relationship between f1 phosphorylation and S phase, the kinetics of f1 phosphorylation were measured at hourly intervals during G_1 and S traverse and related to the fraction of cells labeled with ^{3}H-thymidine (i.e., the fraction of S-phase cells) (Gurley et al. 1974a,b). It was found that f1 phosphorylation was very low in the early stages after release from G_1-arrest but steadily increased prior to the entry of cells into S phase (Figure 1-11). Extrapolation of the phos-

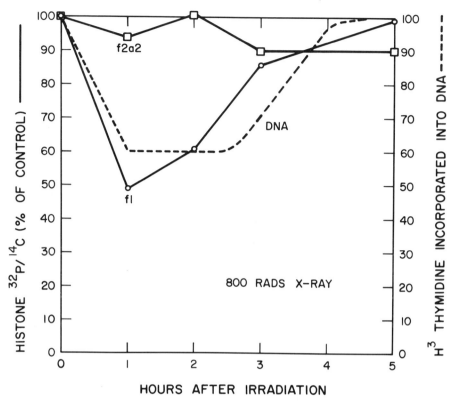

Figure 1-7.
Response of histone phosphorylation to x-irradiation. Exponential cultures of CHO cells prelabeled with ^{14}C-amino acids were irradiated with 800 rads of x-irradiation which prevented cell division for 9 hours. The cells were pulse-labeled with $^{32}PO_4$ at 1-hour intervals after irradiation. The relative specific activity ($^{32}P/^{14}C$) indicates the rate of phosphorylation measured in histone f1 (○) and f2a2 (□) recovered following their purification by preparative gel electrophoresis. The incorporation of ^{3}H-thymidine into DNA is indicated by the dashed line (———). (Reprinted with permission from Gurley and Walters 1971.)

phorylation rate curve to zero indicated that f1 phosphorylation preceded entry into S phase by approximately 2 hours. These results indicated that f1 phosphorylation was not restricted to S phase but was associated with the change of cells from the nonproliferative state to the proliferative state.

When the same measurements were made (Gurley et al. 1974b) in the presence of hydroxyurea (which allowed the cells to traverse G_1 but inhibited entry into S phase), the phosphorylation of f1 began on schedule 2 hours prior to the scheduled entry into S phase (Figure 1-11). During late G_1-arrest, f1 phosphorylation was maintained at a high rate even though both DNA and f1 synthesis were not initiated. These data confirmed our conclusions that f1 phosphorylation was not

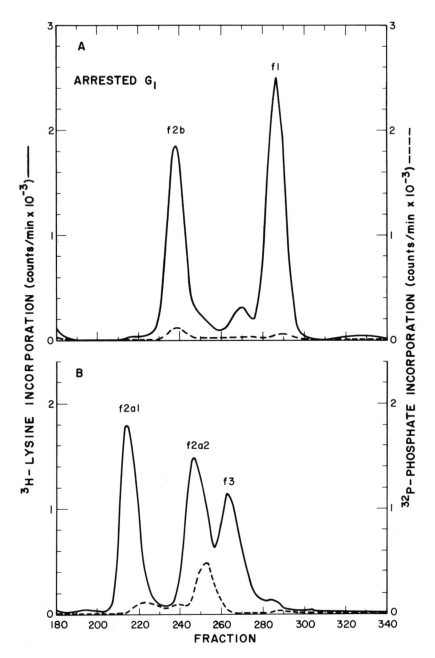

Figure 1-8.
Histone phosphorylation in cells arrested in the G_1 state by isoleucine-deprivation. The lysine-rich histones (A) and arginine-rich histones (B) were separately purified by preparative polyacrylamide gel electrophoresis. The individual histone fractions are indicated by long-term ^{3}H-lysine incorporation (——). The rate of phosphorylation of each fraction is indicated by the ^{32}PO$_4$ 1-hr pulse incorporation (–––). (Reprinted with permission from Gurley et al. 1973b.)

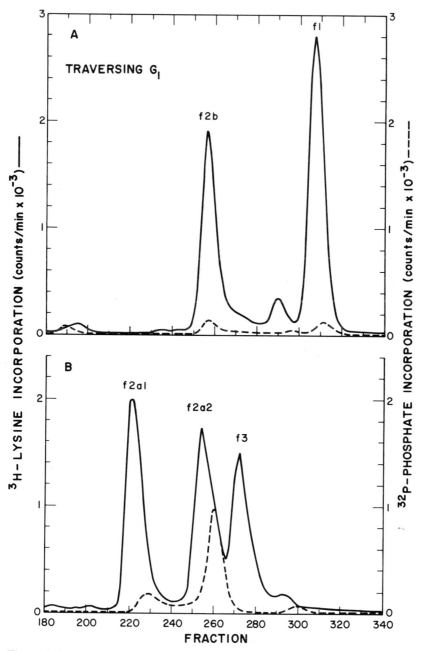

Figure 1-9.
Histone phosphorylation in proliferating cells traversing the G_1 phase 2 hours after release from synchronization by isoleucine-deprivation (explanation of the legend same as in Figure 1-8). (Reprinted with permission from Gurley et al. 1973b.)

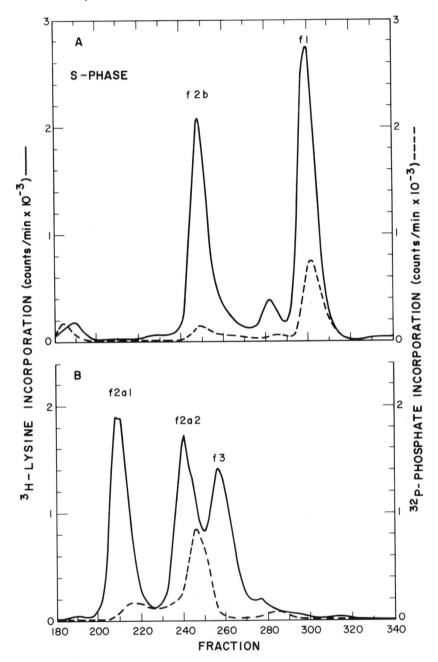

Figure 1-10.
Histone phosphorylation in proliferating cells traversing the S phase 10 hours
after release from synchronization by isoleucine-deprivation (explanation of
the legend same as in Figure 1-8). (Reprinted with permission from Gurley
et al. 1973b.)

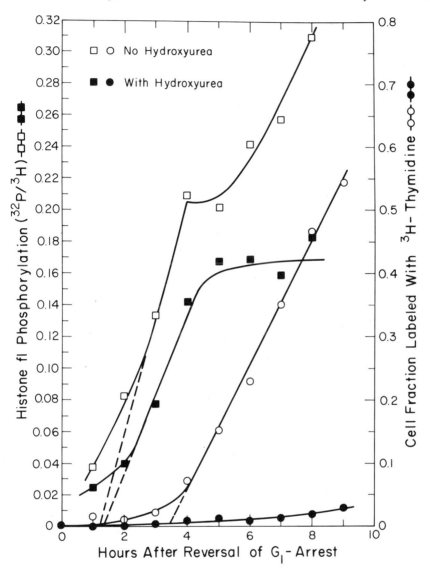

Figure 1-11.
Cell-cycle kinetics of histone f1 phosphorylation. CHO cells prelabeled with
^{3}H-lysine were synchronized in early G_1 by isoleucine-deprivation and then
released to traverse early interphase by adding isoleucine. Histone phosphorylation was measured hourly by a 1-hour incorporation of $^{32}PO_4$. Control cells
received no hydroxyurea (open symbols); treated cells were prevented from
entering S phase by adding 1 mM hydroxyurea (closed symbols). The fraction of
S-phase cells in the culture was determined by a 1-hour incorporation of ^{3}H-
thymidine ($\circ$ or $\bullet$). The relative specific activity ($^{32}P/^3H$) indicates the rate
of f1 phosphorylation measured in f1 recovered following its purification by
preparative polyacrylamide gel electrophoresis ($\square$ or $\blacksquare$). (Reprinted with permission from Gurley et al. 1974b.)

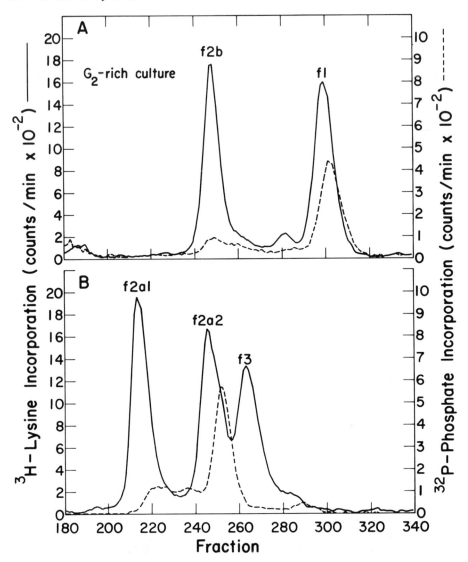

Figure 1-12.
Histone phosphorylation in proliferating cells traversing the G_2 phase 6 hours after release from synchronization by the isoleucine-deprivation/hydroxyurea method (explanation of the legend same as in Figure 1-8). (Reprinted with permission from Gurley et al. 1973a.)

strictly an S-phase event. It was concluded that f1 phosphorylation occurred on "old" presynthesized f1 and that this process did not require either DNA synthesis or histone synthesis (Gurley et al. 1974b).

As cells not treated with hydroxyurea moved into S phase, the rate of f1 phosphorylation increased to twice the rate observed in late G_1-arrested cells (Figure

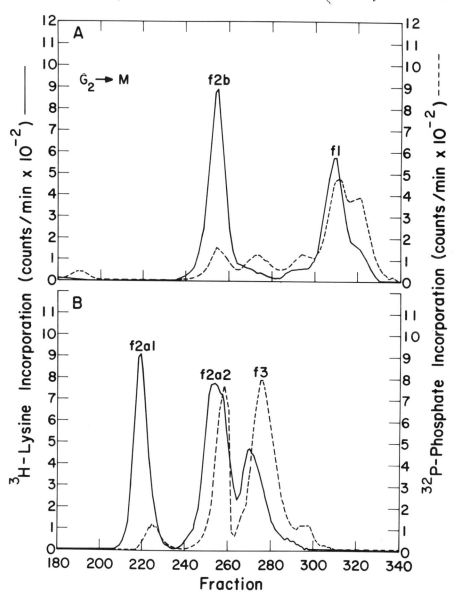

Figure 1-13.
Histone phosphorylation in proliferating cells traversing from G_2 to metaphase. ^{3}H-Lysine prelabeled exponential monolayer cultures were pulse-labeled with $^{32}PO_4$ for 2 hours in the presence of Colcemid. Only the metaphase cells accumulated during this pulse were removed from the monolayer by mitotic selection and collected for histone preparation (explanation of the legend same as in Figure 1-8). (Reprinted with permission from Gurley et al. 1974a.)

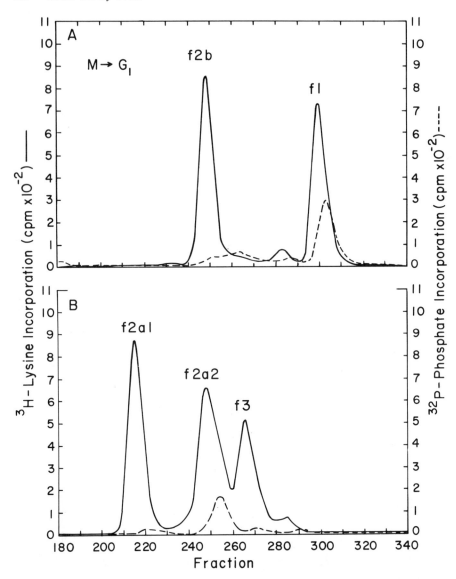

Figure 1-14.
Dephosphorylation of $^{32}PO_4$-prelabeled histones during the transition of metaphase cells to interphase G_1 cells. ^{3}H-Lysine prelabeled exponential monolayer
cultures were pulse-labeled with $^{32}PO_4$ for 2 hours in the presence of Colcemid (as
in Figure 1-13). The metaphase cells were removed by mitotic selection, resuspended in Colcemid-free medium, and allowed to enter G_1 for 2 hours. Histones
were then prepared and subjected to preparative polyacrylamide electrophoresis
(explanation of the legend same as in Figure 1-8). (Reprinted with permission from
Gurley et al. 1974a.)

1-11). This increased rate was not merely due to the phosphorylation of additional newly synthesized f1 but, rather, represented a true increase in f1 phosphorylation rate during S phase (Gurley et al. 1974a,b).

To determine whether this high rate of f1 phosphorylation continued after S phase, we released hydroxyurea resynchronized cultures from blockade at the G_1/S boundary and examined the histones of cells in late interphase (Gurley et al. 1973a). Six hours after release, G_2-rich cultures (Figure 1-12) were found to phosphorylate f1 at twice the rate observed in S-rich cultures 2.5 hours after release (Gurley et al. 1973a). This further confirmed our conclusion that f1 phosphorylation is not restricted to the S phase; rather, f1 phosphorylation appeared to be related to cell proliferation in general, and the rate of this modification increased as cells approached mitosis.

The climactic event of the cell cycle is that change in state a cell makes when preparing to divide: *the transition from interphase to mitosis* (M). During this transition, the chromatin undergoes a conspicuous change in form from the dispersed state in G_2 to the condensed chromosome state in M. To measure changes in histone phosphorylation during this transition, exponential monolayer cultures were labeled with $^{32}PO_4$ for 2 hours in the presence of Colcemid. The metaphase cells accumulated during this period were removed from the monolayer by mitotic selection and then collected by centrifugation. This collection contained only cells labeled with $^{32}PO_4$ during the transition from G_2 to M (i.e., the period of chromosome condensation) (Gurley et al. 1974a). The histones from these cells were subjected to preparative electrophoresis.

As was the case with traversing interphase cells, a high rate of phosphorylation was observed in f2a2 and f1 histones during the $G_2 \rightarrow M$ transition (Figure 1-13). However, two new phosphorylation events specific for mitosis were also observed. Histone f3 phosphorylation was observed for the first time, as was a slow-migrating phosphorylated f1 subfraction. When cells labeled with $^{32}PO_4$ during the $G_2 \rightarrow M$ transition were removed from the Colcemid-containing medium and allowed to reenter G_1 in the absence of $^{32}PO_4$ (i.e., $M \rightarrow G_1$ cells), there was a nearly total dephosphorylation of the prelabeled histone f3 and the slow-migrating f1 subfraction, whereas the $^{32}PO_4$ incorporated into f2a2 and f1 during interphase was lost at a slower rate (Figure 1-14). These data strongly suggest that the phosphorylation of f3 and the slow f1 subfraction is involved in some aspect of the chromatin structural changes observed specifically during this terminal phase of the cell cycle (Gurley et al. 1974a).

SEQUENTIAL f1 PHOSPHORYLATION EVENTS

Kinkade and Cole (1966) have demonstrated that histone f1 is composed of several different subfractions. This raises several questions concerning the nature of cell-cycle-dependent f1 phosphorylation: (1) Does the phosphorylation of f1 occur on all f1 subfractions or preferentially on one subfraction? (2) Does the acceleration of f1 phosphorylation in late interphase result from different f1 subfractions being phosphorylated in S and G_2 from those being phosphorylated in G_1? (3)

Does the acceleration of f1 phosphorylation in S and G_2 result from the phosphorylation of the same sites as those phosphorylated in G_1 but at an increased rate? (4) Are different sites phosphorylated on f1 as cells traverse from G_1 to M? (5) Is there an increased proportion of f1 molecules phosphorylated in late interphase over those phosphorylated in G_1? (6) What is the nature of the slowly migrating form of f1 phosphorylated specifically in mitosis?

We were able to answer some of these questions by modifying the column chromatography method described by Hohmann and Cole (1971). This method enabled us to separate the various f1 subfractions from each other while, at the same time, separating the phosphorylated f1 from its unphosphorylated parent (Gurley et al. 1975). In early G_1, following release from isoleucine-deprivation, no phosphorylation was observed in any of the four parental f1 subfractions (Figure 1-15A). By mid-G_1, phosphorylation had begun in all subfractions (Figure 1-15B). As cells entered S phase (5 and 8 hours after release), a second, more highly phosphorylated form of f1 was observed, appearing as a faster eluting shoulder on the phosphorylated f1 peak (Figure 1-15C and Figure 1-15D). This greater phosphorylated form was found to have a $^{32}P/^3H$ ratio twice that observed in the phosphorylated form first appearing in mid-G_1.

Analysis of these chromatograms (Gurley et al. 1975) demonstrated that some molecules from each f1 subfraction are phosphorylated in G_1 and that phosphorylation precedes S phase by approximately 2 hours in all f1 subfractions (Figure 1-16). Since this chromatographic method separates the phosphorylated and unphosphorylated forms of f1, it was possible to estimate the amount of each phosphorylated form and to relate these to the fraction of S-phase cells in the culture (Figure 1-17). It was found that two different cell-cycle-related phosphorylation events occur in cells traversing early interphase. One event begins in G_1 2 hours before S phase. A second phosphorylation event begins at precisely the time the cells enter S phase. This S-phase-related phosphorylation event appears to occur as the phosphorylation of a new site in addition to the G_1-initiated phosphorylation site. This conclusion is drawn from the observation that the S-related form elutes from the ion exchange column earlier than the G_1-related phosphorylated form and that the $^{32}P/^3H$ ratio of the S-related form is twice that of the G_1-related phosphorylated form (Gurley et al. 1975). While the number of f1 molecules in the G_1-initiated phosphorylated form increased and accumulated in cells traversing interphase, the proportion of f1 molecules in the S-related phosphorylated form remained constant at about 10% of the total f1 (Figure 1-17).

When cells traversing G_1 were treated with hydroxyurea for 8 hours to resynchronize and arrest them in late G_1, only the first phosphorylated form of f1 was observed (Figure 1-18A). When these cells were released to traverse S and G_2, the more highly phosphorylated form was again observed to appear (Figure 1-18B and Figure 1-18C). This confirmed our conclusion that the more highly phosphorylated form of f1 does not appear until the cells have entered S phase (Gurley et al. 1975). Thus, there appears to be distinct phosphorylation events and forms of f1 in G_1 and S.

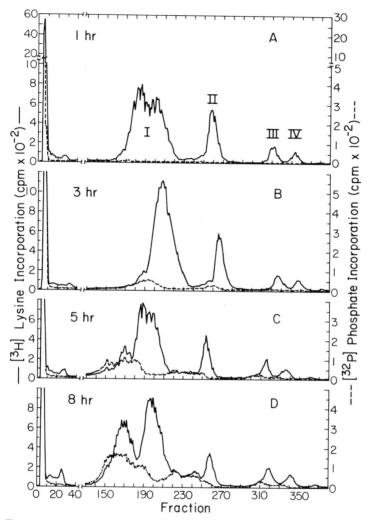

Figure 1-15.
Phosphorylation of histone f1 subfractions in proliferating CHO cells traversing the G_1 and S phases. ^{3}H-Lysine prelabeled cultures synchronized in early G_1 by isoleucine-deprivation were released by adding isoleucine and were pulse-labeled with ^{32}PO$_4$ for 1 hour prior to harvest (A) 1 hour; (B) 3 hours; (C) 5 hours; and (D) 8 hours after release. Histone f1 was subjected to column chromatography which resolved four individual subfractions (I, II, III, and IV) indicated by the ^{3}H-lysine incorporation (———). The phosphorylation of each subfraction is indicated by the ^{32}PO$_4$ incorporation (–––). (Reprinted with permission from Gurley et al. 1975.)

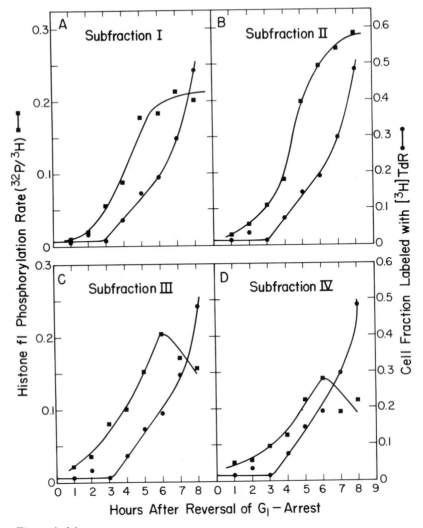

Figure 1-16.
The relationship between the phosphorylation of f1 subfractions and S phase. Analysis of the various f1 subfractions shown in Figure 1-15 are presented in (A) subfraction I; (B) subfraction II; (C) subfraction III; and (D) subfraction IV. The phosphorylation rate of each f1 subfraction is presented as a function of $^{32}PO_4$ incorporation per total ^{3}H-lysine incorporation into the sum of the phosphorylated and unphosphorylated forms of that subfraction (■). The fraction of S-phase cells in the culture was determined by autoradiography following ^{3}H-thymidine incorporation (●). (Reprinted with permission from Gurley et al. 1975.)

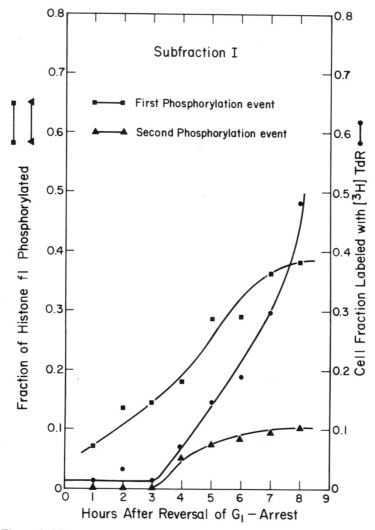

Figure 1-17.
The relationship between entry of cells into S phase and the fraction of f1 subfraction I existing in different phosphorylated forms. The fraction of S-phase cells in the culture after release from isoleucine-deprivation synchronization was determined as in Figure 1-16 (●). The fraction of f1 in the phosphorylated form first occurring in G_1 (■) was estimated from the phosphorylated peak immediately preceding the unphosphorylated form in Figure 1-15. The fraction of f1 in the second phosphorylated from occurring is S phase (▲) was estimated from the more highly phosphorylated shoulder on the left side of the phosphorylated peak in Figure 1-15. (Reprinted with permission from Gurley et al. 1975.)

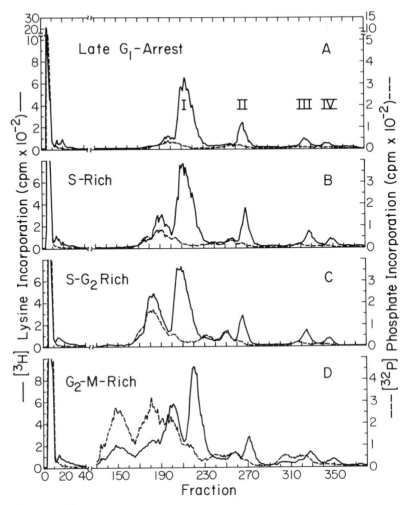

Figure 1-18.
The phosphorylation of histone f1 subfractions in proliferating CHO cells traversing late interphase. ^{3}H-Lysine prelabeled cultures were synchronized in late G_1 by isoleucine-deprivation/hydroxyurea treatment and then released. The cultures were pulse-labeled with [32]PO_4 for 1 hour prior to harvest at the various times indicated by the arrows in Figure 1-19. Histone f1 was subjected to column chromatography which resolved four subfractions (I, II, III, and IV) indicated by the ^{3}H-lysine incorporation (——). The phosphorylation of each subfraction is indicated by the [32]PO_4 incorporation (–––). (A) Culture synchronized in late G_1 by exposure to 1 mM hydroxyurea and harvested without release into S phase. (B) S-rich culture harvested 3.5 hours after release. (C) S-G_2-rich culture harvested 6 hours after release. (D) G_2-M-rich culture harvested 9 hours after release in the presence of Colcemid (to prevent cells from passing through mitosis into G_1). (Reprinted with permission from Gurley et al. 1975.)

When Colcemid was added to cells traversing late interphase to enrich the culture in mitotic cells, chromatographic fractionation revealed two new forms of phosphorylated f1 which eluted far ahead of the G_1-related phosphorylated forms (Figure 1-18D). Both of these forms had $^{32}P/^3H$ ratios four times greater than the G_1-related phosphorylation form, suggesting that the M-related phosphorylation event involves yet additional phosphorylation sites (perhaps four times the number of phosphorylation sites involved in the G_1-related event) (Gurley et al. 1975). Also, all f1 subfractions were observed to be phosphorylated in this manner, indicating that there was no preferential phosphorylation of any of the individual f1 subfractions during the cell cycle.

When the fraction of f1 in each phosphorylated form is plotted in relation to cell-cycle position, it can be seen that the fraction of unphosphorylated f1 steadily decreases as cells traverse late interphase (Figure 1-19). The fraction of f1 existing

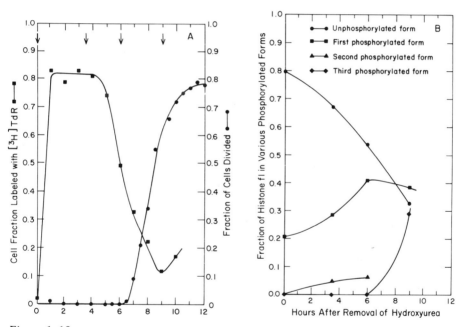

Figure 1-19.
The phosphorylated forms of histone f1 subfraction I existing during late stages of the cell cycle. (A) Cell-cycle analysis of CHO cultures following release from synchronization by the isoleucine-deprivation/hydroxyurea method. The fraction of cells in S phase (■) and the fraction of cells divided (●) were determined from an unlabeled replica culture. (B) The fractions of unphosphorylated f1 (○), the f1 phosphorylated form first occurring in G_1 (■), and the second f1 phosphorylated for occurring in S phase (▲) were determined as described in Figure 1-17. The fraction of f1 in the third phosphorylated form (◆) was estimated from the sum of the two highly phosphorylated peaks (fractions 130–190 in Figure 1-18D) having similar $^{32}P/^3H$ ratios. (Reprinted with permission from Gurley et al. 1975.)

in the first phosphorylated form accumulates through S and G_2, indicating the stable nature of this form. In contrast, the fraction of f1 in the second phosphorylated form did not accumulate to a large proportion, suggesting that this form is more transitory in nature. The rapid increase in the third superphosphorylated forms of f1 at the end of the cell cycle suggests that these forms are associated with mitosis (Gurley et al. 1975).

When f1 was isolated from mitotic cells labeled with $^{32}PO_4$ either during the $G_2 \rightarrow M$ transition or during metaphase, it was found that *all* the f1 existed in the highest phosphorylated forms, eluting rapidly from the ion exchange column (Figure 1-20A and Figure 1-20B). When these cells reentered G_1, the $^{32}PO_4$-prelabeled f1 was rapidly dephosphorylated (Figure 1-20C). Thus, all f1 molecules appear to participate in the "superphosphorylation" related to the condensation and maintenance of chromosomes (Gurley et al. 1975).

MITOTIC PHOSPHORYLATION OF HISTONE f3 SUBFRACTIONS

Zweidler and Cohen (1972) have shown that histone f3 can be subfractionated into more than one protein by polyacrylamide gel electrophoresis in the presence of Triton X-100. This method discriminates among proteins on the basis of their hydrophobic properties. By adapting this method to our preparative electrophoresis system, we have shown that CHO histone f3 could be resolved into two subfractions (Gurley and Walters 1973). Therefore, this method was used to determine whether mitotic f3 phosphorylation occurred preferentially on one f3 subfraction. It was found that both f3 subfractions (f3a and f3b) were phosphorylated to the same extent during the $G_2 \rightarrow M$ transition (Figure 1-21B) and during metaphase arrest in the presence of Colcemid (Figure 1-21C). When Colcemid was removed and the cells reentered G_1, both $^{32}PO_4$-prelabeled f3 subfractions were immediately dephosphorylated (Figure 1-21D).

We conclude, therefore, that the phosphorylation events observed for both f1 and f3 are cell-cycle phenomena which do not discriminate among the various subfractions of these histones (Gurley et al. 1975).

SUMMARY OF CELL-CYCLE-DEPENDENT HISTONE PHOSPHORYLATION IN CHO CELLS

We have summarized our observations on histone phosphorylation by superimposing them on a diagram of the CHO cell cycle (Figure 1-22). This diagram illustrates that histone f2a2 phosphorylation is constitutive in CHO cells, occurring at all times in both G_1-arrested cells as well as in proliferating cells (Gurley et al. 1974a, 1975). Histone f2a1 is also phosphorylated constitutively (not shown), but this occurs only to a limited extent (Gurley et al. 1974a), and histone f2b is not phosphorylated at all (Gurley and Walters 1973).

The phosphorylation of f1 is complex, occurring at three different times in the cell cycle ($f1_{G_1}$, $f1_S$, and $f1_M$ in Figure 1-22). Nonproliferating cells do not phosphorylate f1 at all. When cells begin proliferation, $f1_{G_1}$ phosphorylation begins well

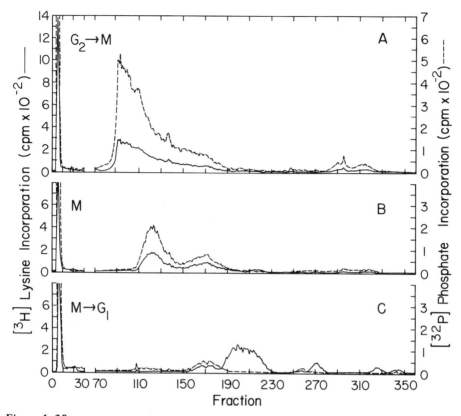

Figure 1-20.
Phosphorylation and dephosphorylation of histone f1 at mitosis and cell division. CHO cells prelabeled with ^{3}H-lysine were grown exponentially in monolayer. (A) Metaphase cells were accumulated in the presence of Colcemid and $^{32}PO_4$ for 2 hours and removed by mitotic selection to produce a culture which had incorporated $^{32}PO_4$ while undergoing the transformation from interphase to metaphase ($G_2 \rightarrow M$). (B) Metaphase cells were accumulated for 2 hours in the presence of Colcemid and removed by mitotic selection, after which they were labeled for 2 hours with $^{32}PO_4$ while arrested in methaphase by Colcemid (M cells). (C) Metaphase cells labeled with $^{32}PO_4$ for 2 hours during the $G_2 \rightarrow M$ transition [as in (A) above] were removed by mitotic selection and resuspended in Colcemid-free medium for 2 hours before harvest, which allowed the cells to divide and enter G_1. Histone f1 was isolated from these three cultures and subjected to column chromatography. Histone f1 is indicated by the ^{3}H-lysine incorporation (——) and its phosphorylation by the $^{32}PO_4$ incorporation (– – –). (Reprinted with permission from Gurley et al. 1975.)

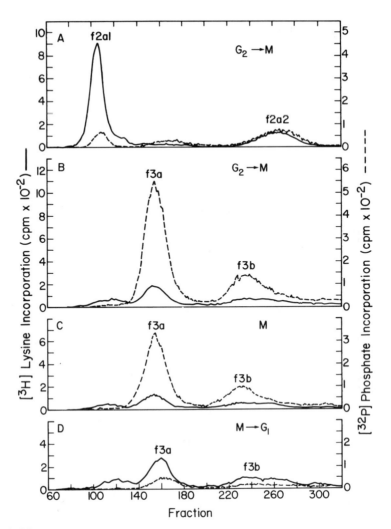

Figure 1-21.
Phosphorylation and dephosphorylation of histone f3 subfractions at mitosis and cell division. CHO cells prelabeled with ^{3}H-lysine were grown exponentially in monolayer. Cells were labeled with ^{32}PO$_4$ for 2 hours exactly as described in Figure 1-20. Histones were isolated and subjected to preparative Triton X-100 polyacrylamide gel electrophoresis. The histones are indicated by the ^{3}H-lysine incorporation (——) and the phosphorylation by the ^{32}PO$_4$ incorporation (– – –). (A) Electrophoresis of histones f2a1 and f2a2 from cells incorporating ^{32}PO$_4$ during the G$_2$ → M transition. (B) Electrophoresis of histone f3 obtained from cells incorporating ^{32}PO$_4$ during the G$_2$ → M transition. (C) Electrophoresis of histone f3 obtained from cells incorporating ^{32}PO$_4$ during metaphase. (D) Electrophoresis of histone f3 obtained from cells which were labeled with ^{32}PO$_4$ during the G$_2$ → M transition and then allowed to divide and enter G$_1$ (M → G$_1$ cells). (Reprinted with permission from Gurley et al. 1975.)

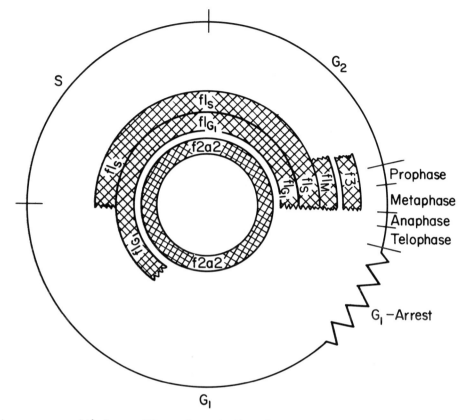

Histone Phosphorylation in the
Chinese Hamster (Line CHO) Cell Cycle

Figure 1-22.
The relationship of histone phosphorylation to the cell cycle of line CHO Chinese
hamster cells. The 16.5-hour generation time of these cells in F-10 medium may be
divided into 9 hours G_1, 4 hours S, 3 hours G_2, and 0.5 hour M (Tobey 1973). The
periods in which histones f1, f2a2, and f3 are phosphorylated are indicated by the
shaded bands. The symbol $f1_{G_1}$ denotes f1 phosphorylation which begins in the G_1
phase; $f1_S$ denotes f1 phosphorylation which begins in the S phase; and $f1_M$
denotes f1 phosphorylation which occurs in mitosis. (Reprinted with permission
from Gurley et al. 1975.)

ahead of S phase and may represent one of the earliest biochemical events involved
with the conversion of cells from the nonproliferative state to the proliferative
state. The $f1_S$ phosphorylation begins exactly when cells begin synthesizing DNA.
While $f1_{G_1}$ phosphorylation is probably involved in prereplicative preparations of
chromatin for DNA synthesis, $f1_S$ phosphorylation may be more directly involved
in DNA synthesis per se, or it may be involved in the deposition of newly synthe-

sized f1 on the newly synthesized DNA. The limited existence of this double-phosphorylated $f1_S$ form throughout S phase opens for consideration the speculation that $f1_S$ phosphorylation may occur only at the site of the DNA replication fork and that $f1_S$ is dephosphorylated when the replication fork has passed that part of the chromatin containing the $f1_S$ molecule (Gurley et al. 1975).

When cells enter mitosis, a third f1 phosphorylation event occurs. This $f1_M$ phosphorylation occurs at a high rate on all f1 molecules and possibly involves four phosphorylation sites. Simultaneously with the phosphorylation of $f1_M$, histone f3 is also phosphorylated. The coincidence of these two phosphorylation events with the condensation of dispersed chromatin into chromosomes (which occurs specifically at this time) strongly suggests that $f1_M$ and f3 phosphorylation are involved with these structural changes in chromatin. From studies on the physical state of chromatin, Bradbury et al. (1973a,b, 1974) have also suggested that f1 may play an integral role in chromosome condensation.

Sadgopal and Bonner (1970) have measures a greater degree of disulfide bond formation within histone f3 from metaphase chromosomes than in the dispersed chromatin from interphase cells. This implies that the oxidation of f3 may be involved in maintaining the condensed state of the chromosome structure. We suggest that the phosphorylation of f3 during mitosis may alter f3 confrontation, resulting in disulfide bond formation to stabilize the condensed chromosome structure (Gurley et al. 1974a).

CELL-CYCLE-DEPENDENT PHOSPHORYLATION OF SERINE AND THREONINE IN HISTONE f1

Previous studies by Lake (1973) demonstrated that different phosphopeptides could be resolved from the f1 histones of interphase and mitotic cells. This suggested that f1 was modified at different sites when the physical state of the chromatin changed from the dispersed to the condensed state during mitosis. Our studies (described above) have led us to suggest that three distinct phosphorylation events occurring on histone f1 are associated with three distinct physical states of chromatin: prereplicative, replicating, and condensed (Gurley et al. 1975). Since phosphorylation-induced changes in the conformation of histones are likely to be involved in these chromatin structural changes, it is of fundamental importance to determine which sites on f1 are phosphorylated at each phase of the cell cycle.

Recent work by Langan and Hohmann (1974) demonstrated that both serine and threonine are phosphorylated in the f1 of growing Erhlich ascites and rat hepatoma cells. Since the phosphorylation of different amino acids is a direct indication of the phosphorylation of different sites, our initial f1 structural studies were concentrated on establishing whether a correlation existed between cell-cycle position and the preferential phosphorylation of serine and threonine (Hohmann et al. 1975).

Histone f1 labeled with $^{32}PO_4$ for 2 hours was isolated from mitotic and exponential cultures and purified by column chromatography (Hohmann et al. 1975). Phosphoserine and phosphothreonine were liberated from the f1 by partial acid hydrolysis and separated by high-voltage paper electrophoresis (Hohmann et al.

1975). Autoradiographic analysis of the paper revealed that in exponential cells most of the phosphorylation occurred on serine, while in mitotic cells more of the phosphorylation occurred on threonine (Figure 1-23). The phosphorylation rates of serine and threonine were determined by radioanalysis. The relative phosphorylation rate of serine-to-threonine in the f1 of exponential cultures was 5.0, as determined from the average of three closely agreeing experiments. In the particular experiment shown in Figure 1-23, the mitotic population was 74% and the relative phosphorylation rate of serine-to-threonine 0.87. In experiments where the mitotic populations approached 99%, the relative phosphorylation rate of serine-to-threonine approached 0.5. Thus, a tenfold difference was observed in the relative rates of phosphorylation of these two amino acids between exponential and mitotic cultures (Hohmann et al. 1975).

We next initiated a study of the phosphorylation of different sites in f1 by examining the extent of phosphorylation of different peptides. Purified f1 histone labeled with $^{32}PO_4$ was subjected to tryptic digestion, followed by high-voltage

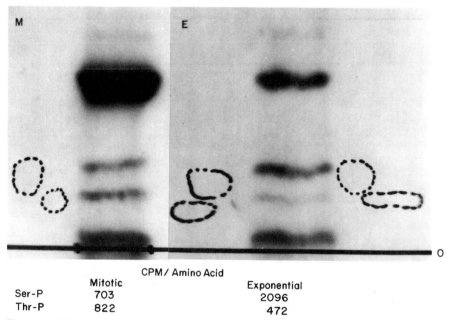

	CPM / Amino Acid	
	Mitotic	Exponential
Ser-P	703	2096
Thr-P	822	472

Figure 1-23.
Autoradiographs of the electrophoretic separation of ^{32}P-phosphoserine and ^{32}P-phosphothreonine isolated from purified f1 of CHO cells. Migration of the standards is indicated by dotted circles on the film, phosphoserine migrating farther away from the origin (O) than phosphothreonine. The counts per minute of ^{32}P in each phosphoamino acid are shown at the bottom of the autoradiographs for both exponential cultures (E) and mitotic cultures (M). The fastest migrating radioactive substance is inorganic ^{32}P-phosphate. The undefined ^{32}P-labeled material at the origin is probably unhydrolyzed protein. (Reprinted with permission from Hohmann et al. 1975.)

paper electrophoresis at pH 7.9 (Hohmann et al. 1975). Autoradiographic analysis of the paper indicated that the phosphopeptide pattern of f1 is more complex in mitotic cells than in exponential cells (Figure 1-24). The bands designated II and III were eluted from the paper and subjected to a second electrophoresis at pH 1.9 (Figure 1-25). Band III appears homogeneous and is common to both mitotic and exponential cells. Further analysis has shown that this band contains only phosphoserine. In contrast, band II is very complex in mitotic cells, containing at least five phosphopeptides. However, in exponential cells, band II contains only one of these phosphopeptides in significant quantity, the other four peptides being either very weak or completely absent. In mitotic cells, the first, strongest labeled band (indicated by the arrow in Figure 1-25) has been shown to contain only phosphothreonine. The second band contains both phosphoserine and phosphothreonine. The third band contains phosphothreonine, the fourth band contains phospho-

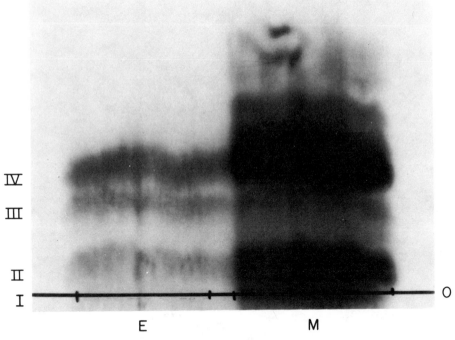

Figure 1-24.
Autoradiographs of the tryptic ^{32}P-phosphopeptides of histone f1 resolved by high-voltage paper electrophoresis at pH 7.9. Histone f1 was isolated from ^{32}PO$_4$-labeled exponential cultures (E) and mitotic cultures (M) of CHO cells. Following purification by column chromatography, the f1 was subjected to tryptic digestion. Electrophoresis of the digest resolved the phosphopeptides between the origin (O) and the anode at the top of the paper. Four distinct peaks (I, II, III, and IV) were resolved from the f1 of exponential cells. (Reprinted with permission from Hohmann et al. 1975.)

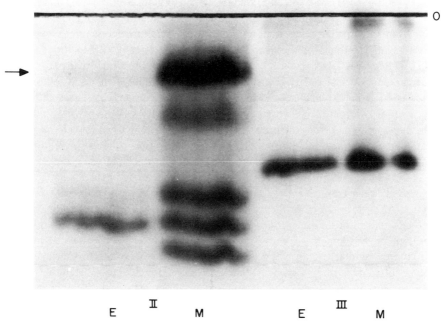

Figure 1-25.

Autoradiographs of ^{32}P-phosphopeptides II and III fractionated by high-voltage paper electrophoresis at pH 1.9. Bands II and III from both exponential cultures (E) and mitotic cultures (M) were eluted from the paper (shown in Figure 1–27) and subjected to electrophoresis at pH 1.9 which resolved these bands into more phosphopeptides between the origin (O) and the cathode at the bottom of the paper. Five distinct bands were resolved from band II of mitotic cells. The arrow indicates the phosphothreonine-containing peptide restricted to M cells. (Reprinted with permission from Hohmann et al. 1975.)

serine, and the fifth band contains phosphothreonine. This observation is a direct demonstration that the phosphorylation of at least one threonine is completely restricted to mitotic cells (Hohmann et al. 1975).

CELL-CYCLE CHANGES IN CHROMATIN ORGANIZATION

Since the superphosphorylation of $f1_M$ and the phosphorylation of f3 occur solely at mitosis, it is easy to correlate these phosphorylation events with the chromatin structural changes observable in the microscope specifically at this time. Unfortunately, correlating f1 phosphorylation with chromatin structural changes in interphase is not so straightforward, because the dispersed chromatin of cells does not change its gross microscopic appearance during cell-cycle progression through interphase. However, we have recently made some progress in measuring chromatin structural changes by probing interphase nuclei with agents whose interaction with chromatin is structure-dependent.

For example, cell-cycle-specific changes in the organization of chromatin have been observed using heparin, a natural polyanion (Hildebrand and Tobey 1975a). It has been shown that, when heparin is added to nuclei, this polyanion specifically interacts with histones in the chromatin (Arnold et al. 1972; Hildebrand et al., 1975). This causes DNA to be released from the nucleus in a highly dispersed state (Kraemer and Coffey 1970). Thus, changes in the amount of heparin required to release DNA from the nucleus can be interpreted as representing changes in the organization of chromatin (Hildebrand and Tobey 1975a).

The cell-cycle kinetics of such changes in chromatin organization were measured in cells traversing G_1 by determining the concentration of heparin required to release 50% of the maximum amount of DNA from nuclei (Figure 1-26). These values showed that the resistance of chromatin to heparin-mediated DNA dispersion increased as cells traversed early interphase (Hildebrand and Tobey 1975b). The timing of this chromatin organization change was similar to the timing of the G_1-specific f1 phosphorylation event (Figure 1-11 and Figure 1-16). When this heparin measurement was made on cells traversing early interphase in the presence of hydroxyurea, similar timing was observed even though DNA and histone synthesis were inhibited (Figure 1-27). This indicates that the increased resistance of chromatin to heparin-mediated dispersion across G_1 does not result from an increase in the histone or DNA components of chromatin (Hildebrand and Tobey 1975b).

When these cells were released from hydroxyurea blockade at the G_1/S boundary, there was a further increase in the resistance of chromatin to heparin-mediated DNA dispersion as the cells traversed S and G_2 (Figure 1-27). Then, coincident with cell division and reentry into G_1, this resistance rapidly decreased.

These data are interpreted as meaning that a continual change in chromatin structure occurs as cells traverse interphase which makes the chromatin more resistant to heparin as the cells approach mitosis (Hildebrand and Tobey 1975b). The close correspondence between the timing of $f1_{G_1}$ phosphorylation and the changes in chromatin organization suggests that these processes may be related. Several hypothetical mechanisms may be offered to explain this correlation: (1) the affinity of histones for DNA could be increased by phosphorylation-induced histone conformational changes. (2) The accessibility of histones to heparin could be decreased by phosphorylation-induced histone conformational changes. (3) The accessibility of histones to heparin could be reduced by an increased association of histones with nonhistone chromosomal proteins induced by f1 phosphorylation. Many more mechanisms could be envisioned.

CELL-CYCLE CHANGES IN DNA ASSOCIATED WITH LIPOPROTEIN

Using a modification of the sarkosyl crystal technique (Hildrebrand and Tobey 1973) [often called the M-band method (Tremblay et al 1969)], we have also demonstrated that cell-cycle-specific changes occur in the association of DNA with lipoprotein during G_1 (Figure 1-28). By specifically absorbing the DNA-lipoprotein complex to magnesium-sarkosyl crystals, the association of DNA with lipoprotein

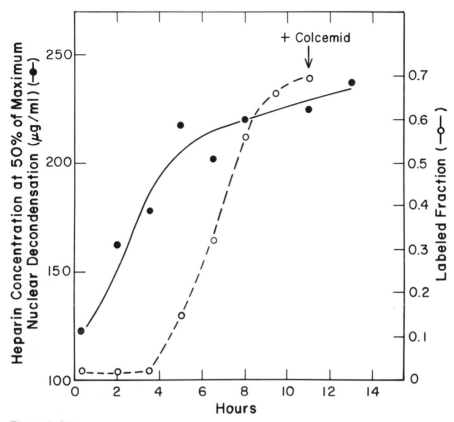

Figure 1-26.
Changes in the state of chromatin condensation in early interphase. CHO cells pre-labeled with ^{14}C-thymidine were synchronized by isoleucine-deprivation and released to traverse early interphase. Changes in the state of chromatin condensation were measured at various times as a function of the amount of heparin necessary to decondense the chromatin in isolated nuclei and to release 50% of the maximum amount of DNA released at large heparin concentrations (●). The fraction of cells in S phase (labeled fraction) was determined by autoradiography of cells pulse-labeled with ^{3}H-thymidine in replica cultures (○). Colcemid was added at 11 hours to prevent the cells from passing through mitosis into G_1.

was found to increase as cells traversed G_1 (Hildebrand and Tobey 1973). This enhanced association commenced approximately 2 hours prior to the initiation of DNA replication (Figure 1-28A). These cell-cycle kinetics are essentially identical with those of fl_{G_1} phosphorylation (Figure 1-11 and Figure 1-16). By inhibiting the entry of cells into S phase with hydroxyurea, the amount of DNA complexed with lipoprotein was limited to approximately 20% of the cellular DNA (Figure 1-28B). However, if cells were allowed to progress through S and G_2, the association of DNA with lipoprotein was further enhanced over that observed in late G_1

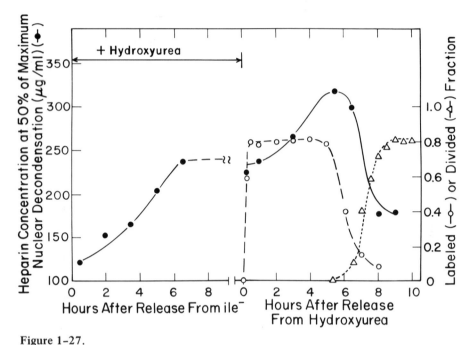

Figure 1-27.
Changes in the state of chromatin condensation in hydroxyurea-treated early interphase cells and in synchronized late interphase cells. CHO cultures prelabeled with ^{14}C-thymidine were synchronized by isoleucine-deprivation and released to traverse early interphase in the presence of 1 mM hydroxyurea. Following resynchronization at the G_1/S boundary, hydroxyurea was removed, allowing the cells to synchronously traverse late interphase. Changes in the state of chromatin condensation were measured at various times as a function of heparin-mediated release of DNA from nuclei, as described in Figure 1-26. (●). The fraction of cells in S phase (labeled fraction) was determined as described in Figure 1-26 (○). Cell division was monitored using an electronic cell counter (△).

(Figure 1-28B). Likewise, an increased fl_{G_1}-type phosphorylation was observed as cells progressed through S and G_2 (Figure 1-19B).

These data, together with the observations on chromatin organizational changes and fl phosphorylation, suggest that fl_{G_1} phosphorylation may alter the structural configuration of chromatin beginning 2 hours prior to entry into the S phase, thus allowing an enhanced association of chromatin with lipoprotein. Why such a change in chromatin structure is necessary in mid-G_1 is unknown. We have speculated that fl phosphorylation may be involved with chromatin structural changes necessary for the orderly separation of daughter DNA molecules during genome replication and cell division (Gurley et al. 1974a, 1975). It is reasonable to expect that such structural changes might be necessary to establish an organized chromatin structure on which DNA could be orderly separated after its replication. Such a segregating process may have to persist and even accelerate in G_2 in preparation for final genome separation during mitosis. This could account for the increased fl

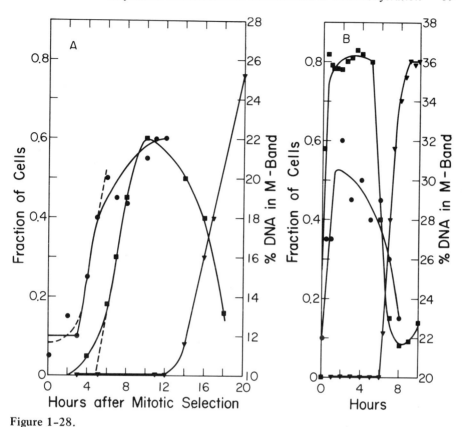

Figure 1-28.
Cell-cycle analysis of DNA-lipoprotein complexes in synchronized CHO cells. (A) Cells prelabeled with ^{14}C-thymidine were synchronized by mitotic selection. While traversing early interphase, the fraction of ^{14}C-labeled DNA-lipoprotein associated with magnesium-sarkosyl crystals (% DNA in M band) was determined (●). The fraction of cells in S phase was determined on a replica culture by autoradiography of ^{3}H-thymidine pulse-labeled cells (■). The fraction of cells divided was determined with an electronic cell counter (▲). (B) ^{14}C-Thymidine prelabeled cells were synchronized by the mitotic selection/hydroxyurea method at the G_1/S boundary and released to traverse later interphase. The same measurements were made as stated above. (Reprinted with permission from Tobey et al. 1974.)

phosphorylation and the continual structural changes in chromatin during late interphase. Experiments by Marks et al. (1973) have resulted in similar speculations.

DISCUSSION: THE CHROMOSOME CYCLE

We interpret our studies on the cell-cycle-dependent phosphorylation of histones as reflecting specific changes in histone structure which are related to changing the structure and function of chromatin. The preferential phosphorylation of at least one threonine residue in f1 of mitotic cells is a clear example of such a *specific* modification. Another example of such a specific phosphorylation has been reported by Hohmann and Langan (1975). They demonstrated that a specific serine residue was phosphorylated on the f1 of H-35 rat hepatoma cells after cyclic AMP stimulation. Thus, the effect of cyclic AMP on the regulation of cellular activity appears, at least in part, to be modulated by the phosphorylation of f1 at a distinct site. Specific modifications of histones have also been correlated with the developmental program of trout testes in extensive studies by Louie and Dixon (1973), and model studies by Adler et al. (1971, 1972) have demonstrated that specific, multiple phosphorylation of f1 can alter the ability of f1 to change the conformation of DNA. In view of these independent studies and the results of our own investigations, we envision an orderly progression of specific histone modifications as necessary for the proper functioning of chromatin as cells progress through their proliferation cycle.

The concepts developed from experimentation with the CHO cell system support and extend a hypothesis proposed earlier by Mazia (1963, 1974; also see discussion by Mitchison 1971). In 1963, Mazia speculated that the changes in chromatin structure observed from prophase to telophase might be only part of a larger chromosome structural cycle which extended from G_1 to mitosis, most of which is unobservable in the microscope. Such a "chromosome cycle" might explain, in terms of chromosome structure, the control of initiation and termination of chromosome replication as well as the ultimate orderly separation of the genome at cell division. Pederson (1972) obtained experimental evidence that such a cycle might be interrupted in nondividing G_0 cells and that reinitiation of chromatin structural changes may lead to a reinitiation of those biochemical events of the cell cycle leading to cell proliferation. Our own experimental data support this general concept by suggesting that $f1_{G_1}$ phosphorylation may alter the structural configuration of chromatin in G_1, thus allowing an enhanced association of chromatin with lipoprotein leading to cell-cycle traverse. Progression toward cell division is then also dependent upon subtle modifications of the chromatin structure induced by other specific f1 phosphorylations (such as $f1_S$ phosphorylation). This "chromosome cycle" would ultimately be culminated by the phosphorylation of f3 and $f1_M$ at specific sites during mitosis, resulting in chromosome condensation.

ACKNOWLEDGMENTS

This work was performed under the auspices of the U.S. Atomic Energy Commission. The authors are grateful for the excellent technical assistance of J. G. Valdez, J. L. Hanners, and P. C. Sanders.

LITERATURE CITED

Adler, A. J., B. Schaffhausen, T. A. Langan, and G. P. Fasman. 1971. Altered conformational effects of phosphorylated lysine-rich histone (f1) in f1-DNA complexes. *Biochemistry* **10**:909–913.

Adler, A. J., T. A. Langan, and G. D. Fasman. 1972. Complexes of DNA with lysine-rich (f1) histone phosphorylated at two separate sites: Circular dichroism studies. *Arch. Biochem. Biophys.* **153**:769–777.

Allfrey, V. G. 1971. Functional and metabolic aspects of DNA-associated proteins. *Histones and Nucleohistones.* Plenum Press, New York, pp. 241–244.

Arnold, E. A., D. H. Yawn, D. G. Wyllie, and D. S. Coffey. 1972. Structural alterations in isolated rat liver nuclei after removal of template restriction by polyanions. *J. Cell Biol.* **53**:737–757.

Balhorn, R., M. Balhorn, H. P. Morris, and R. Chalkley, 1972. Comparative high resolution electrophoresis of tumor histones: Variation in phosphorylation as a function of cell proliferation rate. *Cancer Res.* **32**:1775–1784.

Bradbury, E. M. and C. Crane-Robinson. 1971. Physical and conformational studies of histones and nucleohistones. In *Histones and Nucleohistones.* Plenum Press, New York, pp. 85–134.

Bradbury, E. M., B. G. Carpenter, and H. W. E. Rattle. 1973a. Magnetic resonance studies on deoxyribonucleoprotein. *Nature* **241**:123–126.

Bradbury, E. M., R. J. Inglis, H. R. Matthews, and N. Sarner, 1973b. Phosphorylation of very-lysine-rich histone in *Physarum polycephalum.* Correlation with chromosome condensation. *Eur. J. Biochem.* **33**:131–139.

Bradbury, E. M., R. J. Inglis, and H. R. Matthews. 1974. Control of cell division by very lysine rich histone (f1) phosphorylation. *Nature* **247**:257–261.

Enger, M. D. and R. A. Tobey, 1972. Effects of isoleucine deficiency on nucleic acid and protein metabolism in cultured Chinese hamster cells. Continued RNA and protein synthesis in the absence of DNA synthesis. *Biochemistry* **11**:269–277.

Gurley, L. R. and J. M. Hardin. 1968. The metabolism of histone fractions. I. Synthesis of histone fractions during the life cycle of mammalian cells. *Arch. Biochem. Biophys.* **128**:285–292.

Gurley, L. R. and R. A. Walters. 1971. Response of histone turnover and phosphorylation to x-irradiation. *Biochemistry* **10**:1588–1593.

Gurley, L. R. and R. A. Walters. 1973. Evidence from Triton X-100 polyacrylamide gel electrophoresis that histone f2a2, not f2b, is phosphorylated in Chinese hamster cells. *Biochem. Biophys. Res. Commun.* **55**:697–703.

Gurley, L. R., R. A. Walters, and R. A. Tobey. 1973a. Histone phosphorylation in late interphase and mitosis. *Biochem. Biophys. Res. Commun.* **50**:744–750.

Gurley, L. R., R. A. Walters, and R. A. Tobey. 1973b. The metabolism of histone fractions. VI. Differences in the phosphorylation of histone fractions during the cell cycle. *Arch. Biochem. Biophys.* **154**:212–218.

Gurley, L. R., R. A. Walters, and R. A. Tobey. 1974a. Cell cycle-specific changes in histone phosphorylation associated with cell proliferation and chromosome condensation. *J. Cell Biol.* **60**:356–364.

Gurley, L. R., R. A. Walters, and R. A. Tobey. 1974b. The metabolism of histone fractions. Phosphorylation and synthesis in late G_1-arrest. *Arch. Biochem. Biophys.* **164**:469–477.

Gurley, L. R., R. A. Walters, and R. A. Tobey. 1975. Sequential phosphorylation of histone subfractions in the Chinese hamster cell cycle. *J. Biol. Chem.* **250**:3936–3944.

Hildebrand, C. E. and R. A. Tobey. 1973. Temporal organization of DNA in Chinese hamster cells: Cell-cycle dependent association of DNA with membrane. *Biochim. Biophys. Acta* **331**:165–180.

Hildebrand, C. E., L. R. Gurley, R. A. Tobey, and R. A. Walters. 1975. Fed. Proc. **34**:581.

Hildebrand, C. E. and R. A. Tobey. 1975a. Cell-cycle-specific changes in chromatin organization. *Biochem. Biophys. Res. Commun.* **63**:134–139.

Hildebrand, C. E., and R. A. Tobey. 1975b. (manuscript in preparation).

Hohmann, P. and R. D. Cole. 1971. Hormonal effects on amino acid incorporation into lysine-rich histones in the mouse mammary gland. *J. Mol. Biol.* **58**:533–540.

Hohmann, P. and T. A. Langan. 1975. Phosphorylation of lysine-rich histones: Distinct types of phosphorylation associated with cell replication and cyclic AMP stimulation. *J. Biol. Chem.* (submitted).

Hohmann, P., R. A. Tobey, and L. R. Gurley. 1975. Cell-cycle-dependent phosphorylation of serine and threonine in Chinese hamster cell f1 histones. *Biochem. Biophys. Res. Commun.* **63**:126–133.

Kinkade, J. M., Jr., and R. D. Cole. 1966. The resolution of four lysine-rich histones derived from calf thymus. *J. Biol. Chem.* **241**:5790–5797.

Kraemer, P. M. and R. A. Tobey. 1972. Cell-cycle dependent desquamation of heparin sulfate from the cell surface. *J. Cell Biol.* **55**:713–717.

Kraemer, R. J. and D. S. Coffey. 1970. The interaction of natural and synthetic polyanions with mammalian nuclei. II. Nuclear swelling. *Biochim. Biophys. Acta* **224**:568–578.

Lake, R. S. 1973. Futher characterization of the f1-histone phosphokinase of metaphase-arrested animal cells. *J. Cell Biol.* **58**:317–331.

Langan, T. A. and P. Hohmann. 1974. Phosphorylation of threonine and serine residues of lysine-rich histone in growing cells. *Federation Proc.* **33**:1597.

Louie, A. J. and G. H. Dixon. 1973. Kinetics of phosphorylation and dephosphorylation of testis histones and their possible role in determining chromosomal structure. *Nature* **243**:164–168.

Marks, D. B., W. K. Paik, and T. W. Borun. 1973. The relationship of histone phosphorylation to DNA replication and mitosis during the HeLa S-3 cell cycle. *J. Biol. Chem.* **248**:5660–5667.

Mazia, D. 1963. Synthetic activities leading to mitosis. In Symposium on Macromolecular Aspects of the Cell Cycle. *J. Cell. Comp. Physiol.* **62** (Suppl. 1):123–140.

Mazia, D. 1974. The chromosome cycle in the cell cycle. In *Cell Cycle Controls,* eds. G. M. Padilla, I. L. Cameron, and A. Zimmerman. Academic Press, New York, pp. 265–272.

Mitchison, J. M. 1971. *The Biology of the Cell Cycle.* Cambridge University Press, Cambridge, Great Britain, pp. 89–90.

Pederson, T. 1972. Chromatin structure and the cell cycle. *Proc. Natl. Acad. Sci. U.S.A.* **69**:2224–2228.

Petersen, D. F. and E. C. Anderson. 1964. Quantity production of synchronized mammalian cells in suspension culture. *Nature* **203**:642–643.

Petersen, D. F., R. A. Tobey, and E. C. Anderson. 1969. Essential biosynthetic activity in synchronized mammalian cells. In *The Cell Cycle. Gene-Enzyme Interactions,* eds. G. M. Padilla, G. L. Whitson, and I. L. Cameron. Academic Press, New York, pp. 341–359.

Sadgopal, A. and J. Bonner, 1970. Proteins of interphase and metaphase chromosomes compared. *Biochim. Biophys. Acta* **207**:227–239.

Shepherd, G. R., B. J. Noland, and C. N. Roberts. 1970. Phosphorus in histones. *Biochim. Biophys. Acta* **199**:265–276.

Sinclair, W. K. 1967. Hydroxyurea: Effects of Chinese hamster cells grown in culture. *Cancer Res.* **27** (Part 1):297–308.

Tobey, R. A., E. C. Anderson, and D. F. Petersen. 1967. Properties of mitotic cells prepared by mechanically shaking monolayer cultures of Chinese hamster cells. *J. Cell. Physiol.* **70**:63–68.

Tobey, R. A. and K. D. Ley. 1971. Isoleucine-mediated regulation of genome replication in various mammalian cell lines. *Cancer Res.* **31**:46–51.

Tobey, R. A. and H. A. Crissman. 1972. Preparation of large quantities of synchronized mammalian cells in late G_1 in the pre-DNA replicative phase of the cell cycle. *Exp. Cell Res.* **75**:460–464.

Tobey, R. A., H. A. Crissman, and P. M. Kraemer. 1972. A method for comparing effects of different synchronizing protocols on mammalian cell cycle traverse: The traverse perturbation index. *J. Cell Biol.* **54**:638–645.

Tobey, R. A. 1973. Production and characterization of mammalian cells reversibly arrested in G_1 by growth in isoleucine-deficient medium. *Methods in Cell Biology* **6**:67–112.

Tobey, R. A., L. R. Gurley, C. E. Hildebrand, R. L. Ratliff, and R. A. Walters. 1974. Sequential biochemical events in preparation for DNA replication and mitosis. In *Control of Proliferation in Animal Cells*, Vol. 1, eds. B. Clarkson and R. Baserga. Cold Spring Harbor Laboratory Press, Cold Spring Harbor, New York, pp. 665–679.

Tobey, R. A., L. R. Gurley, C. E. Hildebrand, P. M. Kraemer, R. L. Ratliff, and R. A. Walters. 1975. Sequential biochemical events in the mammalian cell cycle. In *The LASL First Annual Life Sciences Symposium on Mammalian Cells: Probes and Problems.* AEC Symposium Series, Technical Information Center, Oak Ridge, Tenn. pp. 152–167.

Tremblay, G. Y., M. J. Daniels, and M. Schaechter. 1969. Isolation of a cell membrane-DNA-nascent RNA complex from bacteria. *J. Mol. Biol.* **40**:65–76.

Walters, R. A., R. A. Tobey, and R. L. Ratliff. 1973. Cell-cycle-dependent variations of deoxyribonucleoside triphosphate pools in Chinese hamster cells. *Biochim. Biophys. Acta* **319**:336–347.

Zweidler, A. and L. H. Cohen. 1972. A new electrophoretic method revealing multiplicity, tissue specificity and evolutionary variation in histones f2a2 and f2b. *Federation Proc.* **31**:926.

Contractile Proteins in Eukaryotic Chromatin

Angeline S. Douvas and James Bonner   California Institute of Technology

INTRODUCTION

Chromatin, the *in vitro* isolated chromosomal material of eukaryotes, is composed of about 26% by mass of nonhistone protein. The function (more likely functions) of this considerable and heterogeneous mass of proteins is still unknown. Numerous investigations have been undertaken to explore the possible role of nonhistones as regulators of gene expression in eukaryotes (for reviews of some of these results see Spelsberg et al. 1972 and Johnson et al. 1974). As yet, no specific regulatory role has been demonstrated for any of these proteins. Biochemical studies of nonhistones as a class have yielded some fundamental observations that help us to conceptualize their possible function: (1) In a complex eukaryote such as the rat, each cell contains over a hundred nonhistone polypeptide species, representing an average of 2×10^5 copies per haploid genome (Garrard et al. 1974). (2) A small number of polypeptides makes up the bulk of the nonhistone mass. In rat liver, for example, as few as six polypeptides account for 38% of the nonhistone protein mass (Douvas et al. 1975a). (3) Nonhistone proteins show more tissue-specific and organism-specific variation than histones (Elgin and Bonner 1970); however a limited number of nonhistone polypeptides appears to be present in the same tissue of a wide variety of organisms (Wu et al. 1973) and in different tissues of the same organism (Wu et al. 1975). In most cases, however, these homologies have been identified solely on the basis of electrophoretic patterns in gels, and should be accepted with some reservation.

We have aimed to isolate and characterize the few major nonhistone species which account for nearly 40% of the nonhistone mass (Douvas et al. 1975a). Because of the relatively great abundance of some species of nonhistones in chromatin (three of the major polypeptides are present in roughly 10^6 copies per haploid genome) (Garrard et al. 1974) it is likely that their involvement in the control of gene expression occurs at a gross structural level analogous to that of the histones. The identification of six polypeptides as myosin, actin, two tropomyosin subunits, a myosin breakdown product, and tubulin strengthens arguments in favor of a structural role for these major proteins. One must also consider the possibility that they may be cytoplasmic contaminants. Evidence has been presented in an earlier communication that they are probably integral proteins of chromosomes and not contaminants (Douvas et al. 1975a).

The present communication is in part a summary of our efforts to isolate and characterize contractile proteins in chromatin. The possible role of contractile proteins in chromatin is discussed in the light of recent discoveries of contractile proteins, actin in particular, in other nonmuscle cells and organelles.

If nonhistone proteins are instrumental in bringing about activation and deactivation of genes, it is reasonable to expect that the presence or amount of certain nonhistones is correlated with the quantity of genomic DNA in active use in a given cell type. A negative correlation between gene activity and the histone/DNA ratio has been known for some time (Bonner et al. 1968; Marushigi and Ozaki 1967). In the present communication we report that brain chromatin, which is approximately 2.5-fold more active in transcription than liver (Grouse et al. 1972), has about one-half as much chromosomal actin (a 45, 000 dalton polypeptide) as liver. A survey of published comparisons of chromosomal proteins from active and inactive cells has revealed an inverse correlation between the quantity of a 45,000 dalton protein and transcriptional activity.

EXPERIMENTAL SYSTEM

Chromatin preparation. Frozen livers from albino Sprague-Dawley rats were obtained from Pel Freeze. Chromatin was prepared from batches of 16 livers (about 150 grams of frozen tissue) as previously described (Douvas et al. 1975a). The crude chromosomal material obtained from grinding the tissue in saline-EDTA followed by low speed centrifugation was washed once in 10 mM Tris-HCl pH 8.0 then treated with diisopropylfluorophosphate (DFP) to inactivate proteases as described elsewhere (Douvas et al. 1975a; 1976a). After DFP treatment the usual procedure (3 washes 10 mM Tris-HCl pH 8.0 followed by sucrose gradient purification) was followed (Douvas et al. 1976a).

Dissociation of nonhistones from chromatin. Initial fractions for further purification of nonhistones were obtained by dissociating purified chromatin in sodium chloride. NaCl (4 M) dissociates on the average 50% of the nonhistone mass along with 85% of the histones (Douvas et al. 1976b). The NaCl-dissociable nonhistones will be hereafter referred to as the soluble nonhistone fraction. DNA-protein complexes resistant to NaCl were dissociated for gel electrophoresis by dialyzing in a sodium dodecyl sulfate (SDS) containing buffer (2.5% SDS, 65 mM Tris-HCl pH 6.8). The DNA was then removed by centrifugation for 18 hr at room temperature in a Beckman Ti 50 rotor at 47,000 rpm.

Column chromatography of nonhistones. Sodium chloride dissociable proteins were separated into histone and nonhistone fractions by cation exchange chromatography on Bio-rex 70 (Bio-Rad laboratories) as described previously (Douvas et al. 1975a). For separation of the contractile proteins from other nonhistones under non-denaturing conditions the nonhistone proteins were dialyzed to 0.06 M phosphate buffer pH 6.8 and applied to a hydroxyapatite (HAP) column (100 μg protein per 1 c.c resin) equilibrated with the same buffer. The contractile proteins can be eluted from the column in 0.06 M phosphate buffer while the majority of other nonhistone species are retained by the resin. Further purification of contractile nonhistones under denaturing conditions was accomplieshed by gel filtration on Sephadex G-200 superfine or G-75 medium in 2.5% SDS, 65 mM Tris-HCl ph 6.8, applying 0.5 mg concentrated contractile proteins at room temperature to columns of dimensions 110 X 1.5 cm (G-200) and 62.5 X 0.8 cm (G-75).

Polyacrylamide gel electrophoresis. Ten percent polyacrylamide disc gels were run as described in detail elsewhere (Garrard et al. 1974). Samples were prepared by dialysis to 2.5% SDS, 65 mM Tris-Hcl pH 6.8, then boiled for 1 1/2 minutes with 5% (v/v) β-merceptoethanol. After electrophoresis at 2 milliamps per gel, the gels were fixed in 50% trichloroacetic acid and 7% acetic acid, and stained with Coomassie brilliant blue. Gels were scanned on a Gilford 2000 spectrophotometer at 600 mM.

RESULTS AND DISCUSSION

Isolation and Characterization of Some Major Nonhistone Chromatin Proteins

Of the 115 to 120 nonhistone chromatin proteins which have been resolved on one-dimensional gels, fewer than 12 make up the majority of the mass (Garrard et al. 1974). Our goal was to isolate and characterize as many of these major proteins as possible. In the present report we describe progress in characterizing eight polypeptides which together account for 49.5% of the mass of chromosomal nonhistones. We summarize evidence that five of the eight are contractile or structural proteins already known to exist in specialized cells (muscle or in other cell organelles (microtubules). These are: myosin (200,000 daltons), actin (45,000 daltons), two tropomyosin subunits (34,000 and 32,000 daltons), and tubulin (50,000 daltons).

Dissociation of up to 70% of the chromosomal proteins from DNA can be accomplished by extracting chromatin with 4 M NaCl. The extract is then fractionated by Bio-Rex chromatography to separate the bulk of the nonhistones from the histones. The histone fraction contains a small quantity of basic nonhistone protein. Polyacrylamide gels of these fractions are shown in Figure 2-1. The electrophoretic profile of gels of the salt-extractable nonhistones is compared to that of the unextractable fraction in Figure 2-2. The gels were scanned, and the curves resolved to show the eight polypeptides of interest. The distribution, by mass, of the eight polypeptides in the NaCl-soluble and insoluble fractions, respectively, is given in Table 2-1. The proteins are listed by molecular weight. Two polypeptides (37,500 and 28,000 daltons) are entirely salt extractable, while the remaining six are preferentially associated with the DNA. Some of the six remaining proteins can be selectively dissociated from chromatin with urea (Douvas et al. 1976a). All isolation procedures described here, however, employ the NaCl-extracted proteins.

"Actin" and "myosin" (200,000 and 45,000 dalton polypeptides) were isolated from the soluble nonhistone fraction by the method outlined in Figure 2-3. This method yields as by-products a 65,000 dalton polypeptide and a means for isolating the two "tropomyosin" subunits (34,000 and 32,000 daltons). The 200,000 and 45,000 dalton proteins are precipitated from the soluble nonhistone fraction by a procedure similar to that used to precipitate actomyosin from muscle extracts (for details see Douvas et al. 1975a). In order to efficiently precipitate all of the 45,000 dalton protein it was necessary to add muscle myosin before dialysis of the nonhistones to 0.025 M KCl. The "actomyosin" precipitate was dissolved in 1 M KCl and the 45,000 dalton protein was further purified by either gel filtration or by repeated polymerization with 5 mM ATP, 5 mM $MgCl_2$ (the procedure used to polymerize muscle actin from the globular to the fibrous form) (Drabikowski and Gergely 1962; Kuehl and Gergely 1969). The bulk of the nonhistones, which remain soluble on dialysis to 0.025 M KCl, were applied to a hydroxyapatite column in 0.06 M phosphate buffer, pH 6.8. Eluting from the column in 0.06 M phosphate were three polypeptides of molecular weights 65,000, 34,000, and 32,000, and some low molecular weight material ($\leqslant$17,000) which can not be resolved on a 10% acrylamide gel (Figure 2-3). This eluate when subjected to gel filtration on Sephadex G-75 yields three fractions containing the 65,000 dalton protein, the 34,000 and 32,000 dalton polypeptides, and the low molecular weight material, respectively.

When chromosomal proteins are separated into histones and nonhistones on Bio-rex 70, less than 5% of the nonhistone mass is recovered in the histone fraction. The major nonhistone found in the histone fraction is a 37,500 dalton polypeptide, shown in Figure 2-1. The method for purifying this protein is outlined in Figure 2-4. The apparent basic nature of this protein led us to suspect that it may be related to basic proteins found in the nucleus which are involved in packaging heterogeneous nuclear RNA (Martin et al. 1973). The RNA packaging proteins range in molecular weight from 34,000 to 40,000 daltons, and have been found in chromatin (Bhorjee and Pederson 1973), as well as in the nucleoplasm in ribonucleoprotein particles (RNP). Disc polyacrylamide gels of the 37,500 dalton chromatin protein and the basic polypeptides from mouse ascites RNP are compared in Figure 2-5. Identical molecular weights are calculated for one of the major bands in the RNP fraction and the basic chromosomal protein. Further evidence that the 37,500 dalton chromatin is involved in packaging RNA transcripts is seen in comparing the proteins of chromatin that has been fractionated into "active" and "inactive"

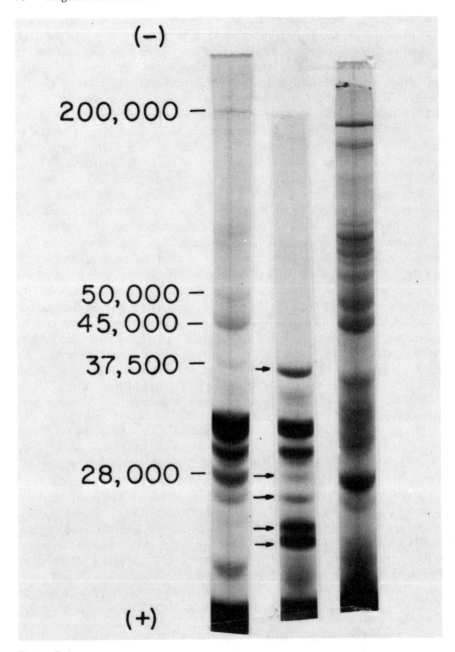

Figure 2-1.
SDS-polyacrylamide gel patterns of sodium chloride dissociated chromosomal proteins. From left to right: the 4 M NaCl dissociated chromatin proteins; the proteins retained on Bio-rex 70 in 0.4 M NaCl, including all of the histones and some nonhistones (indicated by arrows); the NaCl dissociated nonhistones.

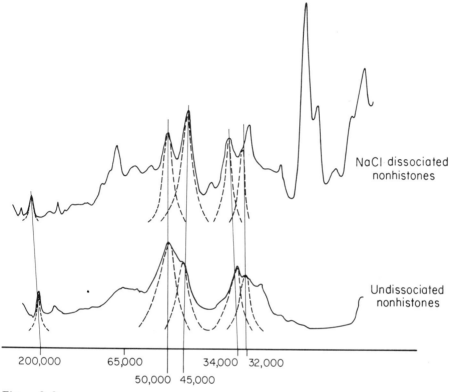

Figure 2-2.
Gel scans, from top to bottom, of sodium chloride dissociated and undissociated nonhistones. Broken lines indicate the resolved areas corresponding to polypeptides of molecular weight 200,000, 50,000, 45,000, 34,000, and 34,000 daltons, respectively.

portions by DNase II digestion. A study with radioactive ribonucleotides shows that 85% of the new transcipts are found in the active fraction. The 37,500 dalton protein is very selectively associated with this active fraction (Gottesfeld et al. 1974).

A nonhistone polypeptide of molecular weight 28,000 is present in the nucleus in 1.7×10^6 copies per haploid genome—more copies than any other chromosomal nonhistone (Garrard et al. 1974). In attempting to purify this protein we have noted a number of unusual properties: (1) it is quantitatively dissociated from chromatin by 0.5 M NaCl at pH 8.0 (Douvas et al. 1976b), a property that it shares with histone I; (2) it is completely resistant to digestion by the nuclear protease (Chong et al. 1974); (3) it is selectively associated with the template inactive fraction of DNase II-digested chromatin (Gottesfeld et al. 1974). Its resistance to protease digestion raised the possibility that it is already a limit-digest of a larger protein (or proteins). Although our chromatin is treated with DFP in early stages of

Table 2–1

Distribution of major nonhistone proteins in NaCl extractable and unextractable fractions of chromatin

Protein (by molecular weight)	% of total nonhistone mass	% NaCl extractable	% unextractable
200,000	1.12	20	80
65,000	5.1	48	52
50,000	12.84	35	65
45,000	7.72	42	58
37,500	1.8	100	0
34,000	6.3	35	64
32,000	4.7	33	67
28,000	10.0	100	0
	49.58		

These data were calculated both by computor, and manually. Computor calculation involved resolution of gel scans into component peaks and estimation of the areas under the peaks of interest relative to the total area. Alternatively the curves were fitted manually, and the area estimated by cutting out and weighing the paper. The two methods were in agreement within 0–3%.

isolation to prevent proteolysis, it is nonetheless possible that breakdown of protein occurs in the tissue before chromatin isolation. Figure 2–6 shows disc polyacrylamide gels of partially purified 28,000 dalton protein from chromatin and muscle myosin which has been incubated with a small amount of non-DFP treated nonhistones (which contain an active protease). All of the myosin, which has a molecular weight of approximately 200,000, has been converted to small fragments, one of which electrophoreses with the same apparent molecular weight as the 28,000 dalton chromatin protein. Besides the correspondence in molecular weights, however, we have as yet no other evidence that indicates that the 28,000 dalton protein is in whole or partly a breakdown product of myosin. Comparisons of the purified protein with myosin are now in progress.

A summary of the properties of the seven polypeptides we have isolated or are in the process of purifying is given in Table 2–2. Details on the characterization of some of these proteins appear elsewhere (Douvas et al. 1975a). An eighth polypeptide, of molecular weight 50,000 daltons, which we tentatively identified as tubulin, is also listed although we have not attempted to purify it. We have reported elsewhere that there is a correlation between the colchicine binding capacity of chromatin and the amount of 50,000 dalton polypeptide (Douvas et al. 1975a). The identification of the 200,000 and 45,000 dalton proteins as myosin and actin, respectively, is based on a number of demonstrated similarities between the chromosomal and muscle proteins (Table 2–2). We have now to consider the possible function of these proteins in chromatin.

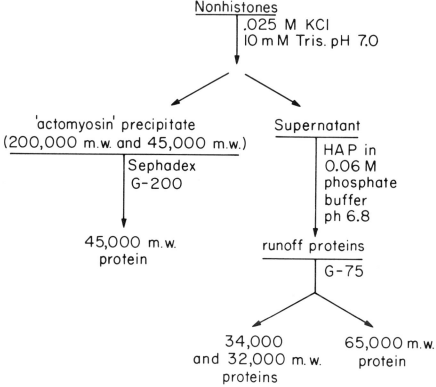

Figure 2-3.
Outline of method for purifying chromosomal actin and three other polypeptides (65,000, 34,000, and 32,000 daltons).

Contractile Proteins and Gene Activity

Contractile proteins have been identified in a number of non-muscle cell structures and organelles in eukaryotes, including the nucleus. Since its discovery in slime mold (Loewy 1952; Ts'o et al. 1956; Hatano and Tazawa 1968), the presence of actin has been reported in a variety of non-muscle cell types (Yang and Perdue 1972; Lazarides and Weber 1974; Sanger 1975). Recently, by use of fluorescent antibodies to actin, Lazarides and Weber (1974) have demonstrated its presence in cytoplasmic microfilaments. Non-muscle cells also contain a significant quantity of myosin (Adelstein et al. 1972; Ostlund et al. 1974; Weber and Groeschel-Stewart 1974). Like actin, the bulk of the myosin appears to be found in the cytoplasm (Ostlund et al. 1974). However, both of these contractile proteins were found in nuclei as early as 1963 (Ohnishi et al. 1963) and very recently in chromatin (Douvas et al. 1975a; LeStourgeon et al. 1975). Actin has been extracted from isolated nuclei of *Physarum polycephalum* (Jockusch et al. 1971), and actin and myosin filaments have been seen in mitotic nuclei of this organism by microscopy (Jockusch et al. 1970; Jockusch et al. 1973). It was postulated that actomyosin

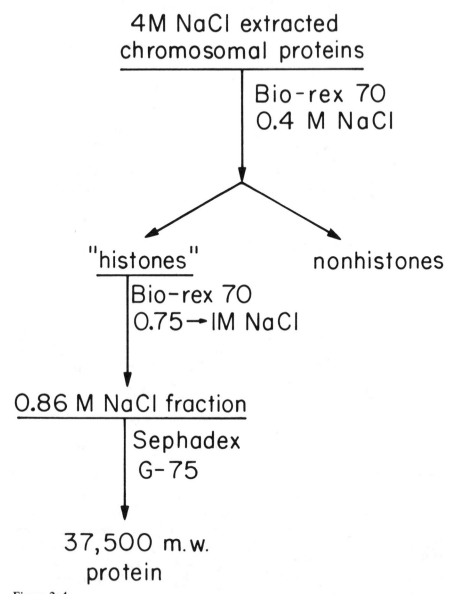

Figure 2-4.
Outline of method for separating NaC1 dissociated proteins into histone and non-histones, and for purifying 37,500 dalton protein.

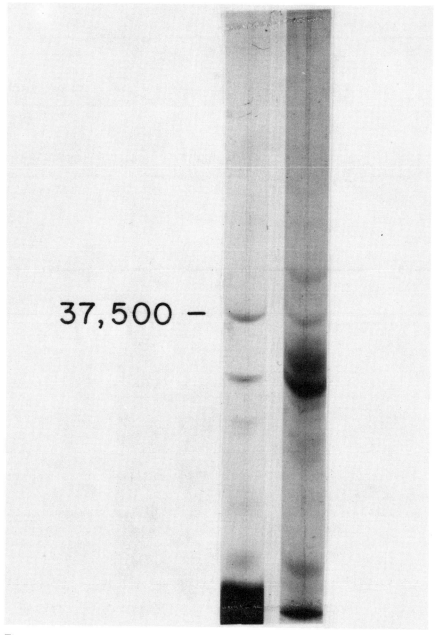

37,500 —

Figure 2-5.
SDS-polyacrylamide gells, from left to right, of partially purified 37,500 dalton protein (0.86 M fraction, Figure 2-4) and mouse ascites RNP preparation.

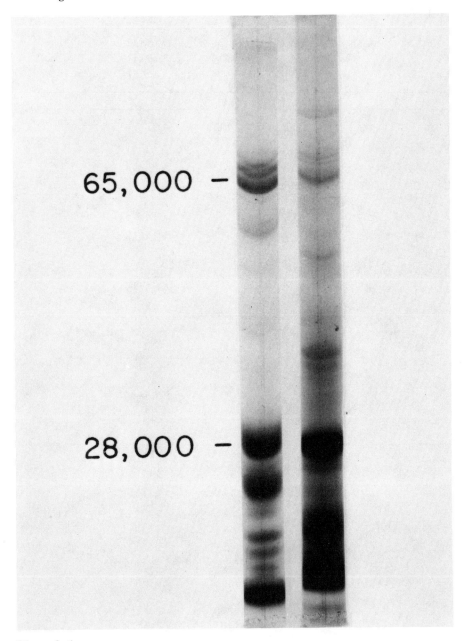

Figure 2-6.
SDS-polyacrylamide gels, from left to right, or protease-digested muscle myosin and partially purified 28,000 dalton polypeptide from chromatin. Digestion of the myosin resulted from incubation with 5% by weight non-DFP treated nonhistones at 4° for 20 hours.

Table 2-2
Properties of major nonhistones

Protein (by molecular weight)	Identity	Properties	Reference
200,000	myosin	(a) coelectrophoreses with muscle myosin (b) forms "actomyosin" precipitate with 45,000 dalton protein. The precipitate has ATPase activity	Douvas et al. 1975(a)
65,000	unknown		Douvas et al. 1975(a)
50,000	tubulin	(a) same molecular weight as tubulin (b) The amount of this protein is directly related to the colchicine binding activity of chromatin	
45,000	actin	(a) coelectrophoreses with muscle actin (b) forms an "actomyosin" precipitate with either muscle myosin or with the 200,000 dalton chromatin polypeptide (c) can be polymerized and depolymerized under conditions that interconvert globular and fibrous actin	Douvas et al. 1975(a) LeStourgeon et al. 1975
37,500	RNP protein	(a) basic protein which elutes from Bio-rex with the histone fraction (b) associated only with the template active fraction of chromatin (c) coelectrophoreses with RNA packaging proteins from purified RNP particles	this report; Gottesfeld et al. 1974 this report

(Continued)

Table 2-2 (continued)

Protein (by molecular weight)	Identity	Properties	Reference
34,000 32,000	tropomyosin subunits	(a) coelectrophorese with tropomyosin subunits from muscle (b) co-purifies with chromosomal actin	Douvas et al. 1975 (a)
28,000	possible breakdown of myosin and other proteins	(a) resistant to nuclear protease (b) quantitatively dissociated from chromatin in 0.5 M NaCl (c) coelectrophoreses with myosin breakdown products	Jockusch et al. 1970 Fleisher-Lambropoulos et al. 1970 this report

may function to constrict the nuclear membrane of *Physarum* during telophase, and/or was possibly involved in chromosome movement. The involvement of contractile protein in chromosome movement during mitosis is further suggested by the recent demonstration of actin in the kinetochores, centrioles, and mitotic spindles of the kangaroo rat (Sanger 1975).

The abundance of actin and myosin in non-muscle cells, and particularly in structures in and around the nucleus, is a condition that favors contamination of chromatin by these proteins during isolation. While we cannot exclude the possibility of some extrachromosomal contamination, there is evidence that both actin and myosin are bonafide chromosomal proteins (Douvas et al. 1975a; LeStourgeon et al. 1975). It is possible that actin and myosin as genuine chromosomal proteins function exclusively as part of the mitotic apparatus. Neither in this case nor in the case of extrachromosomal contamination would their presence be specifically related to the expression of gene activity per se. One might expect to see either increased synthesis or greater absolute amounts of actin and myosin in actively dividing cells if their function were exclusively limited to mitosis. However, comparisons of dividing and nondividing cells, which will be described in detail below, indicate that chromatin from nondividing cells contains the larger amount of actin.

Chromosomal nonhistones are a metabolically active class of proteins which in response to stimuli undergo changes in synthesis, phosphorylation, and quantitative shifts in proportion to the genomic DNA (see Spelsberg et al. 1972 and Johnson et al. 1974 for review). In general, stimuli which induce non-proliferating cells into a state of active growth and division result in more drastic changes in chromatin nonhistone patterns than those which induce new or increased synthesis of specific gene products (hormones are typical of the latter). To study the involvement, if any, of contractile proteins in gene expression an optimal experimental system might be a hormonally responsive cell culture in which the stimulus induces the expression of a few genes. These experiments have not been done. However, it is known that some genes are expressed only in one tissue of a given organism; for example, in mammals the protein S100 is produced only in the brain. In addition to the fact that different genes are expressed, brain-liver comparisons reveal that more of the genomic DNA is expressed in the brain (11% vs 4%, respectively) (Grouse et al. 1972), although it is not known if more of the DNA in any one cell type is expressed in the brain.

We surmised that if the contractile proteins in brain and liver chromatin are involved in maintaining either quantitative or qualitiative differences in gene expression, there might be detectable differences in their banding patterns on polyacrylamide gels. Comparisons of the chromosomal nonhistones from these two tissues are published elsewhere (Douvas et al. 1976b). Several quantitative differences are readily apparent. Of particular interest are the peaks of 200,000 and 45,000 daltons which in liver chromatin correspond to myosin and actin, respectively. As there appeared to be relatively less of these two peaks in brain chromatin, a quantitative comparison was undertaken to estimate what fraction they comprise of the chromosomal proteins applied to the gels in brain and liver respectively. We found that liver chromatin contains approximately 2.3-fold more of the 45,000 dalton protein and 1.8-fold more of the 200,000 dalton protein than brain. A 28,000 dalton component also represents a higher proportion of the total in liver chromatin, but the difference is not as great. Estimations of the proportions of 34,000 and 32,000 dalton proteins was difficult due to the presence of other peaks of nearly the same molecular weight.

Given the nonhistone/DNA ratios are nearly the same in brain and liver (Elgin and Bonner 1970; Bonner et al. 1968; Shaw and Huang 1970; Fleischer-Lambropoulos et al. 1974), the differences in 200,000 and 45,000 dalton proteins imply a twofold greater mass ratio of actin and myosin to genomic DNA in liver (assuming that the two proteins are homologous in brain and liver). This observation prompted us to search for other possible correlations between transcriptional activity and contractile protein content in chromatin. Unfortunately there is a paucity of reported studies in which such a comparison can be made because often proteins are not identified by molecular weights, or because there is no basis on which to assume there exists a homology between actin and myosin and any of the proteins under study. Three examples were found, however, in which changes in a chromosomal polypeptide of molecular weight 43,000–46,000 daltons are correlated with changes in cellular activity. In two of the three examples, a positive identification of actin has been made (LeStourgeon et al. 1975). Myosin was not included in this survey because of its extreme susceptibility to nuclear protease (Douvas et al. 1975b). The individual cases are detailed below:

(1) The slime mold *Physarum polycephalum* is a lower eukaryote with a simple developmental pattern. There is an actively growing plasmodial phase characterized by active synchronous growth and mitosis, and a metabolically quiescent encyst-

ment phase which can be induced by starvation and ultraviolet irradiation (LeStourgeon and Rusch 1973; LeStourgeon et al 1973). Major changes in four nonhistone chomatin proteins have been observed as cells pass from active growth into encystment back to active growth. Three polypeptides (41,000, 37,000, and 34,000) increase when encysted cells reestablish active growth, and one polypeptide of molecular weight 46,000 decreases severalfold. The latter protein was subsequently identified on the basis of several biochemical criteria as actin (LeStourgeon et al. 1975). When dormancy is induced in actively growing cells the 46,000 dalton polypeptide increases to its maximal concentration while the other three polypeptides decrease.

(2) In HeLa cells a dormant state can be induced by allowing the exponentially growing cells to deplete their medium. Growth and mitosis cease upon depletion of the medium, but the cells can be retained in culture for at least 48 hours and reactivated by dilution into fresh medium. Comparisons of nonhistone chromatin proteins between actively growing and medium-depleted cells show that a 46,000 dalton polypeptide is very prominent in the depleted cells and disappears during active growth (LeStourgeon et al. 1973).

(3) Mature goose erythrocytes are nucleated cells which make little protein or RNA and do not divide. Reticulocytes can be induced in response to injection of phenylhydrazine and such blood contains reticulocytes as well as mature erythrocytes. A comparison of nonhistone chromatin fractions in normal and regenerating blood reveals a diminution in the proportion of a 43,000 dalton polypeptide in the regenerating blood cell population (Shelton and Neelin 1971). While the identification of this polypeptide as actin is tenuous, it is, like actin in rat liver, HeLa cells, and slime mold chromatin, a major nonhistone protein.

These few examples point to an inverse correlation between chromosomal actin content and transcriptional activity. Because more data are not available yet, it is not possible to judge how general a phenomenon, and therefore how significant, this relation might be. However, given that actin is found on chromatin, the above observations suggest that this protein may be involved in a negative kind of control, i.e., in rendering some part of the genome unavailable for transcription. There are about 10^6 copies of actin per haploid rat genome as compared to 134×10^6 copies of the sum of all of the histones (Garrard et al. 1974). Thus, there is one actin molecule for every 2,800 base pairs of DNA as compared to one histone molecule for every 21 base pairs (the rat haploid genome size is 2.8×10^9 base pairs). Because of this stoichiometry and its moderately acidic character, it is very unlikely that actin masks a significant fraction of the DNA by direct protein-DNA interactions. However, actin and other contractile proteins may affect the availability of template for transcription more indirectly as structural proteins influencing the conformation of chromatin.

It is generally assumed that transcriptionally active regions of chromatin are relatively more extended and inactive regions more condensed. There is cytological evidence supporting this view in the lampbrush chromosome of *Triturus* in which transcription complexes are seen only in the extended loops of the chromosomes (Miller and Beatty 1969). The molecular mechanisms by which regions of the chromosome are maintained in a relatively condensed state are not understood. Recent evidence suggests that histones and DNA are packed into relatively dense structures know as "ν bodies" (Olins and Olins, 1974). These structures, which alternate with regions of extended DNA in a "beads on a string" arrangement have a higher packing ratio than the extended DNA (Griffith 1975). However, even higher orders of coiling or folding of nucleoprotein are necessary to account for the observed fiber diameters and packing ratio of both interphase and metaphase chromosomes (Olins and Olins 1974; DuPraw 1970). Actin and myosin, with their ability to reversibly form contractile fibers, may be ideally suited for maintaining folds or coils in the primary fibers of chromatin.

Our efforts to isolate and chracterize the major nonhistone proteins of chromatin are directed toward in depth examinations of the functions of individual proteins. The apparent contractile nature of the proteins described here suggests that a large proportion of the nonhistone mass may serve a structural purpose.

LITERATURE CITED

Adelstein, R. S., M. A. Conti, G. S. Johnson, I. Pastan, and T. D. Pollard, 1972. Isolation and characterization of myosin from cloned mouse fibroblasts. *Proc. Natl. Acad. Sci.* **69**:3693–3697.

Bhorjee, J. S. and T. Pederson. 1973. Chromatin: its isolation from cultured mammalian cells with particular reference to contamination by nuclear ribonucleoprotein particles. *Biochemistry* **12**:2766–2773.

Bonner, J., G. R. Chalkley, M. Dahmus, D. Fambrough, F. Fujimura, R. C. Huang, J. Huberman, R. Jensen, K. Marushigi, H. Ohlenbusch, B. M. Olivera, and J. Widholm. 1968. Isolation and characterization of chromosomal nucleoproteins. *Methods Enzymol.* **12**:3–97.

Chong, M. T., W. T. Garrard, and J. Bonner. 1974. Purification and properties of a neutral protease from rat liver chromatin. *Biochemistry* **13**:5128–5134.

Douvas, A. S., C. A. Harrington, and J. Bonner. 1975a. Major nonhistone proteins of rat liver chromatin: Preliminary identification of myosin, actin, tubulin and tropomyosin. *Proc. Natl. Acad. Sci.* **72**:3902–3906.

Douvas, A. S., C. A. Harrington, and J. Bonner. 1975b. Stabilization of nonhistone chromatin proteins against proteolysis. Manuscript in preparation.

Douvas, A. S., C. H. Harrington, and J. Bonner. 1976a. Selective dissociation of rat liver chromatin with sodium chloride and urea. Manuscript in preparation.

Douvas, A. S., A. Bakke, and J. Bonner. 1976b. Evidence for the presence of contractile-like nonhistone proteins in the nuclei and chromatin of eukaryotes. In *Proc. of the Internat. Symp. on Mol. Biol. of the Mammalian Genetic Apparatus.* Associated Scientific Publishers, in press.

Drabikowski, W. and J. Gergely. 1962. The effect of the temperature of extraction on the tropomyosin content in actin. *J. Biol. Chem.* **237**:3412–3417.

DuPraw, E. J. 1970. *DNA and Chromosomes*. Holt, Rinehart and Winston, Inc., New York.

Elgin, S. C. R. and J. Bonner. 1970. Limited heterogeneity of the major nonhistone chromosomal proteins. *Biochemistry* **9**:4440–4447.

Fleischer-Lambropoulos, H., H.-I. Sarkander, and W. P. Brade. 1974. Phosphorylation of nonhistone chromatin proteins from neuronal and glial nuclei-enriched fractions of rat brain. *FEBS Letters* **45**:329–332.

Garrard, W. T., W. R. Pearson, S. K. Wake, and J. Bonner. 1974. Stoichiometry of chromatin proteins. *Biophys. Res. Comm.* **58**:50–57.

Gottesfeld, J. M., W. T. Garrard, G. Bagi, R. F. Wilson, and J. Bonner. 1974. Partial purification of the template-active fraction of chromatin: A preliminary report. *Proc. Natl. Acad. Sci.* **71**:2193–2197.

Griffith, J. 1975. Chromatin structure: Deduced from a minichromosome. *Science* **187**:1202–1203.

Grouse, L., M. D. Chilton, and B. J. McCarthy. 1972. Hybridization of ribonucleic acid with unique sequences of mouse deoxyribonucleic acid. *Biochemistry* **11**: 798–805.

Hatano, S. and M. Tazawa. 1968. Isolation, purification, and characterization of myosin B from myxomycete plasmodium. *Biochim. Biophys. Acta* **154**:507–519.

Jockusch, B. M., D. F. Brown, and H. P. Rusch. 1970. Synthesis of a nuclear protein in G_2-phase. *Biochem. Biophys. Res. Comm.* **38**:279–283.

Jockusch, B. M., D. F. Brown, and H. P. Rusch. 1971. Synthesis and some properties of an actin-like nuclear protein in the slime mold physarum polycephalum. *J. Bacteriol.* **180**:705–714.

Jockusch, B. M., U. Ryser, and O. Behnke. 1973. Myosin-like protein in physarum nuclei. *Exper. Cell Res.* **76**:464–466.

Johnson, J. D., A. S. Douvas, and J. Bonner. 1974. Chromosomal proteins. *Int. Rev. Cytol. Supp.* **4**:273–361.

Kuehl, W. M. and J. Gergely. 1969. The kinetics of exchange of adenosine triphosphate and calcium with G-actin. *J. Biol. Chem.* **244**:4720–4729.

Lazarides, E. and K. Weber. 1974. Actin antibody: The specific visualization of actin filaments in non-muscle cells. *Proc. Natl. Acad. Sci.* **71**:2268–2272.

LeStourgeon, W. and H. P. Rusch. 1973. Localization of nucleolar and chromatin residual acidic protein changes during differentiation in physarum polycephalum. *Arch Biochem. Biophys.* **155**:144–158.

LeStrougeon, W., W. Wray, and H. P. Rusch. 1973. Functional homologies of acidic chromatin proteins in higher and lower eukaryotes. *Exper. Cell Res.* **79**:487–492.

LeStourgeon, W., A. Forer, Y. Z. Yang, J. S. Bertram, and H. Rusch. 1975. Contractile proteins, major components of nuclear and chromosome nonhistone proteins. *Biochim. Biophys. Acta* **379**:529–552.

Loewy, A. G. 1952. An actomysin-like substance from the plasmodium of a myxomycete. *J. Cell Comp. Phsyiol.* **40**:127–156.

Martin, T., P. Billings, A. Levey, S. Ozarslan, T. Quinlan, H. Swift, and L. Urbas. 1973. Some properties of RNA: Protein complexes from the nucleus of eukaryotic cells. *Cold Spring Harbor Symp. Quant. Biol.* **38**:921–932.

Marushigi, K. and H. Ozaki. 1967. Properties of isolated chromatin from sea urchin embryos. *Develop. Biol.* **16**:474–488.

Miller, O. L. and B. Beatty. 1969. Portrait of a gene. *J. Cell Physiol.* **74**:Supp. 225–232.

Ohnishi, T., H. Kawamura, and T. Yamamoto. 1963. Extraction eines dem aktin ähnleichen proteins aus dem zellkern des kalbsthymus. *J. Biochem.* (Tokyo) **54**:298–300.

Olins, A. L. and D. E. Olins. 1974. Spheroid chromatin units (υ Bodies). *Science* **183**:330–332.

Ostlund, R. E., I. Pastan, and R. S. Adelstein. 1974. Myosin in cultured fibroblasts. *J. Biol. Chem.* **249**:3903–3907.

Sanger, J. W. 1975. Presence of actin during chromosomal movement. *Proc. Natl. Acad. Sci.* **72**:2451–2455.

Shaw, L. M. J. and R. C. Huang. 1970. A description of two procedures which avoid the use of extreme pH conditions for the resolution of components isolated from chromatins prepared from pig cerebellar and pituitary nuclei. *Biochem.* **9**:4530–4542.

Shelton, K. R. and J. M. Neelin. 1971. Nuclear residual proteins from goose erythroid cells and liver. *Biochem.* **10**:2342–2348.

Spelsberg, T. C., J. A. Wilhelm, and L. S. Hnilica. 1972. Nuclear proteins in genetic restriction II. The nonhistone proteins in chromatin. *Subcell. Biochem.* **1**:107–145.

Ts'o, P. O. P., J. Bonner, L. Eggman, and J. Vinograd. 1956. Observations on an ATP-sensitive protein system from the plasmodia of a myxomycete. *J. Gen. Physiol.* **39**:325–347.

Weber, K. and U. Groeschel-Stewart. 1974. Antibody to myosin: The specific visualization of myosin-containing filaments in nonmuscle cells. *Proc. Natl. Acad. Sci.* **71**:4561–4564.

Wu, F. C., S. C. R. Elgin, and L. Hood. 1973. Nonhistone chromosomal proteins of rat tissues. A comparative study of gel electrophoresis. *Biochemistry* **12**:2792–2797.

Wu, F. C., S. C. R. Elgin, and L. Hood. 1975. Nonhistone chromosomal proteins of vertebrate kidney and liver: A comparative study by gel electrophoresis. *J. Mol. Evol.*, in press.

Yang, Y. Z. and J. F. Perdue. 1972. Contractile proteins of cultured cells. *J. Biol. Chem.* **247**:4503–4509.

The Regulation of the Plant Cell Cycle by Cytokinin

D. E. Fosket   University of California, Irvine

INTRODUCTION

The cytokinins are a group of naturally occurring plant growth regulators. Although they have been implicated in the regulation of a diverse array of morphological, physiological, and biochemical processes in higher plants (cf. Kende 1971; Skoog and Armstrong 1970; Hall 1973), they are characterized by their ability to initiate cell division in auxin-treated tobacco pith tissue (Skoog et al. 1965). The naturally occurring cytokinins are N^6-substituted adenine derivatives, and zeatin [6-(4-hydroxy-3-methyl-trans-2-butenylamino)purine] is found in a wide variety of plant tissues and organs (Skoog and Armstrong 1970). Although structures exhibiting high mitotic activity have proven to be a particularly rich source of free cytokinins for extraction (Miller 1961; Letham 1963; Zwar et al. 1963; Weiss and Vaadia 1965, Letham and Williams 1969; Blumenfeld and Gazit 1970; Short and Torrey 1972), there is little direct evidence that cytokinins regulate the cell cycle in meristematic tissues of the intact plant.

Nevertheless, there are numerous studies which have demonstrated that cytokinins may play a unique and highly specific role in the regulation of the plant cell cycle. All of these studies have dealt with cultured tissues and cells which require cytokinin for cell proliferation *in vitro*. Two types of experimental systems have been employed for these investigations: (1) mature, differentiated, nondividing tissues in which hormone treatment initiates cell proliferation, and (2) continuously dividing cell cultures in which cytokinin is necessary to maintain cell division activity. Although the results of these studies are not in complete agreement, the most frequent finding is that cytokinin triggers essential events in G_2, or in the transition from G_2 to mitosis in the plant cell cycle. Furthermore, recent work has shown that cytokinin regulates protein synthesis at the translational level and that protein synthesis is necessary for the cytokinin-triggered cell cycle events. This suggests that a cell's progress through its division cycle may be determined, at least in part, through translational control of protein synthesis.

INITIATION OF CELL DIVISION BY CYTOKININ

The cytokinins were discovered by Skoog and his coworkers in the course of their work on the growth of excised, cultured tobacco pith tissue (Miller et al. 1956). As a consequence, considerable effort has been devoted to elucidating the sequence of cellular events that occur during hormone-initiated cell proliferation in this system.

The pith tissue in fully elongated internodes of tobacco consists of large, highly vacuolate, mitotically inactive cells (Nitsch and Lance-Nougaréde 1967). It could be argued that pith cells are terminally, although obviously not irreversibly, differentiated. Normally they stop dividing before internode elongation is complete and they would not divide again during the life of the plant. Feulgen microspectrophotometry has demonstrated that 5% of the pith nuclei have 2C, 47% have 4C, and 48% have 8C levels of DNA (Patau et al. 1957). When pith tissue explants are placed in culture on a medium lacking auxin and cytokinin, about 20% of the nuclei undergo DNA synthesis, as shown by ^{3}H-thymidine labeling and autoradiography, but the cells do not go on to divide. When the explants are cultured on a medium containing auxin alone, cytokinin alone, or both auxin and cytokinin, a considerably higher percentage of the nuclei undergo DNA synthesis during the first 2 days of the culture period. However, neither auxin nor cytokinin treatments alone are sufficient to bring about sustained cell division. The initiation and maintenance of mitotic activity requires the presence of both auxin and cytokinin in the culture medium (Das et al. 1956, 1958; Patau et al. 1957).

Which Cells Respond to the Combined Hormone Treatment and Initiate Cell Division?

Treatment with both auxin and cytokinin for several days leads to the formation of meristematic clumps of small cells in which the nuclear DNA content is within the 2C–4C range expected of dividing, diploid cells. These nests of meristematic cells are separated by large, vacuolate pith cells which rarely divide, but when mitotic figures are observed they are usually polyploid. Only diploid cells are continually meristematic. The data of Patau et al. (1957) and Nitch and Lance-Nougaréde (1967) show that cells stimulated to enter the mitotic cycle by auxin-cytokinin treatment were either the small fraction of the pith cell population which had left the cell cycle from G_1 (the 5% of the cells which initially had 2 C levels of DNA), or some fraction of the 4C population which was in prolonged G_2, or both. That at least some of the 4C cells represented a subpopulation in prolonged diploid G_2 was strongly suggested by an experiment of Patau and Das (1961) in which tobacco pith explants were continuously exposed to ^{3}H-thymidine during treatment with both auxin and cytokinin. After 2 days in culture they observed that nearly 50% of the cells which were entering mitosis for the first time in culture were unlabeled after autoradiography. Since only diploid cells continue to divide, we can conclude that the combined hormone treatment was ineffective in either initiating or sustaining mitotic activity from the nearly 50% of the pith cell population with an 8C level of DNA, and possibly also from any fraction of the 4C cell population which represented G_1 tetraploids.

These observations suggest that a combined auxin and cytokinin treatment of tobacco pith tissue stimulates diploid cells to enter the division cycle by two pathways; (1) induction of mitosis and cytokinesis in the diploid G_2 population, without an intervening round of DNA replication, and (2) the initiation of DNA replication followed by mitosis and cytokinesis in the G_1 diploid cell population. This implies that the hormone treatment overcomes fairly simple cell blocks in G_1 and G_2. That this may be an oversimplification is suggested by the fact that the combined hormone treatment induces profound ultrastructural alterations of tobacco pith cells which both precede and accompany the initiation of cell division (Nitsch and Lance-Nougaréde 1967; Gifford and Nitsch 1969). Thus, it could be argued that the hormones are acting as permissive factors, altering the physiological state of the cells (bringing about "dedifferentiation") so that cell division can occur, rather than as specific regulators of the cell cycle. However, it has not been shown that these ultrastructural changes are necessary for the initiation of cell division. Furthermore, these arguments overlook an important characteristic of this hormone-induced cell division response. Tobacco pith cells* require auxin and cytokinin not only for the initiation of mitotic activity, but also for continued cell proliferation in culture. Furthermore, a number of continuous cell lines of cultured tobacco and soybean cells have been isolated which have an absolute cytokinin requirement for cell proliferation (Miller 1961; Fosket and Torrey 1969; Tandeau de Marsac and Jouanneau 1972). In several cases, variant cell lines no longer requiring cytokinin have been isolated from hormone-dependent strains and these variants have been shown to have regained the capacity to synthesize their own cytokinins (Miura and Miller 1969; Dyson and Hall 1972; Einset and Skoog 1973).

Cytokinin-induced DNA Synthesis

Patau et al. (1957) concluded that cytokinins were required during three phases of the cell cycle: the initiation of DNA synthesis, the initiation of mitosis, and the completion of cytokinesis. Their evidence, which suggested that cytokinin may regulate events leading to the initiation of DNA synthesis, was that treatments of tobacco pith explants with cytokinin alone increased the average nuclear DNA con-

*It should be noted that pith tissue derived from *Nicotiana glauca* does not require cytokinin for growth in culture (Hagen and Marcus 1975). The behavior described above pertains to pith tissue obtained from varieties of *Nicotiana tobacum* (Skoog et al. 1965; Simard 1971).

tent, as compared to controls receiving no hormones, during the first 2 days of the culture period. Simard (1971) has shown that the DNA synthesis which occurs in control pith explants is a wound response, and that cytokinin augments this response. He observed a substantial reduction in the percentage of labeled nuclei when the excised pith segments were aged for 3 days before they were placed in culture. Using aged tissues, he found that cytokinin alone had no effect on DNA synthesis, whereas the combined auxin-cytokinin treatment, or auxin alone, stimulated the incorporation of ^{3}H-thymidine into a high percentage of the nuclei. Recent work of Wardell (1975) has confirmed that auxin treatment alone rapidly induces DNA synthesis in tobacco stem tissue.

Thus, cytokinin does not appear to regulate the events which lead to DNA synthesis in tobacco. However, there are experimental systems in which cytokinin has been shown to initiate cell division by activating DNA synthesis. The best example is Torrey's careful investigation of the response of cultured pea root explants to cytokinin. Torrey (1961) found that cytokinin specifically stimulated polyploid divisions from the cortical cells of a 1 mm explant taken from the region 10-11 mm from the pea root tip. When these segments were cultured on a medium containing auxin, but lacking cytokinin, proliferation occurred from cells in the stele and pericycle, but not the cortex. Virtually all of these mitotic figures were diploid. Libbenga and Torrey (1973) demonstrated that 65% of the cortical cell nuclei had a 2C level of DNA, while the remainder were at the 4C level at the time of excision. Culturing cortical explants on medium containing auxin, but no cytokinin, did not alter this distribution and no mitoses were observed. Thus, auxin alone did not induce DNA synthesis in these cells. When the cortical explants were cultured for 3 days on medium containing both auxin and cytokinin, the 2C population disappeared and the bulk of the nuclei were found to have an 8C level of DNA, with a few at the 16C level. Earlier, Matthysse and Torrey (1967) demonstrated that the potent inhibitor of DNA synthesis, 5-fluorodeoxyuridine (FUdR), completely blocked cytokinin-induced cell division. Thus, cytokinin treatment in the presence of auxin led to the initiation of two rounds of DNA replication before the cortical cells appeared as polyploid mitotic figures. Cytokinin was unable to trigger the division of diploid G_2 cells, should any of the cortical cells with a 4C level of DNA represent such a population. Torrey has not tested the effect of cytokinin alone on DNA replication, but clearly auxin alone did not initiate DNA replication in the cortical explants, as it has been found to do in tobacco. It is apparent that cytokinin is necessary for the initiation of DNA replication in pea root cortical tissue. However, it remains to be shown that cytokinin treatment specifically overcomes a G_1 block, as opposed to activating a series of physiological and metabolic changes which are necessary for cortical cells to enter the mitotic cycle and initiate DNA synthesis.

THE ROLE OF CYTOKININ IN CELL CYCLE REGULATION

The action of cytokinin in the initiation of cell division from mature, differentiated plant tissues appears to be complex. From the evidence presented above it would be difficult to state with certainty that the hormone plays any particular role in cell cycle regulation, although the fact that the hormone is required for continued cell proliferation argues that cytokinin does more than simply initiate "dedifferentiation." A clearer picture of the possible significance of cytokinin in cell cycle regulation has been obtained from studies on the action to the hormone in maintaining cell division in cultured cells which require cytokinin for proliferation. These studies specifically implicate the hormone in regulating the transition from G_2 to mitosis.

Cytokinin-regulated Cell Proliferation in Cultured Soybean Cells

We have isolated callus tissue from the cotyledons of a number of soybean varieties. Some of these, such as callus derived from *Glycine max* cv. Funk Delicious, are merely stimulated by cytokinin and growth will persist through several subculturings onto medium lacking cytokinin. Others, such as callus derived from *G. max* cv. Sodifuri and cv. Biloxi, have an absolute cytokinin requirement for growth (Tepfer and Fosket unpublished data—see Figure 3-1). In all cases so far examined, the cultures cease to produce differentiated cells and lose any semblance of organization after a variable number of subculturings. They grow on the surface of an agar-solidified medium as a loose aggregate of cells and in liquid cultures as a suspension of single cells and small cell clumps. The medium employed is the simple, chemically defined soybean callus medium devised by Miller (1961), as modified by Fosket

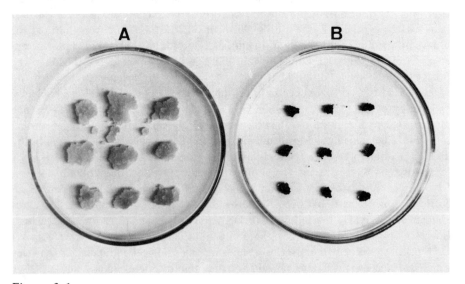

Figures 3-1.
The effect of cytokinin on the growth of cultured soybean cells (*Glycine max* cv. Sodifuri). Twenty-one-day-old soybean tissue was subcultured to freshly prepared medium which either contained (A) or lacked (B) the cytokinin zeatin at a concentration of 5×10^{-7}M. The cells were cultured for 6 days on their respective medium before being photographed.

and Torrey (1969). It contains, in addition to inorganic salts and sucrose, an auxin (indoleacetic acid), a cytokinin (zeatin), myo-inositol, and vitamins.

While the deletion of many components from the medium eventually will halt the growth of Sodifuri or Biloxi cells, cytokinin plays a particularly dramatic role in regulating the proliferation of these cells (Table 3-1). With Biloxi, the cell population doubling time was found to be a direct function of the cytokinin concentration. At 10^{-6} M zeatin, the cell population doubling time was 25 hours, while cells grown in the presence of 10^{-8} M zeatin had a doubling time of 64 hours during the log phase of growth. No growth occurred with concentrations of zeatin below 10^{-9} M (Fosket, unpublished results).

When Biloxi cells were transferred to medium lacking cytokinin, cell proliferation ceased within 48 hours. However, the cytokinin-deprived cells continued to incorporate ^{3}H-thymidine into DNA. Even after 6 days in culture the specific activity of extracted DNA after ^{3}H-thymidine labeling was considerably higher in cells cultured on medium lacking cytokinin than in the cytokinin-treated cells (Table 3-2). That

Table 3-1

Changes in the mitotic index of soybean cells with time in culture

	Mitotic Index	
Time after innoculation, days	*medium plus cytokinin*	*medium lacking cytokinin*
1	6.91	0.19
2	4.10	0.11
3	3.28	0
4	3.20	0
5	3.43	—
6	1.49	—
7	1.13	—
8	1.16	—
9	0.71	—
10	0.21	—
12	0.08	—
14	0	—

Note: Five-day-old log phase cells were subcultured to freshly prepared medium which either contained or lacked 5×10^{-7} M zeatin. At intervals thereafter, some of the cells were fixed, Feulgen squashes prepared, and the mitotic index determined [Mitotic index = (number of cells in mitosis)/(total number of cells) $\times$ 100] (from Short, Tepfer, and Fosket 1974).

Table 3-2

The effect of cytokinin on ^{3}H-thymidine and incorporation into DNA

Zeatin M	Uptake		Incorporation			
	Radioactivity soluble in 80% ETOH, cpm/explant		cpm/explant		cpm/µg DNA	
	Phenol	PCA	Phenol	PCA	Phenol	PCA
0	37,916	39,073	4,173	3,814	3,102	1,624
10^{-7}. . . .	29,117	30,457	4,541	3,397	2,310	1,125
10^{-5}. . . .	24,343	26,935	2,170	3,062	839	852

Note: Twenty-one-day-old Biloxi soybean tissue was subcultured to freshly prepared media containing or lacking the cytokinin zeatin. After 132 hours the cells
were transferred to conditioned medium containing 2µCi/ml ^{3}H-thymidine.
Twelve hours later the tissues were harvested. Hemogenization in 80% ethanol was
used to determine uptake, while incorporation was determined by extracting
nucleic acids with either phenol-sodium dodecyl sulfate or hot perchloric acid
(see Fosket and Short, 1973, for detailed methods).

this ^{3}H-thymidine incorporation in cytokinin-deprived cells actually represented
nuclear DNA synthesis was shown by the fact that the bulk of the label cosedi-
mented with nuclear DNA upon CsCl density gradient centrifugation, and it was
found to be localized over the nucleus upon autoradiography (Fosket and Short
1973). The inverse relationship between the rate of cell proliferation and the specific activity of the DNA after ^{3}H-thymidine labeling in cells grown in the presence
of different cytokinin concentrations appears to be the result of two factors: the
expansion of the intracellular pools of soluble nucleotides, and the inhibition of
^{3}H-thymidine uptake by the hormone (Fosket and Short 1973).

We have determined the DNA content of soybean cells grown in the presence and
absence of cytokinin by two different chemical methods. While the two methods
gave somewhat different quantitative results, with a given procedure cells grown in
the presence of cytokinin had similar nuclear DNA contents, while cells which had
been deprived of cytokinin for 144 hours had nearly twice as much DNA per cell
(Table 3-3). Clearly the cytokinin-requiring soybean cells did not accumulate
behind a block at the initiation DNA synthesis when they left the mitotic cycle
after being deprived of the hormone.

Similar conclusions have been reached by Jouanneau and Tandeau de Marsac
(1973). Jouanneau (1971) succeeded in synchronizing the division of cytokinin-
requiring tobacco cell suspensions by withholding cytokinin from the cells for a
period of time and then supplying the cells with it again. Jouanneau and Tandeau
de Marsac (1973) found that the inhibitor of DNA synthesis, FUdR, did not prevent the cytokinin-induced events which culminated in the first synchronous wave
of cell division in this system, although the inhibitor was highly effective in block-

Table 3–3.
Quantitative determination of the DNA content of soybean cells cultured with and without cytokinin

		Total extractable DNA µg/24 explants		DNA µg/10^6 cells	
Zeatin M	*Final cell number/ explants*	*Phenol-SDS extraction procedure*	*Hot 1 M perchloric acid extraction procedure*	*Phenol-SDS extraction procedure*	*Hot 1 M perchloric acid extraction procedure*
0	63,578	32.29	56.35	2.24	3.90
10^{-7}	140,695	47.19	72.48	1.40	2.15
10^{-5}	181,432	62.10	86.25	1.44	1.99

Note: Twenty-one-day-old Biloxi callus was transferred to freshly prepared media. Six days after inoculation, the tissues were harvested. Cell numbers and the DNA content of the tissues were determined (see Fosket and Short, 1973, for specific methods).

ing both DNA synthesis and mitosis in exponentially growing tobacco cell suspensions. Furthermore, deletion of cytokinin from the culture medium did not markedly affect the amount of ^{3}H-thymidine incorporated into nuclear DNA, at least for the duration of one cell cycle, while omitting auxin from the medium prevented incorporation.

Thus, the cytokinin-regulated events which are necessary for continued progress of cells through their division cycle seem to be localized in G_2, or in the transition from G_2 to mitosis, in both tobacco and soybean cell cultures. One peculiarity of the soybean system, however, is that while cytokinin appears to regulate events which are necessary for the transition from G_2 to mitosis, the cells do not appear to accumulate in G_2 after a prolonged absence of the hormone. Soybean cells deprived of the cytokinin continue to incorporate ^{3}H-thymidine into nuclear DNA for up to 144 hours after the hormone was withdrawn. After 132 hours in the absence of the hormone, 1.4% of the cells still incorporated ^{3}H-thymidine into DNA during a 5-hour exposure to the isotope, as shown by autoradiography (Table 3-4). Since the cell population doubling time was approximately 30 hours in the presence of 5×10^{-7}M zeatin, one would predict that, if a G_2 block occurred in this randomly dividing cell population, one would continue to observe some cells entering the S phase of the cell cycle for perhaps another 24 hours after cytokinin withdrawal, but not for 144 hours.

It was also apparent that the average number of silver grains per labeled nucleus after ^{3}H-thymidine labeling and autoradiography was over three times as great in the cytokinin-deprived, as compared to the cytokinin-treated cells (Table 3-4). A number of studies have shown that the duration of the S phase of the cell cycle is very similar in diploid and polyploid cells of the same species (Troy and Wimber

Table 3–4

The effect of cytokinin on ^{3}H-thymidine incorporation into soybean nuclei, as revealed by autoradiography

Zeatin M	Percent labeled nuclei after 5h exposure to ^{3}H-thymidine	Average number of silver grain/labeled nucleus (harvested immediately)	Average number of silver grain/labeled nucleus (harvested 24 hr after labeling)
0	1.4 ± 0.3	68 ± 27	62 ± 31
5 × 10^{-7}	3.7 ± 0.3	21 ± 6	8 ± 3

Note: Twenty-ond-day-old Biloxi tissue was transferred to medium containing or lacking cytokinin and cultured for 132 hours, after which the cells were exposed to ^{3}H-thymidine (2μCi/ml) for 5 hours. Subsequently tissues were either harvested immediately or they were transferred to unlabeled medium for an additional 24 hours before they were harvested and fixed for autoradiography. The mean of each determination is given, together with the standard deviation of the mean (from Fosket and Short 1973).

1968; Alfert and Das 1969; Yang and Dodson 1970; Friedberg and Davidson 1970). If this is the case, polyploid cells could be expected to incorporate more ^{3}H-thymidine into DNA/nucleus/unit of time than diploid nuclei. Thus, our findings that nuclei were considerably more heavily labeled in the cytokinin-deprived cells argues that they have become more highly polyploid than the cytokinin-grown cells.

The simplest explanation for these observations is that cytokinin specifically regulates events which are necessary for the transition from G_2 to mitosis. Cytokinin-requiring cells stop dividing in the absence of the hormone because they either cannot condense their chromatin, or organize a mitotic spindle, or both. However, they do not necessarily leave the cell cycle. Instead, they will continue to replicate their nuclear DNA, perhaps with the same cyclic periodicity as before, but they will not undergo mitosis between cycles of DNA replication.

THE MODE OF ACTION OF CYTOKININ

Jouanneau (1975) demonstrated that an inhibitor of protein synthesis, 5-methyl-tryptophane, could block cytokinin-induced synchronous mitotic activity in tobacco cell suspensions. He observed that, when cytokinin was given to the cells for an 18-hour period, beginning 18 hours after transfer to fresh medium, a partially synchronous wave of mitoses occurred, the peak of which appeared approximately 30 hours after the end of the cytokinin treatment. When 5-methyltryptophane was added to the culture medium (at a concentration which he showed to be almost 75% effective in blocking S^{35} incorporation into protein) during the cytokinin treatment, it abolished the subsequent synchronous mitotic wave, even though the inhibition of protein synthesis was reversed by adding tryptophane to the culture medium at the end of the cytokinin treatment. However, the addition of cytokinin during the recovery period resulted in the reappearance of the mitotic wave, demonstrating that the temporary inhibition of protein synthesis did not destroy the capacity of the cells to respond to the hormone.

These results suggest that the role of cytokinin in the regulation of the cell cycle is mediated by an effect of the hormone on protein synthesis. Other interpretations of Jouanneau's (1975) results are possible. For example, if cytokinin action were dependent upon the presence of a protein hormone receptor which turned over rapidly, inhibition of protein synthesis could be expected to render the cells temporarily insensitive to the hormone. However, Jouanneau's (1975) work, together with several other studies which have demonstrated that cytokinin can induce both quantitative and qualitative changes in protein synthesis (Jouanneau and Péaud-Lenoël 1967; Jouanneau 1970; Trewavas 1972; Short et al. 1974), indicates that cytokinin is likely to regulate the cell cycle through an effect on protein synthesis.

Cytokinin-induced Polyribosome Formation in Soybean

The control by cytokinin of mitotic activity in cultured cells of *Glycine max* var. Sodifuri is correlated with cytokinin-induced polyribosome formation. Cells which have stopped dividing after transfer to medium lacking cytokinin have an abundance of cytoplasmic ribosomes, as revealed by transmission electron microscopy of thin sections (Figure 3-2). However, sucrose density gradient analysis of ribosomes extracted from such cells has demonstrated that comparatively few of these ribosomes are in polysomal aggregates, engaged in protein synthesis. Cells deprived of cytokinin for 24 hours contained a low level of polysomes and a large pool of free monoribosomes, whereas cytokinin-treated cells exhibited a high level of polyribosomes, with comparatively few monoribosomes (Figure 3-3). The correlation between mitotic activity and polyribosome levels is apparent when one compares the data in Tables 3-1 and 3-5. Transfer of old stationary phase cells (21 days in culture) to fresh cytokinin-containing medium brought about a dramatic increase in the polyribosome levels of the cells and induced a high level of mitotic activity. The polyribosome levels gradually declined over the subsequent 8 days, as did the mitotic index. Cells transferred to medium lacking cytokinin never achieved the polyribosome levels observed in the cytokinin-treated cells and they exhibited little, if any mitotic activity. Cells transferred to medium lacking cytokinin did exhibit an increase in polyribosome content upon transfer to fresh medium, however. This indicated that transfer alone was a significant factor in the response. To better assess the effect of cytokinin on polyribosome formation, stationary phase cells were first transferred to medium lacking cytokinin. Twenty-four hours later cytokinin was added and the level of polyribosomes was determined at intervals over the subsequent 24-hour period. Over a fourfold increase in polyribosomes was observed within the first 6 hours after the addition of the hormone, with no evidence of a lag period (Figure 3-4). This increase in polyribosomes was followed by a partially synchronous wave of mitosis 16 hours after the addition of the hormone

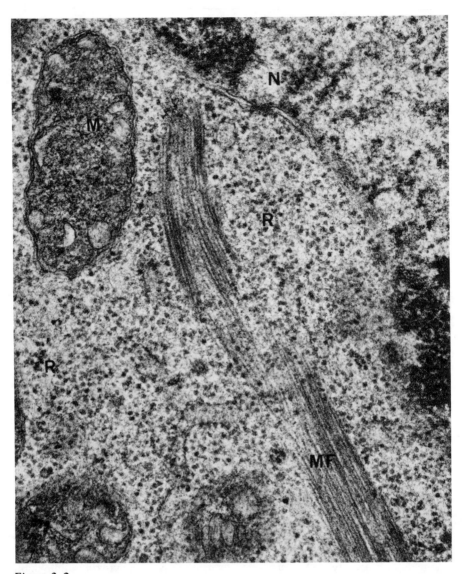

Figure 3-2.
An electron micrograph of a soybean cell after culture for 24 hours in medium
lacking cytokinin. Stationary phase Sodifuri cells were transferred to freshly prepared medium lacking cytokinin. Twenty-four hours later these cells were fixed in
3% glutaraldehyde in 0.2 M phosphate buffer. The fixed tissue was dehydrated
through a graduated ethanol concentration series and then infiltrated and embedded in Araldite-Epon. After the resin had polymerized, silver-gray sections were cut
with a Porter-Blum MT-1 Ultramicrotome. N = nucleus, M = mitochondria, R = ribosomes, MF = microfilaments (Tepfer, unpublished data).

Table 3-5

Time dependent changes in the percent polysomes in cultured Sodifuri cells

	Percent Polyribosomes	
Day of Culture	*5 X 10⁻⁷ M Zeatin*	*No Zeatin*
0	17	17
1	75	36
2	60	37
3	64	30
4	61	22
6	67	21
8	53	12

Note: Ribosomal extracts were obtained from soybean cells at intervals after culture on medium containing or lacking zeatin. The ribosomal material was analyzed by sucrose density gradient centrifugation and the percentage of ribosomes present as polyribosomes was determined by comparing the area under the monoribosome and polyribosome portions of the polysome profile (see Short, Tepfer, and Fosket, 1974, for detailed methods).

(Short et al. 1974). Thus, the ability of the cells to complete the mitotic cycle appears to be correlated with this cytokinin-induced stimulation of polyribosome formation.

Effect of Cytokinin on RNA Synthesis

We next attempted to see if cytokinin exerted its effect on protein synthesis at the translational or the transcriptional level. That is, does cytokinin regulate polysome formation by determining the rate and perhaps the kinds of messenger RNA synthesized by the cells, or does it primarily influence the process by which the information contained in the messenger RNA is utilized for protein synthesis.

Stationary phase cells were transferred to fresh medium lacking cytokinin. Twenty-four hours later, the culture was divided into two aliquots, one of which was treated with cytokinin, while both received $5\mu\text{Ci/ml}$ ³H-uridine. Over a 3-hour period, samples were taken to determine the rate of RNA synthesis, as reflected by the rate of ³H-uridine incorporation. As can be seen by the data presented in Table 3-6, the addition of cytokinin to the medium had a comparatively small effect on the rate of RNA synthesis. Over this same time course the polyribosome content of the tissue more than doubled.

Incorporation data alone tell us nothing about the nature of the RNA synthesized by the cells. In order to obtain some information aobut this, we took advantage of the fact that most eukaryotic messenger RNAs contain a polyadenylic acid

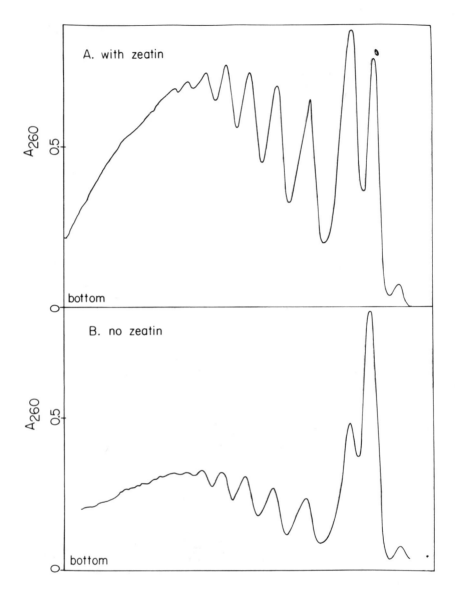

Figure 3-3.
Polyribosome profiles prepared by sucrose density gradient centrifugation of ribosomes extracted from cultured cells of *Glycine max* cv. Sodifuri. Stationary phase cells were transferred to freshly prepared medium which either contained 5×10^{-7} M zeatin or lacked a cytokinin. Twenty-four hours later the cells were homogenized in ice cold extraction medium consisting of 0.25 M sucrose, 0.4 M KCl, 10 mM Mg acetate, 7 mM β-mercaptoethanol, and 1 mM dithiothreitol in a 50 mM Tris buffer (pH 8.5). Diethylpyrocarbonate was added to the extraction beffer just before homogenization at a final concentration of 0.1% to inhibit ribonuclease activity (Weeks and Marcus 1969). After centrifuging at 13,000 $\times$ g for 15 minutes at 4°, a ribosomal pellet was obtained from the supernatant by layering it over a discontinuous sucrose gradient consisting of 1.0 ml of 0.5 M and 1.0 ml of 1.5 M sucrose made up in extraction buffer and centrifuging it at 165,000 $\times$ g in

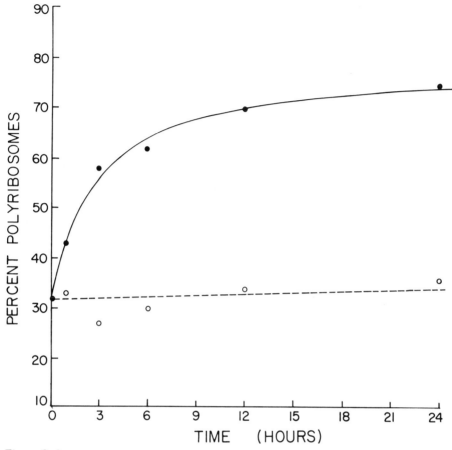

Figure 3-4.
The time course of polyribosome formation after cytokinin stimulation of hormone-deprived cells. Stationary phase cells were transferred to freshly prepared medium lacking cytokinin. Twenty-four hours later, cytokining was added (5 × 10^{-7} M zeatin) and 1.5 to 3.0 gm samples were taken at intervals over the subsequent twenty-four hour period. Ribosomes were extracted and subjected to sucrose density gradient centrifugation, as described in the legend to Fig. 3, and the percentage polyribosomes was calculated by comparing the areas under the monoribosome and polyribosome portions of the gradients.
○ − − − − − ○ = cells which received no cytokinin
● ———— ● = cytokinin-treated cells (from Short, Tepfer, and Fosket, 1974)

---

a Beckman titanium 50 rotor for 3 hours at 4°. The pelleted ribosomes were resuspended in 1 ml of 50 mM Tris buffer (pH 7.8) containing 0.2 M KCL and 10 mM Mg acetate. Aliquots of the resuspended ribosomes were layered over continuous, linear sucrose gradients (5–35%) made up in 50 mM Tris (pH 7.8) and containing 0.2 M KCl and 10 mM Mg acetate. The gradients were centrifuged at 190,000 × g for 35 minutes in an SW 50 rotor at 4°. After centrifugation, 45% sucrose was pumped into the tubes, forcing the gradients out through a tube at the top and into a flow cell where absorbance at 260 nm was monitored with a Beckman Acta III recording spectrophotometer.

Table 3–6

The effect of cytokinin on RNA synthesis

Time (min)	Total acid precipitable counts[1] (CPM/μg RNA)		Counts retained by poly(U) filters[2] (CPM/μg RNA)		% of total CPM bound to poly(U) filters	
	no zeatin	+zeatin	no zeatin	+zeatin	no zeatin	+zeatin
30	1337	1290	185	218	13.8	16.9
60	1972	2053	342	360	17.3	17.5
90	3768	3538	743	710	19.7	20.1
180	7276	7541	1474	1582	20.3	20.9

Note: Stationary Phase cells were transferred to fresh medium lacking cytokinin. The culture was split into two aliquots, one of which was given cytokinin (5×10^{-7} M zeatin). At the same time, both aliquots were given 5μCi/ml ^{3}H-uridine. The cultures were sampled at the times indicated to determine the amount of isotope incorporated into RNA (Fosket, unpublished data).

1. Whole cell RNA was obtained by homogenizing the cells in 0.05 M Tris-HCl (pH 8.0) containing 0.05 M NaCl and 1% sodium dodecyl sulfate. The homogenate was shaken with an equal volume of a phenol-chloroform mixture (2:1) containing 4% isoamyl alcohol. After centrifuging, the lower phenol phase was removed and the aqueous phase + interphase was shaken again with fresh phenol-chloroform. After again centrifuging, the aqueous phase was removed and extracted two additional times with phenol chloroform. The RNA was precipitated with two volumes of ethanol at $-20°$ and collected by centrifugation. The RNA was dissolved in 0.01 M Tris-HCl (pH 7.5) containing 0.1 M NaCl. An aliquot of the dissolved RNA was used to determine the total amount of RNA in the sample. Another aliquot was made 5% in trichloroacetic acid at $4°$ and the precipitated nucleic acids were collected on Millipore filter discs, washed with ice cold 5% TCA and 95% ethanol and the filters were placed in scintillation vials. After they had dried, 0.5 ml of NCS solubilizer was added to each vial. Thirty minutes later, 10 mls of a toluene-based scintillation fluid was added and the samples were counted.

2. An aliquot of each RNA sample, prepared as described above, was passed through a poly (U) glassfiber filter. The filters were prepared as described by Sheldon et al. (1972). After washing the filter sequentially with 0.01 M Tris-HCl (pH 7.5) containing 0.1 M NaCl, 5% trichloroacetic acid, and 95% ethanol, the filters were placed in scintillation vials. After they had dried, the RNA was dissolved with 0.5 ml of NCS, scintillation fluid was added, and the samples were counted.

sequence at the 3' end of the molecule. Thus, they will bind to immobilized poly (U) on fiberglass filters, while other kinds of RNA will pass through the filters without binding (Edmonds and Caramela 1969; Lee et al. 1971a; Darnell et al. 1971). When the labeled RNA, extracted from cytokinin-treated and cytokinin-deprived cells was passed through the poly (U)-fiberglass filters, between 14% and 20% of the counts were retained by the filters. There was no evidence that cytokinin stimulated the synthesis of this RNA fraction at the expense of ribosomal RNA.

Thus, cytokinin treatment appears to bring about a substantial increase in the polyribosome content of the cells, without increasing the rate of messenger RNA synthesis, at least during the initial phase of cytokinin-induced polyribosome formation. This conclusion is supported by work with the potent inhibitor of RNA synthesis, actinomycin D. When this drug was given to the cells, either at the time of hormone stimulation, or 30 minutes preceding it, the drug did not prevent the initial cytokinin-induced stimulation of polyribosome formation, although it led ultimately to polyribosome degradation (Short et al. 1974). Klämbt (1974) also found that cytokinin had no effect on the entry of ^{32}P-labeled RNA into polyribosomes in cultured tobacco pith tissue. Our data and those of Klämbt (1974) suggest that the initial stimulation of polyribosome formation is a result of an effect of cytokinin upon the translational process.

Regulation of Polyribosome Levels in Growing Cells

There is an important difference between our results and those obtained in many other cultured cell systems where a rapid increase in polyribosomes has been shown to occur upon the transfer of stationary phase cells to fresh medium (Levine et al. 1965; Leaver and Key 1967; Hodgson and Fisher 1971; Verma and Marcus 1973, 1974). Specifically, we are not dealing with nutritionally-deprived cells, but rather with cells whose growth is limited by a specific, naturally-occurring hormone known to be required for the division of these cells. These cells are capable of active protein synthesis in the absence of the hormone, and even show a net accumulation of protein over a period of several days (Fosket and Short 1973). However, they neither· divide nor achieve maximal rates of protein synthesis without cytokinins. On the other hand, our observations that cytokinin exerts its effect by regulating protein synthesis at the translational level is not unique in the sense that other investigators have implicated undetermined translational control factors in the stimulation of polyribosome formation and cell division in stationary phase cells following their transfer to fresh medium (van Venrooij et al. 1970; Lee et al. 1971b).

Our results also differ considerably from other reports of hormone-induced polyribosome formation in plants. Two other plant hormones, auxin and gibberellin, have been shown to stimulate polysome formation (Trewavas 1968; Evins and Varner 1972; Travis et al. 1973; Ho and Varner 1974). Auxin treatment produces a relatively small increase in polyribosomes in excised, non-growing tissues. This increase is dependent upon RNA synthesis, and it may result from an auxin-induced stimulation of ribosomal RNA synthesis (Trewavas, 1968), or hormone-induced messenger RNA synthesis (Travis et al. 1973). Gibberellin application to isolated barley aleurone layers brings about a substantial increase in polyribosome formation, but only after a considerable lag period, and this stimulation is dependent upon RNA synthesis (Evins 1971; Evins and Varner 1972). In contrast, cytokinin treatment of hormone-deficient soybean cells produces a rapid stimulation of polyribosome formation with no detectable lag period, which is independent of RNA synthesis.

CYTOKININ AND THE TRANSLATION CONTROL
OF PROTEIN SYNTHESIS

Data from the literature suggest three mechanisms by which cytokinin might regulate protein synthesis at the translational level. These are: (1) they could control the availability of certain isoaccepting species of transfer RNA; (2) they could be involved in the "activation" of cytoplasmic ribosomes; and (3) they could stabilize messenger RNA. While evidence can be marshalled in support of all three of these hypotheses, there is no convincing demonstration that any of these proposed mechanisms represents the mode of action of the hormone.

Cytokinins in Transfer RNA

Cytokinins are known to be constituents of certain transfer RNAs (Skoog et al. 1966; Hall et al. 1967; Letham and Ralph 1967; Burrows et al. 1968, 1969, 1971; Hecht et al. 1969; Verman et al. 1972). These special bases with cytokinin activity include N^6 - ($\triangle^2$-isopentenyl) adenosine (IPA) and ribosylzeatin. Armstrong et al. (1969) have shown that IPA is present only in those t-RNA species responding to codons beginning with U, where it is located next to the $3'$ end of the anticodon. Furthermore, there is evidence that these hypermodified bases are necessary for proper binding of ribosomes to those t-RNA species that contain them (Fittler and Hall. 1966; Gefter and Russell 1969).

However, it is unlikely that cytokinins are incorporated directly into t-RNA. Label from cytokinin has been shown to be incorporated into t-RNA upon fairly long exposure of cultured plant tissues to medium containing labeled hormone (Fox 1966; Fox and Chen 1967), but in senescing leaves, which also respond to the hormone, labeled cytokinin was not incorporated into 4S RNA (Richmond et al. 1970). Rigorous proof that the label incorporated into t-RNA from cytokinin remained associated with the hormone would be necessary to make such demonstrations convincing since there is good evidence that the t-RNAs are synthesized from conventional bases, some of which are modified subsequently at the macromolecular level (Söll 1971). Furthermore, Chen and Hall (1969) found that cytokinin-requiring tobacco cells possessed the enzymatic machinery for attaching the isopentenyl group to adenine residues in t-RNA. Burrows et al. (1971) found that the t-RNA from a cytokinin-requiring strain of tobacco cells contained 6-benzylamino-purine when they were cultured on a medium containing this synthetic cytokinin. However, they also isolated three naturally-occurring cytokinins, 6-(4-hydroxy-3-methyl-2-butenylamino)-9-β-D-ribofuranosylpurine (zeatin), 6-(3-methyl--2-butenylamino)-9-β-D-ribofuranosylpurine (IPA), and 6-(4-hydroxy-3-methyl-2-butenylamino)-2-methylthio-9-β-D-ribofuranosylpurine, despite the fact that the medium did not contain these cytokinins, nor could the 6-benzylaminopurine act as a precursor for these molecules.

While the relationship between the free, hormonal cytokinins and the structural cytokinins in t-RNA remains obscure, the hypothesis that the hormone acts by regulating the processing of those t-RNA species that contain cytokinin is an attractive one since it would explain the diverse effects of the hormone in molecular terms. For example, it is quite possible that the messenger RNA for constituitive enzymes would use one codon for a particular amino acid, while the m-RNA for a specialized cell protein would use a different codon for the same amino acid. The translation of the m-RNA for the specialized protein then would depend upon the availability of a particular t-RNA (Sueoka and Kano-Sueoka 1970).

There are a number of indications that the availability of certain isoaccepting t-RNA species may play a significant role in the regulation of protein synthesis. In several cases it has been shown that the relative abundance of the different isoaccepting t-RNAs changes during development (Anderson and Cherry 1969; Cherry and Osborne 1970; White et al. 1973). Furthermore, it is known that certain isoaccepting t-RNAs are used only for the synthesis of specific polypeptides (Stewart et al. 1971), that the levels of minor t-RNA species can markedly influence the rate of messenger RNA translation *in vitro* (Anderson and Gilbert 1969), and that specific messenger RNAs are translated more efficiently when the t-RNAs supplied in a cell-free system are obtained from the same tissue as the messenger RNA (Gerlinger et al. 1975). There is even some evidence that cytokinin treatment of plant tissues changes the relative abundance of certain isoaccepting t-RNAs (Pillay and Cherry 1974). However, a mechanism by which cytokinin might regulate t-RNA processing or degradation remains to be demonstrated.

Modification of Ribosomes

Cytokinins also might regulate protein synthesis at the translational level by modifying the structure of the ribosome. Lin et al. (1973) have demonstrated that the monomeric ribosomes from growing plant tissues are more active in protein synthesis *in vitro* than are the ribosomes of mature tissues. Furthermore, they found that this difference in ribosomal activity was correlated with differences in ribosomal proteins. Berridge et al. (1970) have shown that cytokinin binds to the ribosomes of Chinese cabbage leaves, where the hormone appears to inhibit the phosphorylation of ribosomal proteins (Ralph et al. 1972). We have obtained data which support these observations. When log phase soybean cells were transferred to medium lacking a cytokinin, a threefold increase in the incorporation of ^{32}P into ribosomal protein occurred over the subsequent 24-hour period. This increase in ribosomal protein phosphorylation was correlated with a marked decline in the percentage of ribosomes bound as polyribosomes in the hormone-deprived cells. Polyacrylamide gel electrophoresis of the proteins extracted from 0.4 M KCl-washed, ^{32}P-labeled ribosomes revealed five peaks of radioactivity, three of which exhibited enhanced ^{32}P incorporation in the cytokinin-deprived cells (Tepfer and Fosket 1975).

The significance of ribosomal protein phosphorylation remains unknown in any biological system. Bitte and Kabat (1972), working with mouse sarcoma 180 cells and rabbit reticulocytes, observed a correlation between the phosphorylation of ribosomal proteins and the inactivation of the ribosome. From our knowledge of ribosome structure, one would expect the phosphorylation of serine and threonine residues of ribosomal proteins to profoundly affect the conformation of the ribosome, and hence its ability to function in protein synthesis (Kurland 1972). Although Eil and Wool (1973) could demonstrate no significant difference in the activity of *in vitro* phosphorylated and non-phosphorylated ribosomes in a cell-free system, they did not examine the possibility that phosphorylation might confer message specificity upon the ribosomes. That is, phosphorylation of ribosomal proteins might restrict the diversity of messenger RNAs that could be translated by the ribosomes.

Messenger RNA Longevity

A third mechanism by which cytokinins may exert translational control over protein synthesis is through an effect on messenger RNA half-life. Cytokinin treatment of various plant tissues has been shown to inhibit RNase activity (Srivastava and Ware 1965; Srivastava 1968). Birmingham and Maclachlan (1972) demonstrated the inhibition by cytokinin of ribonuclease activity associated with the polyribosomes extracted from the apical region of *Pisum* epicotyls. Because of this potentially protective effect of cytokinin on extracted ribosomes, we have been careful to eliminate the possible contribution of endogenous RNase activity during cell homogenization to the polyribosome profiles we have observed in cytokinin-treated and -deprived cells. The cells were homogenized in an extraction buffer known to inhibit plant RNase activity (Davies et al. 1972). Also, a potent inhibitor of RNase activity, diethyl pyrocarbonate, was added to the homogenization medium (Weeks and Marcus 1969; Anderson and Key 1971). Finally, although we found an increase in RNase activity associated with the transfer of soybean cells to fresh medium, cytokinin had little effect on the observed activity of this enzyme (Short et al. 1974).

Although we could find no effect of cytokinin on general RNase activity in soybean cell extracts, this result does not eliminate the possibility that the hormone could stabilize cytoplasmic messenger RNA, thereby increasing its half-life and the probability that it will be translated. Since cytokinin treatment not only increases the polyribosome content of the cells, but also produces qualitative changes in the nature of the proteins synthesized, one must propose that the hormone only stabilizes certain messengers, or a certain class of messengers.

The nature of the factors responsible for messenger RNA degradation in a functioning cell are not known (Greenberg 1975). In prokaryotes, the half-life of a particular messenger RNA appears to be an inherent property of the molecule in that a given mRNA exhibits a characteristic longevity even when present in very different cellular environments (Marrs and Yanofsky 1971). However, in eukaryotic cells there is good evidence that not only do different messengers have markedly different half-lives (Kafatos and Rich 1968; Craig et al. 1971; Murphy and Attardi 1973; Singer and Penman 1973), but also that the half-life of a specific mRNA can change radically during cell development (Buckingham et al. 1974). The basis for such changes in the half-life of mRNA are not known. Sheiness et al. (1975) and

Sheiness and Darnell (1973) have demonstrated that the poly (A) segment of messenger RNA decreases in length with age. They have some evidence that messenger RNA becomes less stable as its poly (A) segment shortens.

PROTEIN SYNTHESIS IN RELATION TO CELL DIVISION

A number of studies have shown that protein synthesis is necessary for the continued progress of a cell through its division cycle (Hotta and Stern 1963; Cummins and Rusch 1966; Jones et al. 1968; Kim et al. 1968; Peterson et al. 1969; Rose 1970; Everhart and Prescott 1972; Highfield and Dewey 1972; Webster and Van't Hof 1973; Jouanneau 1975). However, all stages of the cell cycle are not equally dependent upon on-going protein synthesis for their completion. For example, protein synthesis plays a particularly critical role in the initiation of DNA synthesis (Kim et al. 1968; Everhart and Prescott 1972; Highfield and Dewey 1972), while it does not appear to be required for mitosis iteslf, or for the events of late G_2 in mammalian cells (Peterson et al. 1969). In plants, however, the transition from G_2 to mitosis and possibly even mitosis itself, is strongly dependent upon simultaneous protein synthesis (Rose 1970; Webster 1973).

The nature of the relationship between protein synthesis and progress through the cell cycle is not at all clear. There is little doubt that the division cycle ultimately is under genetic control. Hartwell and his group have isolated a number of temperature-sensitive mutants of yeast which are blocked at specific points in their division cycle when they are placed at the restrictive temperature (Hartwell et al. 1970; Hartwell 1971a,b; Hartwell et al. 1974). Also there is evidence for qualitative differences in the proteins synthesized during different stages of the cell cycle (Kolodny and Gross 1969; Jouanneau 1970; Fox and Pardee 1971). At least some of these stage-specific proteins are nuclear proteins which may be involved in the regulation of particular events in the cell cycle (Jockusch et al. 1970; Salas and Green 1971).

These studies suggest that a cell's progress through its mitotic cycle is dependent upon the sequential synthesis of a number of specific proteins, some of which are labile and do not persist at effective levels into subsequent cycles. Although cells may synthesize and accumulate total protein at a constant rate throughout interphase (Zetterberg 1966), there is evidence that a substantial number of enzymes are synthesized only during a limited part of the cell cycle (Mitchison 1969; Klevecz and Ruddle 1968; Martin et al. 1969). The factors responsible for the cyclic or periodic synthesis of enzymes during the cell cycle are, for the most part, unknown. Nevertheless, they are sufficiently regular to serve as biochemical markers for a given phase of the cell cycle. Furthermore, some of them, such as thymidylate synthetase and DNA synthetase, play a role in the cell cycle itself.

Translational Control of Protein Synthesis During the Cell Cycle

There is good evidence that protein synthesis is controlled at the translational level to some degree during the cell cycle. It is well known that the rate of protein synthesis and the level of polyribosomes declines rapidly to about 30% of the interphase level as many eukaryotic cells move into mitosis (Prescott and Bender 1962; Steward et al. 1968; Martin et al. 1969, Fan and Penman 1970). This reduction in the rate of protein synthesis is not due to a lack of messenger RNA since polyribosomes will reform after mitosis without *de novo* synthesis of messenger RNA (Steward et al. 1968). These studies were confirmed by Fan and Penman (1970), who also showed that the collapse of polysomes during mitosis was not due to a decrease in the amount of functional messenger RNA, to a change in the rate of protein chain elongation, or to the premature release of the ribosomes from the message. They concluded that the reduction in polyribosomes during mitosis is brought about by a deficiency in the initiation phase of polypeptide synthesis. Translational control of protein synthesis at the level of initiation appears to be relatively common in eukaryotic systems (McCormick and Penman 1969; MacKintosh and Bell 1969; Palmiter 1972).

There is evidence that the translational control exercised in mitotic cells not only modulates the rate of protein synthesis but exercises some degree of messenger selectivity as well. Martin et al. (1969) found that the synthesis of tyrosine aminotransferase declined by only 25% during mitosis, whereas the level of general protein synthesis declined by nearly 75%. Similarly, Stein and Baserga (1970) showed that the synthesis of non-histone chromosomal proteins continued at the interphase rate during mitosis while general protein synthesis decreased by 70% to 90%.

The early stages of animal embryogenesis represent a particularly striking example of the correlation of cell division with enhanced protein synthesis activity. In fertilized sea urchin embryos the onset of DNA synthesis and cell division is preceded by a tenfold to twentyfold increase in the rate of protein synthesis (Hultin 1961). Meeker (1970) has provided evidence that at least some of the proteins made after fertilization are proteins required for cell division.

The significance of this to our present discussion is that the activation of protein synthesis upon fertilization is mediated at the translational level. Neither the physical enucleation of the eggs (Denny and Tyler 1964) nor the inhibition of RNA synthesis by actinomycin D (Gross et al. 1964) will prevent this rapid increase in protein synthesis. The messenger RNA required for post-fertilization protein synthesis, containing the information necessary for the early events of development, including cell division, is present in the egg cytoplasm prior to fertilization.

Apparently there is some translational block in the unfertilized egg. The nature of block is not known, but there is evidence to suggest that the messenger RNA is somehow sequestered, or "masked," so that it cannot be used as a template for protein synthesis. Humphreys (1969) demonstrated that there was no difference in the average polysome aggregate size, in the average size of the proteins being synthesized, or in the average time required to complete the synthesis of a protein before and after fertilization. Furthermore, the ribosomes, t-RNAs and aminoacyl t-RNA synthetases of fertilized and unfertilized sea urchin eggs appear to be equivalent in their ability to support amino acid incorporation in a cell-free system, using a synthetic messenger RNA as a template (Gross 1967, 1968; Kedes and Stavy 1969). The bulk of the cytoplasmic messenger RNA in unfertilized eggs appears to be unavailable for translation by an otherwise competent protein synthetic machinery.

Spirin (1969) and Infante and Nemer (1968) have shown that messenger RNA enters the cytoplasm of embryos as an RNA-protein complex (RNP particle). Similar RNP particles have been observed in the cytoplasm of a number of different systems, including mouse L-cells (Perry and Kelly 1966, 1968), ascites tumor cells (Lee et al. 1971a), rat liver (Henshaw 1968), insects (Kafatos 1968), and wheat seeds (Chen et al. 1968; Weeks and Marcus 1971; Schultz et al. 1972). Recently, Gross et al. (1973) have shown that a particular fraction of the cytoplasmic RNP from sea urchin eggs codes for histone proteins. Since histones represent a class of proteins that is not synthesized until after fertilization, there can be no question that at least some of the eggs cytoplasmic RNP particles represent "masked" messenger RNA.

Although some information is available on the nature of the protein bound to cytoplasmic messenger RNA (Bryan and Hayashi 1973), the significance of this association is obscure at the present time. Spirin (1969) has presented considerable evidence to show that newly synthesized messenger RNA normally enters the cytoplasm complexed with protein. Nevertheless, it is possible that the "unmasking" of messenger RNA which occurs upon sea urchin egg fertilization could represent something as simple as the removal of a protein from the initiator site on the mRNA.

CONCLUSION

From the data discussed above we can draw the following conclusions:

1. In actively dividing cells which require cytokinin for growth, the hormone specifically regulates processes either in G_2 or in the transition from G_2 to mitosis. In the absence of cytokinin, such cells either accumulate in G_2 for a period of time and then leave the cell cycle entirely, or they enter an endoreduplication cycle in which successive cycles of DNA replication occur without mitosis or cytokinesis.

2. The events of the cell cycle are the result of a genetically determined program. Cytokinin is not involved in the transcription of this program. Other factors, possibly auxin, initiate the reading of the genetic information for the cell cycle. However, cytokinin is necessary for the translation of certain messages into effector molecules (proteins) which must act in G_2 for the cycle to be completed.

LITERATURE CITED

Alfert, M. and N. K. Das. 1969. Evidence for control of the rate of nuclear DNA synthesis by the nuclear membrane in eukaryotic cells. *Proc. Natl. Acad. Sci.* **63**:123–128.

Anderson, J. M. and J. L. Key. 1971. The effects of diethyl pyrocarbonate on the stability and activity of plant polyribosomes. *Plant Physiol.* **48**:801–805.

Anderson, M. B. and J. H. Cherry. 1969. Differences in leucyl-transfer RNAs and synthetase in soybean seedlings. *Proc. Natl. Acad. Sci.* **62**:202–209.

Anderson, W. F. and J. F. Gilbert. 1969. tRNA-dependent translational control if *in vitro* hemoglobin synthesis. *Biochem. Biophys. Res Comm.* **36**:456–462.

Armstrong, D. J., F. Skoog, L. H. Kirkegaard, A. D. Hampel, R. M. Bock, I. Gillam, and G. M. Tener. 1969. Cytokinins: Distribution in species of yeast transfer RNA. *Proc. Natl. Acad. Sci.* **63**:504–511.

Berridge, M. V., R. K. Ralph and D. J. Letham. 1970. The binding of kinetin to plant ribosomes. *Biochem.* **119**:75–83.

Birmingham, B. C. and G. A. Maclachlan. 1972. Generation and suppression of microsomal ribonuclease activity after treatments with auxin and cytokinin. *Plant Physiol.* **49**:371–375.

Bitte, L. and D. Kabat. 1972. Phosphorylation of ribosomal proteins in Sarcoma 180 tumor cells. *J. Biol. Chem.* **247**:5345–5350.

Blumenfeld, A. and S. Gazit. 1970. Cytokinin activity in avocado seeds during fruit development. *Plant Physiol.* **46**:331–333.

Bryan, R. N. and M. Hayashi. 1973. Two proteins are bound to most species of polysomal mRNA. *Nature New Biology* **244**:271–274.

Buckingham, M. E., D. Caput, A. Cohen, R. F. Whalen and F. Gros. 1974. The synthesis and stability of cytoplasmic messenger RNA during myoblast differentiation in culture. *Proc. Natl. Acad. Sci.* **71**:1466–1470.

Burrows, W. J., D. J. Armstrong, F. Skoog, S. M. Hecht, J. T. A. Boyle, N. J. Leonard, and J. Occolowitz. 1968. Cytokinin from soluble RNA of *Escherichia coli*: 6-(3-Methyl-2-butenylamino)-2-methythio-9-β-D-ribofuranosylpurine. *Science* **161**:691–693.

Burrows, W. J., D. J. Armstrong, F. Skoog, S. M. Hecht, J. T. A. Boyle, N. J. Leonard, and J. Occolowitz. 1969. The isolation and identification of two cytokinins from *Escherichia coli* transfer ribonucleic acid. *Biochemistry* **8**:3071–3076.

Burrows, W. J., F. Skoog, and N. J. Leonard. 1971. Isolation and identification of cytokinins located in the transfer ribonucleic acid of tobacco callus grown in the presence of 6-benzylaminopurine. *Biochemistry* **10**:2189–2194.

Chen, C. M. and R. H. Hall. 1969. Biosynthesis of N^6-(Δ^2-isopentenyl)adenosine in the transfer ribonucleic acid of cultured tobacco pith tissue. *Phytochemistry* **8**:1687–1695.

Chen, D., S. Sarid and E. Katchalski. 1968. Studies on the nature of messenger RNA in germinating wheat embryos. *Proc. Natl. Acad. Sci.* **60**:902–909.

Cherry, J. H. and D. J. Osborne. 1970. Specificity of leucyl-tRNA and synthetase in plants. *Biochem. Biophys. Res. Comm.* **40**:763–769.

Craig, N., D. E. Kelley and R. P. Perry. 1971. Lifetime of the messenger RNAs which code for ribosomal proteins in L-cells. *Biochim. Biophys. Acta.* **246**:493–498.

Cummins, J. E. and H. P. Rusch. 1966. Limited DNA synthesis in the absence of protein synthesis in *Physarum polycephalum*. *J. Cell Biol.* **31**:577–583.

Darnell, J. E., R. Wall, and R. J. Tushinski. 1971. An adenylic acid-rich sequence in messenger RNA of HeLa cells and its possible relationship to reiterated sites in DNA. *Proc. Natl. Acad. Sci.* **68**:1321–1325.

Das, N., K. Patau and F. Skoog. 1956. Initiation of mitosis and cell division by kinetin and indoleacetic acid in excised tobacco pith tissue. *Physiol. Plant* **9**:640–651.

Das, K. N., K. Patau, and F. Skoog. 1958. Autoradiographic and microspectrophotometric studies of DNA synthesis in excised tobacco pith tissue. *Chromosoma* **9**:606–617.

Davies, E., B. A. Larkins and R. H. Knight. 1972. Polyribosomes from peas. An improved method for their isolation in the absence of ribonuclease inhibitors. *Plant Physiol* **50**:581–584.

Denny, P. and A. Tyler. 1964. Activation of protein biosynthesis in non-nucleate fragments of sea urchin eggs. *Biochem. Biophys. Res. Comm.* **14**:245–249.

Dyson. W. H. and R. H. Hall. 1972. N^6 -(Δ^2 -Isopentenyl) adenosine: Its occurrence as a free nucleoside in an autonomous strain of tobacco tissue. *Plant Physiol.* **50**:616–621.

Edmonds, M. and M. G. Caramela. 1969. The isolation and characterization of adenosine monophosphate-rich polynucleotide, synthesized by Ehrlich ascites cells. *J. Biol. Chem.* **244**:1314–1324.

Eil, Charles, and Ira G. Wool. 1973. Functions of phosphorylated ribosomes. The activity of ribosomal subunits phosphorylated *in vitro* by protein kinase. *J. Biol. Chem.* **248**:5130–5136.

Einset, J. W. and F. Skoog. 1973. Biosynthesis of cytokinins in cytokinin-autotrophic tobacco callus. *Proc. Natl. Acad. Sci.* **70**:658–660.

Everhart, L. P. and D. M. Prescott. 1972. Reversible arrest of Chinese hamster cells in G_1 by partial deprivation of leucine. *Exptl. Cell Res.* **75**:170–174.

Evins, W. H. 1971. Enhancement of polyribosome formation and induction of tryptophane-rich proteins by gibberellic acid. *Biochemistry* **10**:4295–4303.

Evins, W. H. and J. E. Varner. 1972. Hormonal control of polyribosome formation in barley aleurone layers. *Plant Physiol.* **49**:348–352.

Fan, H. and S. Penman. 1970. Regulation of protein synthesis in mammalian cells. II. Inhibition of protein synthesis at the level of initiation during mitosis. *J. Mol. Biol.* **50**:655–670.

Fittler, F. and R. H. Hall. 1966. Selective modification of yeast seryl-t-RNA and its effect on the acceptance and binding functions. *Biochem. Biophys. Res. Comm.* **25**:441–446.

Fosket, D. E. and J. G. Torrey. 1969. Hormonal control of cell proliferation and xylem differentiation in cultured tissues of *Glycine max* var. Biloxi. *Plant Physiol.* **44**:871–880.

Fosket, D. E. and K. C. Short. 1973. The role of cytokinin in the regulation of growth, DNA synthesis and cell proliferation in cultured soybean tissues (*Glycine max* var. Biloxi). *Physiol. Plant* **28**:14–23.

Fox, J. E. 1966. Incorporation of kinin, N, 6-benzyladenine into soluble RNA. *Plant Physiol.* **41**:75–82.

Fox, J. E. and C. Chen. 1967. Characterization of labeled ribonucleic acid from tissue grown on ^{14}C-containing cytokinins. *J. Biol. Chem.* **242**:4490–4494.

Fox, T. O. and A. B. Pardee. 1971. Proteins made in the mammalian cell cycle. *J. Biol. Chem.* **246**:6159–6165.

Friedberg, S. H. and D. Davidson. 1970. Duration of S phase and cell cycles in diploid and tetraploid cells of mixoploid meristems. *Exptl. Cell Res.* **61**:216–218.

Gefter, M. L. and R. L. Russell. 1969. Role of modification in tyrosine transfer RNA: a modified base affecting ribosome binding. *J. Mol. Biol.* **39**:145–157.

Gerlinger, P., M. A. LeMeur and J. P. Ebel. 1975. Messenger RNA translation in a t-RNA dependent cell-free system: Role of tRNA from different sources. *FEBS Letters* **49**:376–379.

Gifford, E. M., Jr. and J. P. Nitsch. 1969. Responses of tobacco pith nuclei to growth substances. *Planta* **85**:1-10.

Greenberg, J. R. 1975. Messenger RNA metabolism of animal cells. Possible involvement of untranslated sequences and mRNA-associated proteins. *J. Cell Biol.* **64**: 269-288.

Gross, K. W., M. Jacobs-Lorena, C. Baglioni and P. R. Gross. 1973. Cell-free translation of maternal messenger RNA from sea urchin eggs. *Proc. Natl. Acad. Sci.* **70**:2614-2618.

Gross, P. R. 1967. The control of protein synthesis in embryonic development and differentiation. *Curr. Top. Dev. Biol.* **2**:1-46.

Gross, P. R. 1968. Biochemistry of differentiation. *Ann. Rev. Biochem.* **37**:631.

Gross, P. R., L. T. Malkin and W. A. Moyer. 1964. Templates for the first proteins of embryonic development. *Proc. Natl. Acad. Sci.* **51**:407-414.

Hagen, G. L. and A. Marcus. 1975. Cytokinin effects on growth of quiescent tobacco pith cells. *Plant Physiol.* **55**:90-93.

Hall, R. A. 1973. Cytokinins as a probe of developmental processes. *Ann. Rev. Plant Physiol.* **24**:415-444.

Hall, R. H., L. Csonka, H. David and B. McLennon. 1967. Cytokinins in the soluble RNA of plant tissues. *Science* **156**:69-71.

Hartwell, L. H. 1971a. Genetic control of the cell division cycle in yeast. II. Genes controlling DNA replication and its initiation. *J. Mol. Biol.* **59**:183-194.

Hartwell, L. H. 1971b. Genetic control of the cell division cycle in yeast. IV. Genes controlling bud emergence and cytokinesis. *Exptl. Cell Res.* **69**:265-276.

Hartwell, L. H., J. Culotti and R. Reid. 1970. Genetic control of the cell division cycle in yeast. I. Detection of mutants. *Proc. Natl. Acad. Sci.* **66**:352-359.

Hartwell, L. H., J. Culotti, J. R. Pringle and B. J. Reid. 1974. Genetic control of the cell division cycle in yeast. *Science* **183**:46-51.

Hecht, S. M., N. J. Leonard, W. J. Burrows, D. J. Armstrong, F. Skoog and J. Occolowitz. 1969. Cytokinin of wheat germ transfer RNA: 6-(4-Hydroxy-3-methyl-2-butenylamino)-2-methylthio-9-β-D-ribofuranosylpurine. *Science* **166**: 1272-1274.

Henshaw, E. 1968. Messenger RNA in rat liver polyribosomes: Evidence that it exists as ribonucleoprotein particles. *J. Mol. Biol.* **36**:410-411.

Highfield, D. P. and W. C. Dewey. 1972. Inhibition of DNA synthesis in synchronized Chinese hamster cells treated in G_1 or early S phase with cycloheximide of puromycin. *Exptl. Cell Res.* **75**:314-320.

Ho, D. T. H. and J. E. Varner. 1974. Hormonal control of messenger ribonucleic acid metabolism in barley aleurone layers. *Proc. Natl. Acad. Sci.* **71**:4783-4786.

Hodgson, J. R. and H. W. Fisher. 1971. Formation of polyribosomes during recovery from contact inhibition of replication. *J. Cell Biol.* **49**:945-947.

Hotta, Y. and H. Stern. 1963. Inhibition of protein synthesis during meiosis and its bearing on intracellular regulation. *J. Cell. Biol.* **16**:259-279.

Hultin, T. 1961. Activation of ribosomes in sea urchin eggs in response to fertilization. *Exptl. Cell Res.* **25**:405-417.

Humphreys, T. 1969. Efficiency of translation of messenger-RNA before and after fertilization in sea urchins. *Develop. Biol.* **20**:435-458.

Infante, A. and M. Nemer. 1968. Heterogeneous ribonucleoprotein particles in the cytoplasm of sea urchin embryos. *J. Mol. Biol.* **32**:543-565.

Jockusch, B. M., D. F. Brown, and H. P. Rusch. 1970. Synthesis of a nuclear protein in G_2-phase. *Biochem. Biophys. Res. Comm.* **38**:279-283.

Jones, R. F., J. R. Kates and S. J. Keller. 1968. Protein turnover and macromolecular synthesis during growth and genetic differentiation in *Chlamydomonas reinhardtii. Biochim. Biophys. Acta* **157**:589-598.

Jouanneau, J. P. 1970. Renouvellement des protéines et effet spécifque de la kinétine sur des cultures de cellules de Tabac. *Physiol. Plant* **23**:232-244.

Jouanneau, J. P. 1971. Contrôle par les cytokinines de la synchronisation des mitoses dans les cellules de Tabac. *Exptl. Cell Res.* **67**:329-337.

Jouanneau, J. P. 1975. Protein synthesis requirement for the cytokinin effect upon tobacco cell division. *Exptl. Cell Res.* **91**:184-190.

Jouanneau, J. P. and C. Péaud-Lenoël. 1967. Croissance et synthèse des protéines de suspensions cellulaires de Tabac sensibles à la kinetine. *Physiol Plant* **20**:834-950.

Jouanneau, J. P. and N. Tandeau de Marsac. 1973. Stepwise effects of cytokinin activity and DNA synthesis upon mitotic cycle events in partially synchronized tobacco cells. *Exptl. Cell Res.* **77**:167-174.

Kafatos, F. 1968. Cytoplasmic particles carrying rapidly labeled RNA in developing insect epidermis. *Proc. Natl. Acad. Sci.* **59**:1251-1258.

Kafatos, F. C. and J. Rich. 1968. Stability of differentiation-specific and nonspecific messenger RNA in insect cells. *Proc. Natl. Acad. Sci.* **60**:1458-1465.

Kedes, L. H. and L. Stavy. 1969. Structural and functional identity of ribosomes from eggs and embryos of sea urchins. *J. Mol. Biol.* **43**:337-340.

Kende, H. 1971. The cytokinins *Internat. Rev. Cytol.* **31**:301-338.

Kim, J. H., A. S. Gelbard and A. G. Perez 1968. Inhibition of DNA synthesis by actinomycin D and cyclohexamide in synchronized HeLa cells. *Exptl. Cell Res.* **53**:478-487.

Klämbt, D. 1974. Einfluss von auxin und cytokinin auf die RNA-synthese in sterilem Tabakgewebe. *Planta* **118**:7-16.

Klevecz, R. R. and F. H. Ruddle. 1968. Cyclic changes in enzyme activity in synchronized mammalian cell cultures. *Science* **159**:634-636.

Kolodny, G. M. and P. R. Gross. 1969. Changes in patterns of protein synthesis during the mammalian cell cycle. *Exptl. Cell Res.* **56**:117-121.

Kurland, C. G. 1972. Structure and function of the bacterial ribosome. *Ann. Rev. Biochem.* **41**:377-408.

Leaver, C. J. and J. L. Key. 1967. Polyribosome formation and RNA synthesis during aging of carrot root tissue. *Proc. Natl. Acad. Sci.* **57**:1338-1344.

Lee, S. Y., J. Mendecki and G. Brawerman. 1971a. A polynucleotide segment rich in adenylic acid in the rapidly-labeled polyribosomal RNA component of mouse sarcoma 180 ascites cells. *Proc. Natl. Acad. Sci.* **68**:1331-1335.

Lee, S. Y., V. Krsmanovie and G. Brawerman, 1971b. Initiation of polysome formation in mouse sarcoma 180 ascites cells. Utilization of cytoplasmic messenger ribonucleic acid. *Biochemistry* **10**:895-900.

Letham, D. S. 1963. Purification of factors inducing cell division extracted from plum fruitlets. *Life Sci.* **2**:152-157.

Letham, D. S. and R. K. Ralph. 1967. A cytokinin in soluble RNA from a higher plant. *Life Sci.* **6**:387-394.

Letham, D. S. and M. W. Williams. 1969. Regulators of cell division in plant tissues. VIII. The cytokinins of the apple fruit. *Physiol. Plant* **22**:925-936.

Levine, E. M., Y. Becker, C. W. Boone and H. Eagles. 1965. Contact inhibition, macromolecular synthesis, and polyribosomes in cultured human diploid fibroblasts. *Proc. Natl. Acad. Sci.* **53**:350–356.

Libbenga, K. R. and J. G. Torrey. 1973. Hormone-induced endoreduplication prior to mitosis in cultured pea root cortex cells. *Amer. J. Bot.* **60**:293–299.

Lin, C. Y., R. L. Travis, S. Y. Chia and J. L. Key. 1973. Protein synthesis by 80 S ribosomes during plant development. *Phytochemistry* **12**:515–522.

Lukanidin, E. M., G. P. Georgiev and R. Williamson. 1971. A comparative study of the protein components of nuclear and polysomal messenger ribonucleo-protein. *FEBS Letters* **19**:152–156.

MacKintosh, F. R. and E. Bell. 1969. Regulation of protein synthesis in sea urchin eggs. *J. Mol Biol.* **41**:365–380.

Marrs, B. L. and C. Yanofsky. 1971. Host and bacteriophage specific messenger RNA degradation in T7-infected *Escherichia coli. Nature New Biology* **234**:168.

Martin, D., Jr., G. M. Tomkins and D. Granner. 1969. Synthesis and induction of tyrosine aminotransferase in synchronized hepatoma cells in culture. *Proc. Natl. Acad. Sci.* **62**:248–255.

Matthysse, A. G. and J. G. Torrey. 1967. DNA synthesis in relation to polyploid mitoses in excised pea root segments cultured *in vitro. Exptl. Cell. Res.* **48**:484–498.

McCormick, W. and S. Penman. 1969. Regulation of protein synthesis in HeLa cells: Translation at elevated temperatures. *J. Mol. Biol.* **39**:315–333.

Meeker, C. O. 1970. Intracellular potassium requirement for protein synthesis and mitotic apparatus formation in sea urchin eggs. *Exptl. Cell Res.* **63**:165–170.

Miller, C. O. 1961. A kinetin-like compound in maize. *Proc. Nat. Acad. Sci.* **47**:170–174.

Miller, C. O., F. Skoog, H. H. von Saltzn, F. S. Okumura and F. H. Strong. 1956. Isolation, structure, and synthesis of kinetin, a substance promoting cell division. *J. Amer. Chem. Soc.* **78**:1375.

Mitchison, J. M. 1969. Enzyme synthesis in synchronous cultures. *Science* **165**:657–663.

Miura, G. A. and C. O. Miller. 1969. Cytokinins from a varied strain of cultured soybean cells. *Plant Physiol.* **44**:1035–1039.

Murphy, W. and G. Attardi. 1973. Stability of cytoplasmic messenger RNA in HeLa cells. *Proc. Natl. Acad. Sci.* **70**:115–119.

Nitsch, J. P. and A. Lance-Nougaréde. 1967. L'action conjugée des auxines et des cytokinines sur les cellules de moelle de Tabac: Etude physiologique et microscopic électronique. *Bull. Soc. Franç. Physiol. Végét.* **13**:81–118.

Palmiter, R. D. 1972. Regulation of protein synthesis in chick oviduct. II. Modulation of polypeptide elongation and initiation rates by estrogen and progesterone. *J. Biol. Chem.* **247**:6770–6780.

Patau, K., N. K. Das and F. Skoog. 1957. Induction of DNA synthesis by kinetin and indoleacetic acid in excised tobacco pith tissue. *Physiol. Plant* **10**:946–966.

Patau, K. and N. K. Das. 1961. The relationship of DNA synthesis and mitosis in tobacco pith tissue cultured in vitro. *Chromosoma* **11**:553–572.

Perry, R. and D. Kelley. 1966. Bouyant densities of cytoplasmic ribonucleoprotein particles of mammalian cells: Distinctive character of ribosome subunits and the rapidly labeled components. *J. Mol. Biol.* **16**:255–268.

Perry, R. and D. Kelley. 1968. Messenger RNA-protein complexes and newly synthesized ribosomal subunits: Analysis of free particles and components of polyribosomes. *J. Mol. Biol.* **35**:37–59.

Peterson, D. F., R. A. Tobey and E. C. Anderson. 1969. Essential biosynthetic activity in synchronized mammalian cells. In *The Cell Cycle,* eds. G. W. Paeilla, G. L. Whitson, and I. L. Cameron. Academic Press, New York.

Pillay, D. T. N. and J. H. Cherry. 1974. Changes in leucyl, seryl, and tyrosyl tRNAs in aging soybean cotyledons. *Canad. J. Bot.* **52**:2499–2504.

Prescott, D. M. and M. A. Bender. 1962. Synthesis of RNA and protein during mitosis in mammalian tissue culture cells. *Exptl. Cell Res.* **26**:260–268.

Ralph, R. K., P. J. A. McCombs, G. Tener and S. J. Wojcik. 1972. Evidence for modification of protein phosphorylation by cytokinin. *Biochem. J.* **130**:901–911.

Richmond, A., A. Back and B. Sachs. 1970. A study of the hypothetical role of cytokinins in completion of tRNA. *Planta* **90**:57–65.

Rose, R. J. 1970. The effect of cyclohexamide on cell division in partially synchronized plant cells. *Aust. J. Biol. Sci.* **23**:573–583.

Salas, J. and H. Green. 1971. Proteins binding to DNA and their relation to growth in cultured mammalian cells. *Nature New Biol.* **229**:165–169.

Schultz, G. A., D. Chen, and E. Katchalski. 1972. Localization of a messenger RNA in a ribosomal fraction from ungerminated wheat embryos. *J. Mol. Biol.* **66**:379–390.

Sheiness, D. and J. E. Darnell. 1973. Polyadenylic acid segment in mRNA becomes shorter with age. *Nature New Biology* **241**:265–268.

Sheiness, D., L. Puckett and J. E. Darnell. 1975. Possible relationship of poly (A) shortening to mRNA turnover. *Proc. Natl. Acad. Sci.* **72**:1077–1081.

Sheldon, R., C. Jurale and J. Kates. 1972. Detection of polyadenylic acid sequences in viral and eukaryotic RNA. *Proc. Natl. Acad. Sci.* **69**:417–421.

Short, K. C. and J. G. Torrey. 1972. Cytokinins in seedlings roots of pea. *Plant Physiol.* **49**:115–160.

Short, K. C., D. A. Tepfer and D. E. Fosket. 1974. Regulation of polyribosome formation and cell division in cultured soybean cells by cytokinin. *J. Cell Sci.* **15**:75–87.

Simard, A. 1971. Initiation of DNA synthesis by kinetin and experimental factors in tobacco pith tissue *in vitro. Canad. J. Bot.* **49**:1541–1549.

Singer, R. H. and S. Penman. 1973. Messenger RNA in HeLa cells: Kinetics of formation and decay. *J. Mol. Biol.* **78**:321–334.

Skoog, F., F. M. Strong and C. O. Miller. 1965. Cytokinins. *Science* **148**:532.

Skoog, F., D. J. Armstrong, J. D. Cherayil, A. E. Hampel and R. M. Bock. 1966. Cytokinin activity: Localization in transfer RNA preparation. *Science* **154**:1354–1356.

Skoog, F. and D. J. Armstrong. 1970. Cytokinins. *Ann. Rev. Plant Physiol.* **21**:359–384.

Söll, D. 1971. Enzymatic modification of transfer RNA. *Science* **173**:293–299.

Spirin, A. 1969. Informosomes. *Europ. J. Biochem.* **10**:20–35.

Srivastava, B. I. S. 1968. Increase in chromatin associated nuclease activity of excised barley leaves during senescence and its suppression by kinetin. *Biochem. Biophys. Res. Comm.* **32**:533–538.

Srivastava, B. I. S. and G. Ware. 1965. The effect of kinetin on nucleic acids and nucleases of excised barley leaves. *Plant Physiol.* **40**:62–64.

Stein, G. and R. Baserga. 1970. Continued synthesis of non-histone chromosomal proteins during mitosis. *Biochem. Biophys. Res. Comm.* **41**:715-722.

Steward, D. L., J. R. Shaeffer and R. M. Humphrey. 1968. Breakdown and assembly of polyribosomes in synchronized Chinese hamster cells. *Science* **161**:791-793.

Stewart, T. S., R. J. Roberts, and J. L. Strominger. 1971. Novel species of tRNA. *Nature* **230**:36-38.

Sueoka, N. and T. Kano-Sueoka. 1970. Transfer RNA and cell differentiation. *Prog. Nuc. Acid Res. and Mol. Biol.* **10**:23-55.

Tandeau de Marsac, N. and J. P. Jouanneau. 1972. Variation de l'exigence en cytokinine de lignées clonales de cellules de Tabac. *Physiol. Vég.* **10**:369-380.

Tepfer, D. A. and D. E. Fosket. 1975. Phosphorylation of ribosomal protein in soybean. *Phytochemistry* **14**:1161-1165.

Torrey, J. G. 1961. Kinetin as trigger for mitosis in mature endomitotic plant cells. *Exptl. Cell Res.* **23**:281-299.

Travis, R. L., J. M. Anderson and J. L. Key. 1973. Influence of auxin and incubation on the relative level of polyribosomes in excised soybean hypocotyl. *Plant Physiol.* **52**:608-612.

Trewavas, A. 1968. The effect of 3-indolylacetic acid on the levels of polysomes in etiolated pea tissue. *Phytochemistry* **7**:673-681.

Trewavas, A. 1972. Control of the protein turnover rates in *Lemna minor*. *Plant Physiol.* **49**:47-51.

Troy, M. R. and D. E. Wimber. 1968. Evidence for a constancy of the DNA synthetic period between diploid-polyploid groups in plants. *Exptl. Cell Res.* **53**:145-153.

van Venrooij W. J. W., E. C. Henshaw and C. A. Hirsch. 1970. Nutritional effects on the polyribosome distribution and rate of protein synthesis in Erlich ascites tumor cells in culture. *J. Biol. Chem.* **245**:5947-5953.

Verma, D. P. S. and A. Marcus. 1973. Regulation of RNA synthesis in plant cell culture: delayed synthesis of ribosomal RNA during transition from the stationary phase to active growth. *Develop. Biol.* **30**:104-114.

Verma, D. P. S. and A. Marcus. 1974. Activation of protein synthesis upon dilution of an Arachis cell culture from stationary phase. *Plant Physiol.* **53**:83-87.

Verman, H. J., F. Skoog, C. R. Frihart and N. J. Leonard. 1972. Cytokinins in *Pisum* transfer ribonucleic acid. *Plant Physiol* **49**:848-851.

Wardell, W. L. 1975. Rapid initiation of thymidine incorporation into deoxyribonucleic acid in vegetative tobacco stem segments treated with indole-3-acetic acid. *Plant Physiol.* **56**:171-176.

Webster, P. L. 1973. Effects of cycloheximide on mitosis in *Vicia faba* root-meristem cells. *J. Exptl. Bot.* **24**:239-244.

Webster, P. L. and J. Van't Hof. 1973. Polyribosomes in proliferating and nonproliferating root meristem cells. *Amer. J. Bot.* **60**:117-121.

Weeks, D. P. and A. Marcus. 1969. Polyribosome isolation in the presence of diethyl pyrocarbonate. *Plant Physiol* **44**:1291-1294.

Weeks, D. P. and A. Marcus. 1971. Preformed messenger of quiescent wheat embryos. *Biochem. Biophys. Acta* **232**:671-684.

Weiss, C. and Y. Vaadia. 1965. Kinetin-like activity in root apices of sunflower plants. *Life Sciences* **4**:1223-1326.

White, B. N., G. M. Tener, J. Holden and D. T. Suzuki. 1973. Analysis of tRNAs during the development of Drosophila. *Develop. Biol.* **33**:185–195.

Yang, D. P. and E. O. Dodson. 1970. The amounts of nuclear DNA and the duration of DNA synthetic period (S) in related diploid and autotetroploid species of oats. *Chromosoma* **31**:309–320.

Zetterberg, A. 1966. Synthesis and accumulation of nuclear and cytoplasmic proteins during interphase in mouse fibroblasts *in vitro. Exptl. Cell. Res.* **42**:500–511.

Zwar, J. A., W. Bottomby and N. P. Kefford. 1963. Kinin activity from plant extracts. II. Partial purification and fractionation of kinins in apple extract. *Aust. J. Biol. Sci.* **16**:407–415.

Hormonal Regulation of Cell Division in the Primary Elongating Meristems of Shoots

J. Brent Loy   University of New Hampshire

INTRODUCTION

Although stem elongation per se involves only cell enlargement, proliferation of new cells is necessary for sustained growth, and thereby contributes to the final length of the plant organ. Analysis of hormonal regulation of shoot development has primarily focused on anatomical and physiological aspects of cell enlargement. Whereas cell enlargement can be studied conveniently in excised internodes and hypocotyls, cell division studies are largely restricted to intact shoots. Also, cytological techniques readily amenable to research on root meristems are not always easily applied to complex shoot meristems.

Most studies of hormonal regulation of cell proliferation in "primary elongating" or subapical meristems have dealt with the hormone gibberellin* because this hormone elicits positive stem growth responses in most plants. This report will thus emphasize the effects of gibberellin on cell proliferation, but will also briefly mention its possible interaction with other plant hormones in controlling cell division.

---

*It is recognized that several structurally different gibberellins have been identified, however, unless otherwise indicated, the term gibberellin (GA) will refer to the use of gibberellic acid (GA_3) in the studies cited. Other abbreviations: benzyladenine–BA; indoleacetic acid–IAA; abscisic acid–ABA; deoxyribonucleic acid–DNA.

ANATOMICAL DESCRIPTION OF THE PRIMARY ELONGATING MERISTEM

There is considerable variability in terminology to describe the zonation of the shoot apex (Gifford and Corson 1971). However, the subapical region in dicotyledonous plants is usually defined as the region just below the pith rib meristem, characterized by cells forming distinct longitudinal cell files. As reviewed by Sachs (1965), this is the region chiefly responsible for contributing cells for continued elongation of internodes during stem growth. Sachs invoked the term "primary elongating meristem" as a more descriptive synonym for the subapical meristem. This term is particularly appropriate for discussing stem growth in monocotyledons which have true intercalary meristems localized at nodal plates (Holttum 1955).

GIBBERELLIN PROMOTION OF CELL PROLIFERATION

Relationship of Cell Elongation to Cell Division

When Stowe and Yamaki (1957) reviewed the history and physiology of the gibberellins, there was considerable controversy as to whether the histological basis for gibberellin promotion of stem elongation involved cell division or cell elongation, or both processses. At that time there were few definitive anatomical studies describing the GA response. However, Sachs and Lang (1957) reported that gibberellin stimulation of stem elongation in rosette plants was characterized by marked increases in mitotic activity in the subapical meristem. Greulach and Haesloop (1958) showed that gibberellin treatment of dwarf beans promoted cell division in young internodes. Subsequently, several studies with GA-responsive rosette plants (Bernier et al. 1964; Sachs et al. 1959a, 1959b) genetic dwarfs (Arney and Mancinelli 1966, Basford 1961; Skjedstad 1960), and plants treated with growth retardants (Sachs and Kofranek 1963; Sachs et al. 1960) confirmed gibberellin promotion of cell division. The major effect of gibberellin is to promote cell division in primary elongating meristems, but cell division is often enhanced in the apical meristem as indicated directly by anatomical studies (Bernier et al. 1964; Okuda 1964) and suggested indirectly by increases in rates of leaf initiation in GA-treated plants (Arney and Mancinelli, 1966; Basford, 1961; Liu, 1971; Okuda, 1964).

Mean cell size in meristems of gibberellin treated plants may be similar (Liu 1974), smaller (Bernier et al., 1964; Sachs et al. 1959a) or larger (Okuda 1964) than untreated controls, the response varying among species. In GA-treated intact shoots, increases in cell number are often proportionally greater than increases in length of mature cells (Arney and Mancinelli 1966; Loy and Liu 1974; Okuda 1964). Studies with GA-responsive shoots exhibiting little change in mean cell size and marked increases in cell number have often led to the conclusion that GA has little effect on cell elongation. However, even in species where cell size is diminishing due to GA-stimulation of cell division, GA also increases the rate of stem elongation. This is reflected in faster rates of cell elongation and often a shorter duration of the cell elongation period, so that final cell size may not change appreciably.

The question has thus emerged: which gibberellin response occurs first, an increased rate of cell elongation or the induction of cell division. This question only assumes importance to the extent that one process may directly or indirectly regulate the other. Brotherton and Bartlett (1918) observed that the epicotyl of *Phaseolus multiflorus* elongated 3.6 times more in the dark than in the light, and that 34% of this increase was due to more cell divisions. There appeared to be a specific length of primary epidermal cells, independent of darkness or light, necessary for the initiation of cell division. They hypothesized that for each kind of cell there is a specific size at which division takes place. Cell size does not appear to be a critical factor in the initial stimulation of cell division by gibberellin in primary elongating meristems. As mentioned above, mean cell size is often smaller in GA-treated tissues which are rapidly dividing. In a dwarf strain of *Pharbitis nil* treated with gibberellic acid, Okuda (1964) found most mitotic figures in the upper 500 μm of the shoot after 12 hours. Cell lengths in this region averaged 16.3 μm. Seventy-two hours after gibberellin treatment, most mitotic figures were found 700 μm to 1600 μm below the apex where mean cell size had increased to 27.3 μm. Similar results were found in shoot apex of *Erigeron annuus* treated with gibberellin (Okuda 1964).

Likewise, the fastest rates of cell division do not always occur in regions of the stem exhibiting the fastest rates of cell elongation due to gibberellin treatment. In GA-treated scapes of *Gerbera*, Sachs (1968) found that high rates of pith-cell elongation and division overlapped, but did not coincide for the 5 to 20 cm stages of elongation. Gibberellin-treated dwarf watermelon seedlings behave similarly, exhibiting the highest rates of cell division in the upper 1 to 2 mm of hypocotyls of 5-day-old seedlings, and the highest rates of cell elongation below this region (Loy, unpublished data).

The above examples serve to illustrate that neither cell size nor rates of cell elongation are necessarily correlated directly with rates of cell division. In certain plant tissues or organs cell division or mitosis can occur prior to detectable cell expansion (Haber and Liuppold 1960; Naylor et al. 1954). Thus, although cell elongation always accompanies cell division in primary elongating meristems, it does not appear likely that the former process directly regulates the latter.

Effects of Gibberellin on Rates and Patterns of Cell Proliferation

The general effect of gibberellin on the pattern of cell proliferation in shoots was clearly established by the work of Sachs and coworkers on rosette plants (Sachs et al. 1959a,b) and on caulescent plants treated with GA-antagonistic growth retardants (Sachs and Kofranek 1963; Sachs et al. 1960). The meristematic activity in those plants was determined by examining mitotic figures in longitudinal sections through the axis of the shoot. Gibberellin treatment both extended the length of the subapical region and increased the number of mitotic figures throughout the meristematic zone (Figures 4-1 and 4-2).

Okuda (1964) studied the effects of GA on growth and cell division in a dwarf strain "Kidachi" of *Pharbis nil* Chois. and on a wild growing biennial plant, *Erigeron annuus* L. In the dwarf *Pharbitis*, GA treatment markedly increased the frequency of

A. SAMOLUS

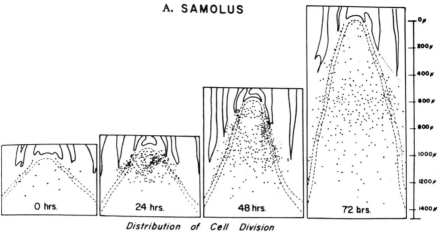

Distribution of Cell Division

B. HYOSCYAMUS

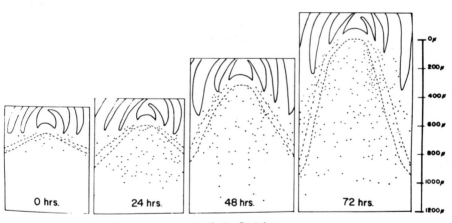

Distribution of Cell Division

Figure 4-1.
Number and position of mitotic figures in the median 64 μm (8 median longisections, 8 μm per section) of the apices of *Samolus*, A, and *Hyoscyamus*, B, following application of GA. Each mitotic figure is indicated by a dot. The pith tissue is bounded by the apical meristem at the top and the vascular tissue on the sides. The boundaries for the vascular tissue and the lower limit of the apical meristem are indicated by dashed lines. Observations for cortical tissue were confined to the area bounded by the outer edge of the vascular tissue and the line connecting the leaf bases (From Sachs et al. 1959a).

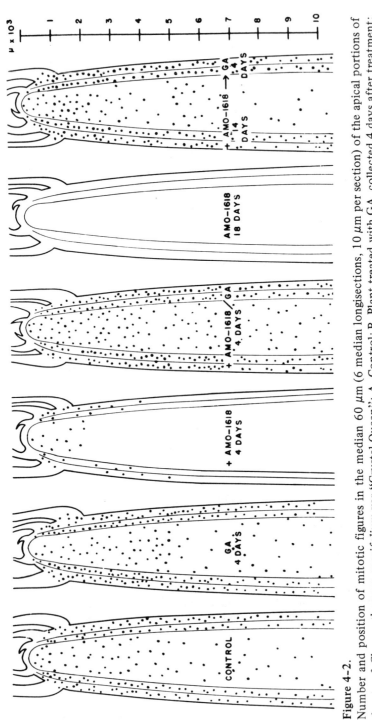

Figure 4-2.
Number and position of mitotic figures in the median 60 μm (6 median longisections, 10 μm per section) of the apical portions of shoots of *Chrysanthemum morifolium* var "Crystal Queen": A. Control; B. Plant treated with GA, collected 4 days after treatment; C. Treated with Amo-1618, observed 4 days after treatment; D. Treated with Amo-1618 plus GA, collected 4 days after treatment; E. Amo-1618, after 18 days; F. Amo-1618, after 14 days followed by GA, observed 4 days after application of GA. Each dot represents a transverse mitotic figure. The pith tissue is bounded by the apical meristem at the top and the vascular tissue on the sides. The boundaries for the vascular tissue (and lower limit of the apical meristem) are indicated by thin lines. Observations for cortical tissue were confined to the area bounded by the outer edge of the vascular tissue and the line connecting the leaf bases. For convenience most of the leaf bases and vascular traces are not shown (From Sachs et al. 1960.).

mitotic figures and also extended the region of mitotic activity as compared to untreated plants. Okuda obtained similar results with gibberellin treatment of *Erigeron*
(Figure 4-3). With both species the meristematic zone lengthened with time and the
region of highest mitotic activity shifted from the distal most region of the sub-
apex to lower regions. This is consistent with Sachs data on *Samolus* (Figure 4-2).

The above reports could be criticized because GA-induced increases in mitotic
figures could reflect increases in the duration of mitosis rather than increased cell
divisions. However, as mentioned previously, it is well documented that gibberellin
can increase the number of cells in an elongating shoot, and GA effects on the cell
cycle suggest that it would shorten and not lengthen the period of mitosis (see
discussion in next section).

Jacqmard (1968) studied both the mitotic activity and the percentage labeled
nuclei following a 4-hour treatment of the shoot apex with ^{3}H-thymidine. Gibberellin stimulated mitotic activity and increased the labeling index in all regions of the
apex, but the largest effect by far was on the subapical region, where the labeling
index increased about 5-fold in GA-treated versus control plants (Figure 4-4). Thus,
in the primary elongating meristems of *Rudbeckia*, gibberellin treatment promotes a
considerable rise in the number of cell nuclei which are synthesizing DNA or are in
S phase of the mitotic cycle.

In the subapical meristem of watermelon seedlings Liu and Loy (1976) made
cell counts in longitudinal cell files throughout the length of hypocotyls. From
these data rates of cell production per cell file per hour were calculated to be 1.20
for dwarf seedlings, 2.09 for normal seedlings, and 3.14 for GA-treated dwarf
seedlings. In a later study (unpublished) dwarf seedlings, plus or minus GA, were
given a 1-hour pulse of ^{3}H-thymidine, and mitotic indices were determined on
squashed pith cells. Labeling indices in the dwarf and GA-treated dwarf were 50%
higher in the upper 0 to 1100 μm region of the subapical meristem than in the
region extending from 1100 to 2200 μm (Table 4-1). In both regions labeling
indices in GA-treated dwarf seedlings were about 40 percent higher than in untreated dwarfs.

By combining the labeling data from Table 4-1 with the cell cycle data from
Table 4-2, it is possible to estimate the proportions of rapidly proliferating cells
using the equation,

$$GF = \frac{LI \times T_c}{T_s + T_1} \tag{4.1}$$

where GF is the growth fraction or the proportion of cells dividing with a short
cycle (see Clowes, 1971, for discussion), T_c is the duration of the cell cycle, LI is
the labeling index, T_s is the duration of S phase, and T_1 is the labeling period.
Using equation 4.1 the proportions of rapidly proliferating cells (GF) in the 0 to
1100 μm region of untreated dwarf and GA-treated dwarf seedlings were calculated
to be 0.32 and 0.49, respectively. Although these values were derived from two
separate experiments, the results indicate that gibberellin induces a substantial
increase in the proportion of proliferating cells.

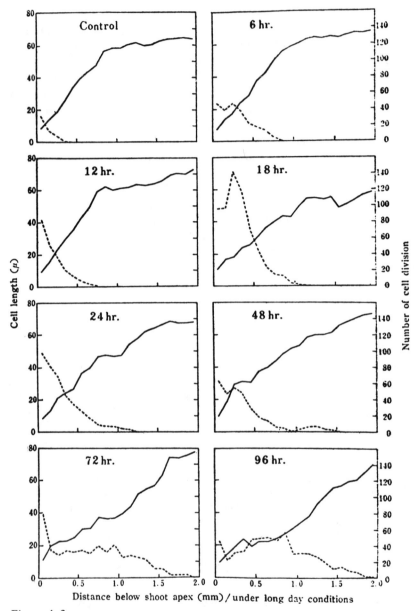

Figure 4-3.
Mean cell length and the number of mitotic figures in median longitudinal sections, 8 μm thick, of the shoot of *Erigeron annuus*. Gibberellin at 1000 ppm was applied as a 0.04 ml drop at the base of the leaves as close to the apex as possible (From Okuda 1964). Dashed Line — number of transverse cell divisions; solid line — mean cell length.

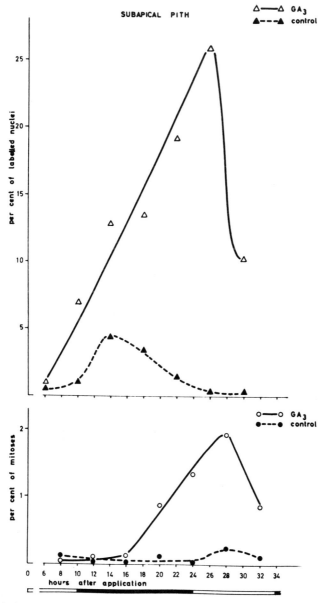

Figure 4-4.
Mitotic index and labeling index for nuclei of subapical pith cells of *Rud-beckia bicolor* at various times after the application of GA$_3$. Partially defoli-ated shoot tips were supplied 4 μC ^{3}H-thymidine (specific activity, 5C/mM) by means of cotton plugs each containing an 0.2 ml aliquot of the labeled solution. The incorporation period was 4 hours. (From Jacqmard 1968.)

Table 4-1
Labeling indices for nuclei of subapical pith cells from longitudinal segments of hypocotyls of dwarf watermelon seedlings untreated or treated with GA_3

| Treatment | *Labeling indices in longitudinal segments below apex* | |
	0–1100 μ	*1100–2200 μ*
WB-2-control	15.1 ± 0.7*	10.2 ± 1.7
WB-2-GA	21.3 ± 1.2	14.2 ± 3.6

Note: For gibberellin treatment seedlings were transferred after 75 hours of germination to Petri dishes containing absorbant wadding saturated with a 3.2×10^{-4} M concentration of GA_3. Forty-five hours after GA treatment ^{3}H-thymidine was administered to the apices of seedlings in 10 μl aqueous droplets containing 4 μC of labeled thymidine (specific activity, 20 C/mM). Following a 1-hour labeling period, the labeled thymidine was washed off, and the hypocotyls were fixed 2-hours later. Squashes were prepared from subapical pith cells and were covered with autoradiographic stripping film (Kodak AR 10). Films were developed after a 20-day exposure period.
*Standard deviation, 5 replications.

The above results emphasize another characteristic of plant meristems often not appreciated; that is, not all cells within an active meristem are dividing even under optimum conditions for growth. This phenomenon has been most extensively studied in root meristems. MacLeod (1968), using cell cycle and labeling index data, found that only 56% of the cells in the apical region of lateral roots of *Vicia faba* were proliferating. Clowes (1971) examined four regions of the primary root of *Zea mays* pulse labeled with ^{3}H-thymidine. He found variations in the percentages of cells dividing with a short cycle ranging from about 17% in the quiescent center to 80% in the root cap initials. Evans and Van't Hof (1975) determined the proportions of fast proliferating cells in root meristems of four species. They reported values of 0.62 for *Helianthus*, 0.62 for *Pisum*, 0.28 for *Triticum*, and 0.56 for *Vicia*. After a long exposure (48 hours) of roots to ^{3}H-thymidine, there were still substantial proportions of unlabeled cells in the meristems of the four species, indicating that some cells were apparently permanently halted in G_1 or G_2.

The apparent distribution of cells into a growth fraction and into a fraction that are temporarily or permanently arrested represents a perplexing problem with respect to understanding the regulation of cell proliferation in plant meristems. Why are some cells triggered to divide by gibberellin or other hormones while other neighboring cells fail to respond?

Effect of Gibberellin on the Cell Cycle

Frequency of mitotic figures has been the most common parameter for determining the timing of the cell division response to gibberellin. In *Pharbitis nil* and *Erigeron*, Okuda (1964) observed a rise in number of mitotic figures within 12 hours after the treatment of apices with gibberellic acid. Since no data are available on cell cycle durations in these species, it is not possible to assess the timing of the GA response. Sachs et al. (1959b) observed a lag period of between 16 to 20 hours prior to gibberellin stimulation of mitosis in *Samolus parviflorus* and *Hyocyamus niger*. In the case of *Samolus*, a cell generation time of 24 hours was estimated from mitotic peaks in pith cell populations partially synchronized in division by gibberellin. Thus, it was inferred that gibberellin must be inducing pith cells in early interphase to divide.

Jacqmard (1968) reported a 20-hour lag period prior to a rise in frequency of mitotic figures in the subapical pith meristem of *Rudbeckia bicolor*, a GA-responsive, long day plant. The lag period for increases in labeled nuclei following a 4-hour pulse of ^{3}H-thymidine was 10 hours for this same region. The evidence, as summarized by Jacqmard (1968), "suggest that the primary action of GA is to induce the release of nuclei from the presynthetic phase G_1 of the mitotic cycle to the phase of DNA synthesis S."

Control nuclei in *Rudbeckia* also exhibited a peak labeling index at 14 hours, indicating a nonrandom distribution of cell divisions over time (Figure 4-4). A peak in per cent mitotic figures at 28 hours is also suggested in Figure 4-4. However, this peak may not be significant since the duration of S and G_2 + M/2 periods together in *Rudbeckia* was later estimated by Jacqmard (1970) to be 20.8 hours, not 14 hours as extrapolated from the labeling and mitotic index curves of Figure 4-4. The closeness of the peaks for labeled nuclei and mitotic figures in GA-treated plants is difficult to interpret, but may indicate that following a lag period, gibberellin treatment shortens either S or G_2, or both phases.

Using the labeled thymidine technique, Liu and Loy (1976) determined the duration of the cell cycles in seedlings of a normal strain of watermelon (Sugar Baby) and in a dwarf strain (WB-2) with and without applied gibberellin. Gibberellin markedly reduced the length of the cell cycle in the genetic dwarf primarily by shortening the S phase from 11.3 to 7.2 hours (Table 4-2). The G_1 and G_2 periods were also shortened in the GA-treated dwarf, G_1 from 6.0 to 4.2 hours and G_2 from 7.1 to 5.8 hours.

The shortening of the S period by gibberellin does not necessarily reflect faster rates of DNA synthesis. In eukaryotes there are numerous DNA replicating sites which are activated at different times during S phase (Mitchison 1971). According to current theory (Taylor 1974), DNA replication occurs by bidirectional movement of DNA replicating forks following initiation of DNA synthesis at the specific replicating sites. Thus, absolute rates of DNA synthesis are determined by rates of replicating fork movement, and this may vary at different times during S phase (Housman and Huberman 1975). Moreover, the rate at which replicating sites are activated can have a greater influence on the length of the S period than differences in rate of fork movement (Painter and Schaefer 1969).

Table 4–2
Duration in hours for cell cycles and their component stages
G_1, S, G_2, and M in subapical pith cells in dwarf (WB-2) and
GA-treated dwarf watermelon seedlings

Treatments	T_c	G_1	S	G_2	M
WB-2	26.2	6.0	11.3	7.1	1.8
WB-2 (GA)	18.9	4.2	7.2	5.8	1.7

Note: Durations of cell cycle phases were estimated from time-
course curves of labeled mitotic figures by the ^{3}H-thymidine
labeling method. Data from Liu and Loy (1976) modified
so as to include estimates of the duration of mitosis (T_m)
from the equation,

$$T_m = \frac{MI \times (T_s + T_1)}{LI}$$

Values for LI used for calculations represent mean labeling
indices for the 0 to 2200 μm region of subapical meristem as
given in Table 4–1.

Embryonic cells of radicles in different species of plants are arrested in G_1 or in
G_1 and G_2 periods (Van't Hof 1973). Likewise, cultured root meristems starved of
a carbohydrate source arrest preferentially in G_1 and G_2 phases (Van't Hof 1966,
1973; Webster and Van't Hof 1970). This has led to the hypothesis that the mitotic
cycle is governed by two principal control points; one in G_1, involving the G_1 to
S transition, and the other in G_2, controlling the G_2 to M transition (Van't Hof and
Kovacs 1972). This hypothesis is relevant in the context of gibberellin regulation of
the cell cycle. Although the main effect of GA on the duration of the mitotic cycle
in dwarf watermelon seedlings was to shorten S phase, the shortening of G_1 may
reflect an equally important physiological function of GA in regulating the principal
control point for G_1 to S transition. This would be consistent with the increased
proportion of proliferating cells observed in subapical meristems of GA-treated
Rudbeckia and dwarf watermelon. Exactly what metabolic processes constitute
the G_1 to S transition are not known, but certainly protein synthesis is needed
(González-Fernández et al. 1974; Van't Hof and Kovacs 1970; Webster and
Van't Hof 1970), presumably for directing synthesis of DNA and other chromosomal components. In mammals and lower eukaryotes several enzymes involved in
DNA replication exhibit periodic peaks of activity which coincide with or just
precede DNA synthesis (Mitchison 1971). However, there is no direct evidence that
levels of these enzymes regulate the onset of DNA synthesis, and it has been suggested that proteins participating in the activation of DNA replicating sites may
control the onset of DNA replication (Cummins and Rusch 1966; Muldoon et al.
1971).

INTERACTION OF GIBBERELLIN WITH AUXINS AND CYTOKININS

The initial discovery of the cytokinin, kinetin, was based on the property of kinetin to stimulate cell division in callus cultures of tobacco (Miller et al. 1956). Indole acetic acid was required in the culture medium to obtain cytokinin promotion of cell division in the pith-cell cultures. A major feature of cytokinin action was an enhancement of cytokinesis in binucleate cells resulting from IAA treatment (Das et al. 1956). Autoradiographic and microspectrophotometric estimates of nuclear DNA synthesis in cultured pith cells of tobacco indicated that in the presence of added IAA, kinetin stimulated DNA synthesis, mitosis, and cytokinesis (Das et al. 1958; Patau et al. 1957).

The early results with tobacco pith-callus cultures have largely been confirmed by more recent studies. Thus, although the effects of cytokinins on cell proliferation in tissue cultures vary depending upon the culture conditions (Simard 1971) and the species of plant studied (Fosket and Short 1973; Phillips and Torrey 1973), cytokinins in concert with auxins usually promote DNA synthesis, mitosis, and cytokinesis.

In contrast, direct evidence for participation of cytokinins in controlling cell division in primary elongating meristems is meager. Exogenously applied cytokinins often have no effect (de Ropp 1956; Kende and Sitton 1967) or inhibit (Sprent 1968; Wittwer and Dedolph 1963) stem elongation. Perhaps exogenous applications result in supra-optimal concentration of cytokinins. Application of 0.1 μg of benzyladenine (BA) to apices of *dw*-2 dwarf watermelon seedlings increases hypocotyl length by 50%, but only slightly increases the number of pith cells per longitudinal file. However, BA inhibits hypocotyl elongation in the non-dwarf watermelon variety, "Sugar Baby," and also retards prolonged gibberellin promotion of growth in dwarf seedlings (Loy, unpublished data).

Since root tips have been shown to be a major source of cytokinins (Kende 1965; Kende and Sitton 1967; Short and Torrey 1972; Weiss and Vaadia 1965), the use of rootless seedlings provides a means of studying cytokinin effects on shoot growth under conditions where cytokinins as well as other hormones produced in the roots might be limiting. In light-grown, rootless pea seedlings, GA but not cytokinins stimulated hypocotyl elongation (Kende and Sitton 1967). Holm and Key (1969) studied the effect of several plant hormones on cell elongation and DNA synthesis in rootless soybean seedlings. In the apical section of the hypocotyl, GA alone or in combination with a cytokinin stimulated growth, whereas cytokinin alone was ineffective. Either GA or cytokinin alone enhanced levels of DNA synthesis above that of controls, and together, these two hormones synergistically promoted DNA synthesis in the apical sections. Since continued growth of the soybean hypocotyl is dependent on cell division in the apical section, GA and cytokinin promotion of DNA synthesis is probably a reflection of the effect of these compounds on cell division in the primary elongating meristem. Rootless soybeans failed to respond to the auxin 2, 4-D over a wide range of concentrations.

Fisher (1970) studied the effect of auxins, gibberellins, and cytokinins on stem elongation and on cell division in a basal intercalary meristem of *Cyperus alterni-*

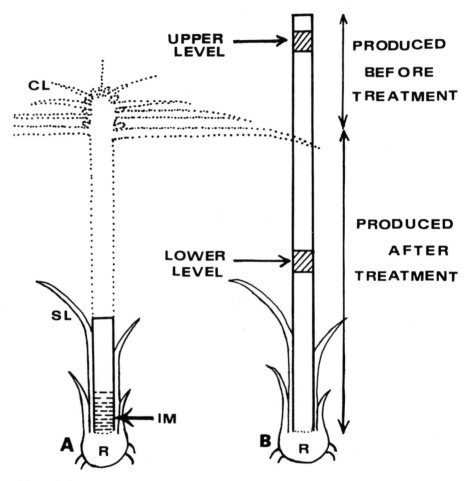

Figure 4-5.
Diagram of growing culm of *Cyperus alternifolius* decapitated at start of treatment (A) and some weeks after treatment (B). Plant hormones were applied in 1% agar solutions in plastic capsules placed on the cut surface of the internode immediately after decapitation. Upper and lower levels for anatomical are indicated. IM, inter- calary meristem; R, rhizome; SL, sheathing basal leaf; CL, crown leaves. (From Fisher 1970.)

folius (Figure 4-5). Decapitation of the apical portion of the stem and crown of the plant completely halted growth and cell division in the intercalary meristem. By removing either the crown, the crown leaves, the crown buds, or the basal buds or leaves, it was shown that the presence of the upper part of the stem and not the other organs were requisite for internodal growth. Both benzyladenine (10 ppm) and gibberellin (7.5 ppm) restored normal stem growth in decapitated plants while indoleacetic acid at 100 ppm was only slightly stimulatory. Anatomically, both BA and GA promoted cell division in the intercalary meristem, and GA also enhanced cell enlargement. Since either BA or BA + GA restored the normal growth pattern in decapitated *Cyperus*, it was concluded that the flow of these hormones from the upper culm regulates stem growth. The presence of rhizomes and roots in decapitated *Cyperus* apparently did not serve as a source of cytokinins for cell division in the intercalary meristem.

In view of the cytokinin requirement for cell division in most tissue cultures and for cell division associated with bud morphogenesis in several species (Fox 1969), it seems probable the cytokinins are also required for cell division in primary elongating meristems. However, more research is needed to substantiate their role in intact plants.

Auxins

Auxins have been implicated as regulators of cell division in several plant growth phenomena such as xylem regeneration (Jacobs 1952), phloem development (Camus 1949; La Motte and Jacobs 1963), root formation (Went and Thimann 1937), and in plant tissue cultures (Gautheret 1955).

The extent of auxin-mediated promotion of cell proliferation varies with the plant organ or tissue under observation, the species of plant being studied, or with the interaction of auxin with other plant hormones or with nutrients. Auxin alone can induce xylem formation in wounded stems or in explants (Jacobs 1952; Wareing, 1958) and stimulate cell division in tissue cultures (Das et al. 1956; Gautheret 1955). However, the presence of a cytokinin is often required (Fosket and Torrey 1969; Phillips and Torrey 1973) or enhances (Das et al. 1956) cell division in tissue cultures. In addition, Wareing (1958) demonstrated in disbudded shoots of three species of woody plants that gibberellin together with auxin elicited more complete xylem development than either compound applied alone.

Auxins are present in elongating shoots, and probably participate in regulating cell division in primary elongating meristems. Unfortunately, there is little direct evidence for such a role for auxin. Most studies on cell division in primary elongating meristems require an intact developing shoot. Exogenously applied auxins usually do not promote growth of intact plants, presumably because such plants are maintaining their own endogenous supplies of auxin (Phillips 1971).

One type of specialized stem occurring in some plant species, the scape or flower stalk, is promoted by exogenously applied auxin (Sachs 1965). Sachs (1968) reported that both IAA and GA promoted cell elongation and cell division in deflowered scapes of *Gerbera jamesonii*. In decapitated scapes (receptacle and bracts removed along with inflorescence) GA and IAA promoted cell elongation without cell division. This indicated that some other substance(s) produced in the bracts or receptacle was necessary for cell division. Kinetin did not effect growth of the scapes either alone or in combination with GA, IAA, or GA + IAA.

To clarify the action of auxin on cell division in primary elongating meristems, test systems are needed in which endogenous auxin is limiting in the target organ. One approach would be to use antiauxins. Another means would be to locate auxin deficient mutants such as has been done in the case of gibberellin responding dwarfs.

TERMINATION OF CELL PROLIFERATION

One pertinent problem which has been largely neglected is the regulation of the termination of cell division in primary elongating meristems. Why does an internode or hypocotyl cease producing new cells for continued elongation?

It could be proposed for an elongating and developing shoot that as internodes are farther removed from the source of auxins and gibberellins (in young expanding leaves), the levels of these hormones drop below concentrations needed to induce cell division. For instance, Jones and Phillips (1966) demonstrated a good correlation between the amount of diffusible gibberellins in internodes of *Helianthus annuus* and the degree of elongation. Young expanding internodes, which would be expected to have active primary elongating meristems, had much higher GA levels than older internodes. A similar relationship has been shown for auxin content and the amount of stem elongation in epicotyls of *Pisum* (Scott and Briggs 1960).

The above correlations of hormone levels to plant growth probably exist in other plant species. Nevertheless, they are probably not the most important factors mediating the termination of cell proliferation. In seedlings such as dwarf watermelon, gibberellin can be supplied continuously, and yet, cell division still ceases after 8 or 9 days of seedling growth (unpublished data). Natural inhibitors could very well be involved in controlling the arrest of cell division. Inhibition of stem growth is a common effect of abscisic acid (ABA). In addition, ABA has been shown to counteract promotions of growth mediated by GA and IAA (Addicott and Lyon 1969). It is thus possible that levels of ABA increase with age of seedlings or internodes, subsequently counteracting the promotive effects of GA and other hormones on cell division in primary elongating meristems. Unfortunately, it may be difficult to test this possible role of ABA because of lack of competitive inhibitors or ABA or of compounds which interfere with synthesis of ABA.

Ethylene is another hormone which can inhibit stem elongation. Burg and Burg (1968) and Burg et al. (1971) have also shown the ethylene can inhibit cell division. However, there is no direct evidence that ethylene plays a natural role in controlling cell arrest in primary elongating meristems.

CONCLUDING REMARKS

There is a surprising lack of research on hormonal regulation of cell proliferation in primary elongating meristems, considering the importance of these meristems in shoot development. It is probable that a number of hormones are involved in the regulation of cell division in primary elongating meristems, but the role of hormones other than gibberellin needs to be clarified. The importance of gibberellin in regulating subapical meristematic activity is well founded, primarily because of the availability of good experimental systems for studying GA effects such as genetic dwarfs and rosette plants. Evidence from *Citrullus* and *Rudbeckia* suggests that gibberellin can induce cells in G_1 to enter S phase and can shorten the period required for DNA replication. These results need to be substantiated in other species of plants. In meristems of *Citrullus* or *Rudbeckia*, temporarily arrested cells may be predominantly in G_1 phase, so that the response to GA is seen in the G_1 to S transition. At least in root meristems, cell arrest usually occurs in both G_1 and G_2 phases (Van't Hof 1973). If this condition occurs commonly in primary elongating meristems, how would cells arrested in G_2 react to gibberellin? Perhaps the cells arrested in G_2 represent the fraction of meristematic cells which are not induced to divide by GA. If indeed, a major effect of gibberellin on cell division is controlling the initiation and subsequent replication of DNA, then experimental systems must be concerned with the effects of gibberellin on nuclei acid pools, on enzymes involved in DNA synthesis, on the number of DNA replicating sites and the rate of their activation during S phase, and finally, on the rate of fork movement during DNA replication.

ACKNOWLEDGEMENTS

The author wishes to thank Dr. Roy Sachs, University of California, Davis, and Dr. Cleon Ross, Colorado State University, Fort Collins, for their critical reviews of the manuscript.

I am especially grateful to Dr. Russell Jones, University of California, Berkeley, for his helpful suggestions and his support of my research on cell proliferation in his laboratory while I was on sabbatical leave in 1974-75.

LITERATURE CITED

Addicott, F. T. and J. L. Lyon. 1969. Physiology of abscisic acid and related substances. *Ann. Rev. Plant Physiol.* **20**:139–164.

Arney, S. E. and P. Mancinelli. 1966. The basic action of gibberellic acid in elongation of 'meteor' pea stems. *New Phytol.* **65**:161–175.

Basford, K. H. 1961. Morphogenetic responses to gibberellic acid of a radiation-induced mutant dwarf in groundsel, *Senecio vulgaris* L. *Ann. Bot.* (N.S.) **25**: 279–302.

Bernier, G., R. Bronchart and A. Jacqmard. 1964. Action of gibberellic acid on mitotic activity of the different zones of the shoot apex of *Rudbeckia bicolor* and *Perilla nankinensis*. *Planta* **61**:236–244.

Brotherton, W. and H. H. Bartlett. 1918. Cell measurement as an aid in the analysis of quantitative variation. *Amer. J. Bot.* **5**:192–206.

Burg, S. P. and E. H. Burg. 1968. Ethylene formation in pea seedlings: its relation to the inhibition of bud growth caused by indole-3-acetic acid. *Plant Physiol.* **43**:1069–1074.

Burg, S. P., A. Apelbaum, W. Eisinger and B. G. Kang. 1971. Physiology and mode of action of ethylene. *Hortscience* **6**:359–364.

Camus, G. 1949. Recherches sur le role des bourgeons dans les phenoménes de morphogénése. *Rev. Cytol. Biol. Veg.* **11**:1–195.

Clowes, F. A. L. 1971. The proportion of cells that divide in root meristems of *zea mays* L. *Ann. Bot.* (N.S.) **35**:249–261.

Cummins, J. E. and H. P. Rusch. 1966. Limited DNA synthesis in the absence of protein synthesis in *Physarum polycephalum*. *J. Cell Biol.* **31**:577–583.

Das, N. K., K. Patau, and F. Skoog. 1956. Initiation of mitosis and cell division by kinetin and indoleacetic acid in excised tobacco pith tissue. *Physiol. Plant.* **9**: 640–651.

Das, N. K., K. Patau, and F. Skoog. 1958. Autoradiographic and microspectrophoto-metric studies of DNA synthesis in excised tobacco pith tissue. *Chromosoma* **9**:606–617.

Evans, L. S. and J. Van't Hof. 1975. The age-distribution of cell cycle populations in plant root meristems. *Exp. Cell Res.* **90**:401–410.

Fisher, J. B. 1970. Control of the internodal intercalary meristem of *Cyperus alternifolius. Amer. J. Bot.* **57**:1017–1026.

Fosket, D. E. and K. C. Short. 1973. The role of cytokinin in the regulation of growth, DNA synthesis and cell proliferation in cultured soybean tissues (*Glycine max* var. Biloxi). *Physiol. Plant.* **28**:14–23.

Fosket, D. E., and J. G. Torrey. 1969. Hormonal control of cell proliferation and xylem differentiation in cultured tissues of *Glycine max* var. Biloxi. *Plant Physiol.* **44**:871–880.

Fox, E. J. 1969. The cytokinins. In *Physiology of Plant Growth and Development*, ed. M. B. Wilkins, McGraw-Hill, New York, pp. 85–123.

Gautheret, R. J. 1955. The nutrition of plant tissue cultures. *Ann. Rev. Plant Physiol.* **6**:433–484.

Gifford, E. M. and G. E. Corson. 1971. The shoot apex in seed plants. *Bot. Rev.* **37**:143–229.

González-Fernández, A. G. Giménez-Martín, M. E. Fernández-Gómez and C. de la Torre. 1974. Protein synthesis requirements at specific points in the interphase of meristematic cells. *Exp. Cell. Res.* **88**:163–170.

Greulach, V. A. and J. G. Haesloop. 1958. The influence of gibberellic acid on cell division and cell elongation in *Phaseolus vulgaris. Amer. J. Bot.* **45**:566–570.

Haber, A. H. and H. J. Liuppold. 1960. Separation of mechanisms initiating cell division and cell expansion in lettuce seed germination. *Plant Physiol.* **35**: 168–173.

Holm, R. E. and J. L. Key. 1969. Hormonal regulation of cell elongation in the hypocotyl of rootless soybean: an evaluation of the role of DNA synthesis. *Plant Physiol.* **44**:1295–1302.

Holttum, R. E. 1955. Growth-habits of monocotyledons, variations on a theme. *Phytomorphology* **5**:399–413.

Housman, D. and J. A. Huberman. 1975. Changes in the rate of DNA fork move-ment during S phase in mammalian cells. *J. Mol. Biol.* **74**:173–181.

Jacobs, W. P. 1952. The role of auxin in differentiation of xylem around a wound. *Amer. J. Bot.* **39**:301–309.

Jacqmard, A. 1968. Early effects of gibberellic acid on mitotic activity and DNA synthesis in the apical bud of *Rudbeckia bicolor. Physiol. Veg.* **6**:409–416.

Jacqmard, A. 1970. Duration of mitotic cycle in apical bud of *Rudbeckia bicolor. New Phytol.* **69**:269–274.

Jones, R. L. and I. D. J. Phillips. 1966. Organs of gibberellin synthesis in light-grown sunflower plants. *Plant Physiol.* **41**:1381–1386.

Kende, H. 1965. Kinetin-like factors in the root exudate of sunflowers. *Proc. Natl. Acad. Sci.* **53**:1302–1307.

Kende, H. and D. Sitton. 1967. The physiological significance of kinetin- and gib-berellin-like root hormones. *Ann. N.Y. Acad. Sci.* **144**:235–243.

La Motte, C. E. and W. P. Jacobs. 1963. A role for auxin in phloem regeneration in *Coleus* internodes. *Devel. Biol.* 8:80–98.

Liu, P. B.-W. 1971. A genetic, morphological and physiological investigation of bush and vine genotypes in *Citrullus lanatus.* M.S. Thesis, University of New Hampshire, Durham.

Liu, P. B.-W. 1974. Cell proliferation and elongation in normal, dwarf and gibberellin-treated dwarf watermelon seedlings. Ph.D. Dissertation, University of New Hampshire, Durham.

Liu, P. B.-W. and J. B. Loy. 1976. Action of gibberellic acid on cell proliferation in the subapical shoot meristem of watermelon seedlings. *Amer. J. Bot.* 63: 700–704.

Loy, J. B. and P. B. W. Liu. 1974. Response of seedlings of a dwarf and a normal strain of watermelon to gibberellins. *Plant Physiol.* 53:325–330.

MacLeod, R. D. 1968. Changes in the mitotic cycle in lateral root meristems of *Vicia faba* following kinetin treatment. *Chromosoma* 24:177–187.

Miller, C. O., F. Skoog, F. S. Okumara, M. H. Von Salta and F. M. Strong. 1956. Isolation, structure and synthesis of kinetin, a substance promoting cell division. *J. Amer. Chem. Soc.* 78:1375–1380.

Mitchison, J. M. 1971. *The Biology of the Cell Cycle.* Cambridge University Press, London.

Muldoon, J. J., T. E. Evans, P. F. Nygaard and H. H. Evans. 1971. Control of DNA replication by protein synthesis at defined times during S period in *Physarum polycephalum. Biochim. Biophys. Acta* 247:310–312.

Naylor, J., G. Sander and F. Skoog. 1954. Mitosis and cell enlargement without cell division in excised tobacco pith tissue. *Physiol. Plant.* 7:25–29.

Okuda, M. 1964. Physiological observation of the gibberellin effects on the development and growth of plants. *Contr. Biol. Lab. Kyoto Univ.* 18:1–36.

Painter, R. B. and A. W. Schaefer. 1969. Rates of synthesis along replicons of different kinds of mammalian cells. *J. Mol. Biol.* 45:467–479.

Patau, K., N. K. Das and F. Skoog. 1957. Induction of DNA synthesis by kinetin and indoleacetic acid in excised tobacco pith tissue. *Physiol. Plant.* 10:949–966.

Phillips, I. D. J. 1971. *Introduction to the Biochemistry and Physiology of Plant Growth Hormones.* McGraw-Hill, New York.

Phillips, R. and J. G. Torrey. 1973. DNA synthesis, cell division and specific cytodifferentiation in cultured pea root explants. *Devel. Biol.* 31:336–347.

deRopp, R. S. 1956. Kinetin and auxin activity. *Plant Physiol.* 31:253–254.

Sachs, R. M. 1965. Stem elongation. *Ann. Rev. Plant Physiol.* 16:73–96.

Sachs, R. M. 1968. Control of intercalary growth in the scape of *Gerbera* by auxin and gibberellic acid. *Amer. J. Bot.* 55:62–68.

Sachs, R. M. and A. M. Kofranek. 1963. Comparative cytohistological studies on inhibition and promotion of stem growth in *Chrysanthemum morifolium. Amer. J. Bot.* 50:772–779.

Sachs, R. M. and A. Lang. 1957. Cell division and gibberellic acid. *Science* 125: 1144–1145.

Sachs, R. M., C. F. Bretz, and A. Lang. 1959a. Shoot histogenesis: the early effects of gibberellin upon stem elongation in two rosette plants. *Amer. J. Bot.* 46:376–384.

Sachs, R. M., C. F. Bretz, and A. Lang. 1959b. Cell division and gibberellic acid. *Exp. Cell Res.* 18:230–244.

Sachs, R. M., C. F. Bretz, A. Lang, and J. Roach. 1960. Shoot histogenesis: subapical meristematic activity in a caulescent plant and the action of gibberellic acid and amo-1618. *Amer. J. Bot.* **47**:260–266.

Scott, T. K. and W. R. Briggs. 1960. Auxin relationships in the Alaska pea (*Pisum sativum*). *Amer. J. Bot.* **47**:492–499.

Short, K. C. and J. G. Torrey. 1972. Cytokinins in seedling roots of pea. *Plant Physiol.* **49**:155–160.

Simard, A. 1971. Initiation of DNA synthesis by kinetin and experimental factors in tobacco pith tissues *in vitro. Can. J. Bot.* **49**:1541–1549.

Skjedstad, K. R. 1960. Dwarfism and the anatomical basis for the gibberellin response in *Zea mays*. Ph.D. Dissertation, Univ. of Calif., Los Angeles.

Sprent, J. 1968. The effects of benzyladenine on the growth and development of peas. *Planta* **78**:17–24.

Stowe, B. B. and T. Yamaki. 1957. The history and physiological action of gibberellins. *Ann. Rev. Plant Physiol.* **8**:181–216.

Taylor, J. H. 1974. Units of DNA replication in eukaryotes. *Intern. Rev. Cytol.* **37**:1–20.

Van't Hof, J. 1966. Experimental control of DNA synthesizing and dividing cells in excised root tips of *Pisum. Amer. J. Bot.* **53**:970–976.

Van't Hof, J. 1973. The regulation of cell division in higher plants. In *Basic Mechanisms in Plant Morphogenesis*. V. 25, Brookhaven Symposia in Biology, pp. 152–165.

Van't Hof, J. and C. J. Kovacs. 1970. Mitotic delay in two biochemically different Q1 cell populations in cultured roots of pea (*Pisum sativum*). *Radiation Res.* **44**:700–712.

Van't Hof, J. and C. J. Kovacs. 1972. Mitotic cycle regulation in the meristem of cultured roots: the principal control point hypothesis. In *The Dynamics of Meristem Cell Populations. V. 18, Advances in Experimental Medicine and Biology*. Plenum Press, New York, pp. 15–30.

Wareing, P. F. 1958. Interaction between indole-acetic acid and gibberellic acid in cambial activity. *Nature* **181**:1744–1745.

Webster, P. L. and J. Van't Hof. 1970. DNA synthesis and mitosis in meristems: requirements for RNA and protein synthesis. *Amer. J. Bot.* **57**:130–139.

Weiss, C. and Y. Vaadia. 1965. Kinetin-like activity in root apices of sunflower plants. *Life Sciences* **4**:1323–1326.

Went, F. W. and K. V. Thimann. 1937. *Phytohormones*. The MacMillan Co., New York.

Wittwer, W. H. and R. R. Dedolph. 1963. Some effects of kinetin on the growth and flowering of intact green plants. *Amer. J. Bot.* **50**:330–336.

Responses of the Plant Cell Cycle to Stress

Thomas L. Rost   University of California, Davis

INTRODUCTION

Plant meristems are composed of cell populations actively progressing through the cell cycle. These cells progress at different rates, and some cells are naturally arrested. Each cell, however, is positioned at some period within the cycle. The cell cycle consists of four stages first named by Howard and Pelc (1953) (Figure 5-1). G_1 is the stage during which metabolic events are completed in preparation for DNA synthesis; the DNA synthesis period itself is called S; G_2 is the stage preceding mitosis (M). All of these stages are related to each other, in the sense that, in order for one stage to be entered the one preceding it must usually be completed. Each of these stages requires certain specific enzymes, proteins, and RNAs in order to progress to the next cycle stage. In addition, all of the metabolic events require a constant source of oxygen, a certain level of hydration, a carbon source, and the means to create chemical energy. Inhibition of any of these requirements by environmental stresses, either by scarcity or excess will affect the cell cycle (Figure 5-1). Figure 5-2, taken in part from Levitt (1972), shows an example of possible sources of stress which may affect the cell cycle. This review will examine some of the literature on various sources of stress on the cell cycle in plant tissues. Readers are referred to several books which have been written on aspects of the subject (Lea 1947; Kihlman 1966; Levitt 1972; Ashton and Crafts 1973).

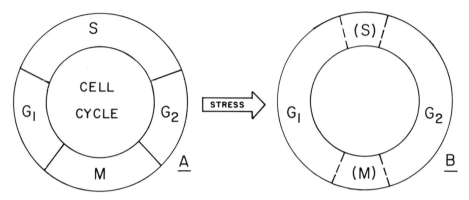

Figure 5-1.
The cell cycle before (A) and after (B) stress. In a proliferating tissue cells are present in all cycle stages G_1, the pre-DNA synthetic period; S, DNA synthesis period; G_2, post-DNA synthetic period; and M, mitosis. After application of environmental stress from a number of sources, the cell cycle distribution will become altered. Figure 5-1B shows this in symbolic form. G_1 and G_2 are usually most sensitive to stress factors so that cell cycle arrest in G_1 and G_2 commonly occurs. S and M are most refractive to stress generally and consequently complete their passage even under stress. Acute stress however, such as extreme oxygen deficiency may arrest cells in S and M as well.

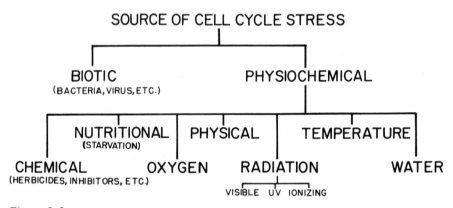

Figure 5-2.
Sources of cell cycle stress. Chart taken in part with permission from Levitt (1972).

NUTRITIONAL STRESS

Brown and Rickless (1949) grew excised pumpkin roots (*Cucurbita pepo*) in medium without sugar. Under these conditions no root extension occurred, no cells divided, and the number of cells in the meristem decreased. If sugars and inorganic salts were added to the medium, division resumed, thus demonstrating the requirement of a carbon source to insure cycle progression.

Van't Hof (1973, 1974) has described in some detail his work since 1965 on the effects of sucrose starvation on root meristems. Since these two recent papers have fully described this work, only a superficial discussion will follow.

Cells in the meristem of excised roots (1 cm) grown in liquid culture medium will proliferate such that all portions of the cell cycle will be present. Roots grown in medium lacking sucrose for 48-72 hours, however, will stop their cycle progression (Van't Hof 1966a). This stationary phase (SP) condition is not random in that each plant species examined had a specific repeatible ratio of cells which stopped selectively in G_1 and/or G_2 (Van't Hof 1973; 1974) (Table 5-1). Cells in SP meristems are reversibly stationary and can be stimulated to resume progression by reculturing in a medium containing sucrose. Addition of tritiated thymidine to the medium makes it possible to follow the progression of cells previously arrested in G_1 as they progress into S; the progression of these same cells into G_2 and M can be followed by scoring for labeled division figures. The progression of previously arrested G_2 cells into mitosis is determined by scoring the numbers of unlabeled division figures. Sample data for this type of experiment is shown in Figure 5-3. This experimental system can be applied readily to experiments designed to examine the cell requirements needed to initiate and facilitate the continued progression of cells through the cycle.

Three species, pea (Webster and Van't Hof 1970), sunflower (Van't Hof and Rost 1972; Rost and Van't Hof 1973), and longpod bean (Van't Hof et al. 1973) have been examined in this way to determine the requirements for RNA and protein synthesis. Protein synthesis inhibitors such as puromycin and cycloheximide inhibit the passage of cells into S or M from their arrested position in G_1 or G_2. González-Fernández et al. (1974) similarly observed in onion roots (*Allium cepa*) that the protein synthesis inhibitor anisomycin inhibits the passage of G_1 and G_2 cells into S and M. These observations indicate the necessity of protein synthesis in order for cycle progression to occur.

Certain specific proteins are necessary for DNA synthesis and mitosis. Yeoman (1974) and Aitchison and Yeoman (1974) point out that specific protein syntheses occur in early and mid-S and during G_2 in Jerusalem artichoke explants. Such enzymes as thymidine kinase and DNA polymerase, for example, increase with the

Table 5-1
Percentage of cells arrested in G_1 and G_2 in carbohydrate-starved stationary phase root meristems

Species	G_1	G_2	S
Glycine max	50	50	0
Helianthus annuus	75	19	2
Pisum sativum	44	56	0
Vicia faba	28	71	0
Triticum aestivum	63	34	0
Zea mays	53	47	0

Data used with permission from Van't Hof 1973.

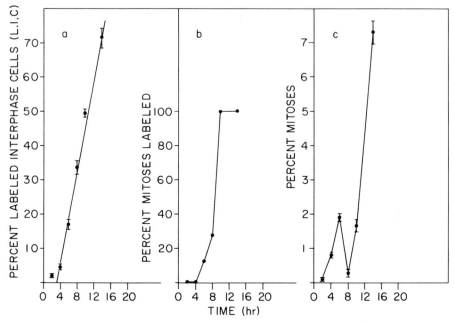

Figure 5–3.
Sample data to show progression of cells from stationary phase meristems of 48-hour starved *Helianthus annuus* roots (from Rost and Van't Hof 1973). Figure 5–3A. Progression of G_1 cells into S as percentage ^{3}H-thymidine labeled interphase cells. Figure 5–3B. Progression of previously arrested G_1 cells into mitosis as percent labeled mitotic figures. Figure 5–3C. This panel shows the passage of G_2 cells into mitosis.

start of the S period (Aitchison and Yeoman 1974). Mazia (1970, 1974) has suggested the requirement of two specific proteins, DNA polymerase and spindle protein (tubulin), which are needed for S and M, respectively. Although the amounts of these proteins increase preceding S and M, their triggered synthesis may not be a regulatory step, but instead an indication that S and M have been "turned-on."

RNA synthesis, or at least the presence of certain RNAs is necessary for S and M to be initiated. Webster and Van't Hof (1970), Rost and Van't Hof (1973), and Van't Hof et al. (1973) transferred starved, stationary phase roots to sucrose medium containing actinomycin D. Interestingly, in all three species, DNA synthesis and mitosis were not inhibited from initiation and progression, at least in part. In pea roots (Webster and Van't Hof 1970) S and M proceeded for 8 hours before being inhibited. In the other two species, sunflower and longpod bean, progression continued unhibited for the sample time duration without new, actinomycin sensitive RNA synthesis. This is interpreted to mean that long-lived RNA, which remains active for the entire starvation period, is present in the roots. Jakob (1972) reported a similar observation in longpod bean roots. In his experiments, roots were treated with actinomycin D during early germination, and in spite of the treatment, DNA synthesis could occur indicating the presence of previously existing RNA.

Additional insight was provided concerning this mechanism of action during starvation by studying polyribosome profiles in starved pea root meristems (Webster and Van't Hof 1973). They noticed that immediately after root excision the number of polyribosomes decreased. If roots were transferred to medium with sucrose, the polyribosome profile increased, but if transferred to medium lacking sucrose it remained low. Even at this low level, however, some protein synthesis did occur. The polyribosome level also resumed when starved roots were transferred to sucrose medium even in the presence of actinomycin D. These experiments demonstrated that protein synthesis was inhibited, although not completely, by sucrose starvation and that polyribosome reformation occurred with the replenishment of sucrose even if actinomycin sensitive RNA synthesis was inhibited.

Principal Control Point Hypothesis

Experiments such as those described above indicated that sucrose starved root tips stopped their cycle progression in G_1 and G_2. DNA synthesis and mitosis, therefore, have a requirement which, if not satisfied, inhibited the entry of cells into these stages. This same observation was also made by D'Amato and his coworkers in seed embryos in response to dehydration (this will be discussed in the next section). Van't Hof and Kovacs (1972) consolidated these observations to formulate the principal control point hypothesis. The hypothesis maintains that under stress cells will stop their cycle progression and will accumulate in G_1 and G_2. Further, this arrested condition is tissue and species specific and is reversible. Additional corroborative work was done by González-Fernández et al. (1974) who demonstrated G_1 and G_2 arrest in onion roots treated with the protein synthesis inhibitor anisomycin. That this control mechanism is a genetic phenomena is suggested from the work of Evans and Van't Hof (1974b). In their study mature unstarved tissue was shown to have cells arrested in G_1 and G_2 with the same ratio as that of starved SP meristems.

The seeds of the principal control point idea can be traced to the work of Gelfant (1962, 1963, 1966), who observed two separate self-perpetuating cell populations in mouse tissues; one which stopped only in G_1 and another which stopped only in G_2. Also, Mazia (1970) discusses two points of regulation—one before DNA synthesis which increases the synthesis of DNA polymerase and a second before mitosis when mitotic proteins must be synthesized. Others, too, (Mitchison 1971) have demonstrated the presence of points in the cell cycle after which progression could not be inhibited, by irradiation for example, but which could be inhibited before these cycle events were completed.

WATER STRESS

During plant growth and development, cells are entirely dependent upon the presence of water for metabolic reactions to take place. A turgid plant organ, such as a root is composed of in excess of 90% water. Any reduction in the degree of hydration must result in altered metabolism and, consequently, in changes of cell cycle behavior.

Seed maturation involves first the development of the embryo and its storage and protective structures. After reaching their structural maturity, most seeds become dehydrated by a natural phenomena. Seed dormancy and self-induced water stress is ideal for the study of cell cycle responses, and D'Amato (1972) and his coworkers have examined a number of different species in this regard. Brunori (1967) studied the changes in distribution of DNA content and mitosis in radicle cells of *Vicia faba* seeds during their ripening period. At 39 days before seed harvest the cotyledons contained 91% water, the radicle mitotic index was 10.5 and, based on cytophotometric measurements, cells were found in G_1, S, and G_2 (2 C, 2 C-4 C, and 4 C relative DNA content, respectively). At 25 days before harvest the cotyledon water content decreased to 83.5%, the radicle mitotic index to 2.7, with cells still distributed in G_1, S, and G_2. At 12 days the reduced water content caused the mitotic index to decrease to 0.4 and the distribution of cells to become accumulated in G_1 and G_2 with no cells in S. At 5 days before harvest the cotyledon water content was reduced to 55%, the mitotic index was zero, and all cells were in G_1 and in G_2 with most cells stopped in the G_1 phase.

Three things are apparent from these data. First, as water content is reduced cells are no longer able to progress through the cell cycle, and they become arrested in G_1 and in G_2. Second, in terms of dehydration, DNA synthesis was most sensitive as no more cells entered S after 75% water content was reached. Third, mitosis was also inhibited but only after cotyledon water content was 55%. Inhibition of the two events would, of course, account for the apparent arrest in G_1 and G_2.

Brunori et al. (1966) x-irradiated dry *Vicia faba* seeds and analyzed the types of chromosome aberrations which occurred during the first post-irradiation mitotic cycle after germination. Chromosome breaks (B'') would occur only if the irradiation induced DNA lesion occurred to a G_1 cell, while chromatid breaks (B') would indicate a break induced while the cell was in G_2. Brunori et al. (1966) first measured the DNA distribution of dry seed radicles cytophotometrically. They found all cells stopped in G_1 and G_2, but with a good deal of variability. Some seeds were irradiated and then grown in water containing ^{3}H-thymidine. In this case the first mitotic figures were observed to be labeled and to have responded to irradiation with B'' breaks; these cells had been previously arrested in G_1. These data strengthen the original observation that cells in dry embryos are arrested in G_1 and G_2.

Cells in all embryonic organs do not become arrested in the same cycle stages, due to water stress. Avanzi et al. (1969) examined the development of *Triticum durum* embryos in terms of cycle position of cells in each embryo organ. During dehydration, DNA synthesis stopped at a higher water content than mitosis. In the shoot apex and leaf primordia all cells contained the G_1 DNA content. The root apex contained both G_1 and G_2 cells, ranging from 61-86% (G_1): 14-39% (G_2). The seminal root also contained a G_1/G_2 mixture. In addition to being organ specific within one species, the ratio of arrested cells due to water stress in embryos is also species specific. For example, *Vicia faba, Triticum durum, Pisum sativum,* and *Hordeum vulgare* radicles all have both G_1 and G_2 arrested cells. *Helianthus annuus, Allium cepa,* and *Pinus pinea,* on the other hand, have arrested radicles with only

G_1 nuclei. Van't Hof (1974) has compiled a list (Table 5-2) of dry seed radicles for which arrested G_1 and G_2 position data are available.

These two considerations, organ and species specificity, and the environmentally induced variability which occurs led Brunori and D'Amato (1967) to the following conclusion: "The variation in relative proportions of 2 C (G_1) and 4 C (G_2) nuclei in mature embryos would then be the resultant of genotypic constitution and environmental factors: an assumption well in line with the experimental evidence available."

Investigations of cell cycle water stress responses in intact plants are not so clear-cut as in embryo studies. Yee (1976) conducted experiments with intact *Vicia faba* seedlings grown in 20% polyethylene glycol in low light. Under stress, roots showed a reduction in the number of cells in S and M for the first 24 hours. After this time cells of treated roots demonstrated a "self-repair" mechanism in that from 48 to 72 hours the number of cells in S and M returned to the control levels. Possibly, the depressed solute potential of the stressed roots caused the transport of water from the cotyledons to the root cells.

Levitt (1972), in his excellent book on plant responses to environmental stresses, has thoroughly reviewed whole mature plant responses to water deficiency. Water stress inhibits photosynthesis and the translocation of already synthesized photosynthates. The resulting starvation could inhibit cell growth and cell cycle progression throughout the plant. Levitt, however, concludes that in most instances intact plants have quantities of storage reserves available so that starvation can be eliminated as a causal factor in water stress effects. Inhibition of other events, however, such as protein synthesis, protein degradation, and enzyme inactivation are also affected by water stress. The inhibition of these processes would affect cell cycle progression.

Chen et al. (1968) reported that dehydration of wheat embryos causes inactivation of RNA and arrested protein synthesis. Kessler (1961) studied the effect in leaves of water stress over long duration. He noted that RNA content decreased as RNase activity increased. In addition, DNA synthesis was also inhibited under these conditions.

Continued cell progression through DNA synthesis and mitosis in plant meristems is also influenced by the presence of critical concentrations of cytokinin (Fosket, this volume). Under water stress, as discussed, cells cease progression and accumulate in G_1 and G_2. One explanation for this arrested state is suggested by the experiments of Itai and Vaadia (1968). They exposed roots of intact plants to increasing osmotic concentrations of salt and concluded that induced water stress reversibly decreased the translocation of cytokinins from the roots to the shoot system. The reduction of cytokinin translocation and possibly its synthesis could contribute to the arrested cycle state under water stress.

TEMPERATURE STRESS

Brown (1951) measured the duration of mitosis and the total mitotic cycle time in pea roots at 15-30°C. Interphase and all mitotic phases were accelerated in their

Table 5-2
Relative percentage of cells arrested in G_1 and G_2 in dry seed embryos

Species	G_1	G_2
Allium cepa	100	0
Helianthus annuus	100	0
Pinus pinea	100	0
Pisum sativum	65	35
Triticum durum	75	25
Vicia faba	83	17

Data used with permission from Van't Hof 1974.

rate of progression with increasing temperature. Evans and Savage (1959) also observed that mitosis was sensitive to temperature changes and reported that metaphase was the most sensitive period. Wimber (1966) compared the mitotic cycle stages as affected by three temperatures, 13°C, 21°C, and 30°C. At 21°C and 30°C the duration of mitotic phases was essentially the same, but the G_1 period was more than doubled. The comparison of effects between 13°C and 21°C was most striking, as the duration of all phases was prolonged. The G_2 and G_1 duration almost tripled over those at 21°C while that of S was doubled.

Van't Hof (1966b) reported essentially no reduction in number of mitotic figures in *Pisum* roots grown at 10°C and 20°C. He also compared the effects of these temperatures after chronic gamma irradiation. At both temperatures DNA and RNA syntheses were reduced, but at 10°C nucleic acid synthesis was almost eliminated, indicating the accumulation of cells in G_1 and G_2. Protein synthesis was unaffected after irradiation in both temperatures.

Bodson (1970) examined the mitotic activity of shoot apex cells in *Sinapsis alba* plants grown at 2-3°C and determined that division was completely inhibited. When the temperature was restored, however, the number of mitotic figures returned to normal after a delay of approximately 16 hours.

Van't Hof and Ying (1964) examined root growth rates and mitotic cycle times in *Pisum* roots at various temperatures from 10°C to 30°C. Rate of root growth increased with increasing temperature from 10°C to 20°C; no further increase was observed at 25°C. The rate of cell population also increased linearly with temperatures ranging from 10-25°C, but there was no further increase at 30°C. Mitotic cycle time decreased with increasing temperature, as did the duration of mitosis.

Murin (1966) used the colchicine metaphase accumulation technique to estimate cell cycle parameters as changed by temperatures ranging from 3°C to 35°C on *Vicia faba* roots. From 8°C to 30°C the duration of the total cycle time, interphase, and mitotic duration decreased with increasing temperatures. At 3°C, 50% of the roots scored had no dividing cells. At the maximum temperature (35°C) the mitotic cycle time was doubled in duration over that at 25°C, while the duration of mitosis itself continued to decrease in response to temperature.

Burholt and Van't Hof (1971) conducted perhaps the most complete analysis of temperature induced cell cycle and root growth responses with *Helianthus* roots grown at temperatures ranging from 10°C to 38°C. The optimal temperature for root growth and cell enlargement was determined to be 25°C to 30°C. Other growth indicators such as increases in cell files and number of cells per file also were maximum at these temperatures. At 30°C to 35°C the cycle time (CT) is reduced to 1.3 hours, while at 38°C the CT was not measureable. At these high temperatures, cell proliferation was reduced to almost zero, the rate of cell differentiation increased, and the size of the meristem decreased. Interestingly, the mitotic index remains constant with temperatures up to 35°C and it does not reflect the decreasing cycle time duration. The relative amount of time spent in G_2 and M remained constant with temperature change, while the relative duration of S increased as the cycle time decreased with warmer temperatures. The duration of G_1, however, decreased with increasing temperature (Figure 5-4). Wimber (1966) observed this same pattern of G_1 and S durations in response to temperature changes.

Burholt and Van't Hof (1971) attempted to explain the shortening G_1 phase and lengthening S phase response in two ways. (1) G_1 reactions are more temperature sensitive than the other stages, and (2) DNA synthesis may be initiated before all G_1 reactions are completed, resulting in a shortened G_1 and lengthened S. The authors also point out the important consideration that even though the MI remains constant with remperature changes, the CT is decreased and the number of proliferating cells is also decreased with increased temperature; MI alone, therefore, does not indicate an accurate temperature response.

Lopez-Saez et al. (1966) and González-Fernández et al. (1971) also investigated cell cycle temperature responses in roots of onion (*Allium cepa*). Their results are somewhat contrary to those studies previously discussed. Lopez-Saez et al. (1966) reported that in onion the relative number and duration of all mitotic stages remained constant without regard to temperature. The cycle time and the absolute number of dividing cells, however, decreased with increasing temperature. González-Fernández et al. (1971) expanded this analysis to include the duration of G_1, S, and G_2 as well. They noted that temperature affects all phases of the cell cycle but that the relative proportion of all phases remained constant (Figure 5-4), even though the absolute numbers of cells in each phase is reduced with increasing temperature. They provided an interesting illustration showing this relationship (see Giménez-Martin et al., this volume).

OXYGEN STRESS

The continued presence of oxygen is a requirement for the production of chemical energy via respiration. Oxygen deficiency is a known cause of plant disease and the direct inhibition of growth (Bergman 1959). Coulomb and Coulomb (1972) report that particular damage occurs to organelles important for respiration and high metabolic activity after oxygen deficiency.

Bullough (1952) investigated the energy requirements for mitosis in mouse ear epidermal cells. He made the observation that once the energy supply has been built up in G_2, mitosis will progress to completion without regard to energy relation-

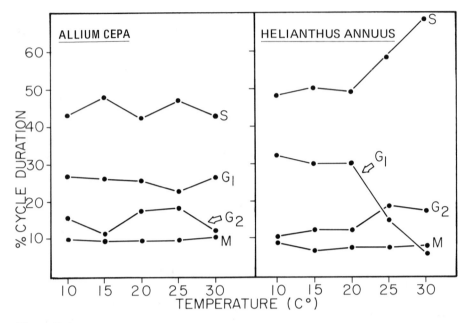

Figure 5-4.
Comparative effect of temperature on percent cycle duration of each cycle phase in two species. In *Allium cepa,* the length of each phase decreases with increasing temperature, but the percent relative duration which each phase contributes to the cycle time remains constant. In *Helianthus annuus* the durations of all phases except S decrease with increasing temperature. The relative percent cycle duration of G_2 and M remain constant while the contribution of G_1 is drastically reduced and S is increased. Data reprinted with permission from González-Fernández, et al. 1971, and Burholt and Van't Hof 1971.

ships. Gelfant (1959) showed that mitotic progression is not quite so simple, since a combination of glucose content and oxygen tension also exists. At 0.002 M glucose he noted that cells were more tolerant to O_2 tension than they were at 0.02 M glucose.

Amoore (1961a) studied the effect of oxygen stress and cyanide poisoning on mitotic progression. He noted that root excision, oxygen deficiency (growing roots in nitrogen), and low (10^{-3} M) concentration of NaCN inhibited cells from entering mitosis, but did not arrest those undergoing division. Application of highly pure nitrogen or 10^{-2} M cyanide also inhibited mitotic entry, but in addition, these treatments arrested or slowed progression of all cells actually in mitosis. This observation is contrary to that of Bullough (1952), who postulated that mitosis is unaffected by oxygen lack. Amoore suggested from his results that mitotic progression requires O_2 in order to complete, and that initiation of mitosis is also an oxygen requiring step. His results suggest that initiation of mitosis is more sensitive to O_2 stress than is actual progression.

Amoore (1961b) expanded his first observations in a more thorough study of oxygen tension in pea root tips. Oxygen concentrations of 0.05% to 0.2% were necessary to inhibit progression of interphase cells into division, but even in oxygen tensions as low as 0.001%, cells were still able to complete division. Tensions as low as 0.0005% were required to arrest mitosis altogether. Amoore postulated that O_2 susceptibility depends upon the cells position in the cycle at the time of oxygen tension; interphase had the highest O_2 requirement followed by prophase, metaphase, anaphase, and telophase. Calculation of the approximate amounts of chemical energy available at low O_2 levels led Amoore to conclude that complete loss of chemical energy did not occur and that this did not account for mitotic arrest.

Kihlman (1966) discussed at length the inhibition of cell entry into division and the factors involved in mitotic arrest. Kihlman pointed out that in addition to oxygen stress and cyanide other oxidative phosphorylation inhibitors such as ozide and dinitrophenol (DNP) may also arrest mitosis as well as stop their initial entry. Amoore (1962a,b and 1963a,b) noted that a greater amount of oxygen tension was required to stop mitosis than was needed to inhibit respiration. Even when respiration was stopped about 1.5% the normal value of ATP was still present in the cells, which is (Amoore 1963a,b) enough ATP for mitosis to continue. Kihlman continued his discussion by a description of Amoore's concept of a "non respiration mitotic ferrous complex (MFH)," which would be a mitotic specific molecule and which assists mitotic progression when oxygenated.

Webster and Van't Hof (1969) starved pea roots of sucrose to accumulate cells in G_1 and G_2 (stationary phase). After formation of the stationary phase condition they released the cells into S and M by adding sucrose to the medium. They investigated the oxygen requirements for these events by growing roots in hypoxic conditions and by adding a respiration inhibitor 2,4-dinitrophenol (DNP). Without sufficient oxygen, sucrose, or in the presence of DNP, no cells were able to leave G_1 or G_2 and progress into S or M. Late G_1 and late G_2, therefore, have a greater O_2 requirement than do S and M themselves. The observation was also made that in experiments where roots were given a short sucrose pulse followed by hypoxia or DNP treatment some cells were either arrested in S and M or greatly slowed down. DNA synthesis and mitosis, therefore, required aerobic conditions plus oxidative phosphorylation for initiation and continued progression.

Van't Hof (1970) investigated the necessity for O_2 aeration in pea root cultures after irradiation with 300 R γ-rays. Roots were starved of sucrose to cause their accumulation in G_1 and G_2, irradiated and further cultured either under N_2 or air for 24 hours. They were then placed in medium containing sucrose plus ^{3}H-thymidine. The time of entry into S by cells of irradiated roots grown with and without O_2 indicated that they had the same recovery time, while their entry into mitosis was quite different. Cells of roots irradiated and placed in aerated medium recovered and divided within 4–5 hours while those irradiated and kept under hypoxic conditions for 24 hours required 10-12 hours. It was clear from the data that G_2 cells have a greater oxygen dependency than do G_1 cells.

BIOTIC STRESS

The subject of biotic induced effects on the cell cycle and its associated biochemical events will not be discussed. Braun (1975) has discussed at length the cell cycle

Table 5-3
Selected list of cell cycle inhibitors

Chemical	Effect	Reference
ABA	Prophase inhibition	Nagl (1972)
Actinomycin D	Cell arrest in G_1 and G_2, blocks early prophase.	Webster & Van't Hof (1969); Rost & Van't Hof (1973); Van't Hof, Hoppin & Yagi (1973); Nagl (1970a).
	RNA synthesis 75% reduced by 3 hrs. Protein synthesis inhibited by 24 hrs. No further cycle progression.	Bal (1970)
Aminopteria	Reduced rate of S and extended its duration. Elongated G_1 but not G_2.	Badr (1972)
Chloramphenicol	Lengthened cycle time and increased duration of G_2.	Benbadis et al. (1971); Benbadis (1970); Benbadis & Levy (1974)
Colchicine	Lengthens cell cycle time and G_1 phase.	Macleod (1972)
	Measured cycle times.	Evans et al. (1957)
	G_1 arrest, metaphase block.	Nagl (1972)
Cycloheximide	Lengthens cycle time and increases G_2 duration.	Benbadis et al. (1971); Benbadis & Levy (1974)
	Blocked all mitotic stages.	Rose (1970)
	Blocks mitosis and arrests some cells in M; arrests other cells in G_2. Suggests CH inhibition of respiration.	Webster (1973); Lytinskaya (1971)
	Stops some cells in G_2 causes mitotic aberrancies.	Levy (1971)
3'-Deoxyadenosine (Cordycepin)	Inhibits DNA synthesis G_1 arrest. Accumulates cells in prophase.	Giménez-Martín et al. (1971)
Histone f_1	Blocks early prophase. Arrests cells in G_1.	Nagl (1970a)
Hydroquinone	Increased G_2 and duration.	Valadaud-Barrieu and Izard (1973)
IAA	G_2 arrest, blocks M.	Nagl (1972)

Table 5-3 (continued)

Chemical	Effect	Reference
Maleic hydrazide	Lengthens S and blocks G_1 and G_2.	Bonaly (1971)
2-mercaptoethanol	G_2 and mitotic arrest.	Nagl (1970b)
Ni^{++}	Antimitotic (C-mitoses)	Constantinesea and Radu (1970)
Nitrous acid	Mitotic arrest.	Badr, Mourad and Nagrib (1972)
2,4,5-trichlorophenoxy-acetic acid	Increases S and G_2 duration.	Macleod (1969)
Tubulosine	G_2 arrest, inhibits protein synthesis.	Pareyre (1970)
Vincaleukoblastine	Lengthens S.	Srivastava and Rao (1972).

information available on tumorigenesis in plants. He pointed out that tumor formation is not one of inducing cycle blocks, but of removing them. This permits abnormal differentiation resulting in reversible tumor development. Braun suggested the activity of auxin and cytokinin in this process. He also cited earlier work (Wood et al. 1972; Wood and Braun 1973; Wood et al. 1974), which suggested cell division regulating roles for 8-bromo-cAMP (3':5') and for cytokinesin.

Most of the other literature in this area concerns teratological growth patterns, conditioning and induction events and the physiology of tissue formation (Block 1953; Kehr and Smith 1953; Kunkel 1953; Pilet et al. 1960; Wood and Braun 1961; Arnold 1969; Braun 1969; Jauffret et al 1970).

CHEMICAL STRESS

Kihlman (1966), in his treatise "Actions of Chemicals on Dividing Cells," presented a definitive survey of the effects of chemicals on mitosis. Kihlman examined the action of numerous chemical compounds on the basis of their mechanisms and sites of action. Deysson (1968) reviewed the effects of numerous anti-mitotic substances, but from the point of view of their modes of effect and not their biochemical mechanisms of action. Ashton and Crafts (1973) also examined the mitotic effects of numerous chemicals, those used as herbicides.

Chemicals act on mitosis in one of three ways (D'Amato 1960). (1) They may be preprophase inhibitors (D'Amato 1949), which are chemical compounds which may arrest cell cycle progression at some stage of interphase (G_1, S, or G_2). These compounds may inhibit entry of cells into mitosis or DNA synthesis. A selected list of such chemicals and their effects is given in Table 5-3. Ideally a compound in this category would cause interphase cell arrest at any concentration, and would not interfere with those cells already in mitosis. In many cases, however, these compounds, at high concentrations, also act as spindle poisons because of their "shotgun" inhibition of many metabolic activities. (2) Compounds which interfere with the synthesis (Pickett-Heaps 1967; Hess 1973, 1975) or the orientation of the mitotic spindle (Hepler and Jackson 1969). These are the so-called "C-mitotic" or "mitoclasic" agents (Deysson 1968) of which colchicine exhibits the classic effect. These compounds cause the controlled movement of mitotic chromosomes to be disrupted. The number and type of chemicals which can produce these effects is very large. Reviews by Hadder and Wilson (1958), Kihlman (1966), Wilson (1966), Deysson (1968, 1970), Hess (1973, 1975), Soifer (1975), and many others should be consulted. (3) The last category of effects are those which inhibit cytokinesis. Chemicals such as caffeine completely inhibit cell plate and cell wall formation between daughter cells resulting in binucleate cells (Giménez-Martín et al. 1966, and this volume).

Preprophase Inhibitors

DNA Synthesis Inhibition

Compounds which are base analogs may be substituted into the DNA strand and inhibit further chain growth. This phenomenon has been capitalized on by several workers as a chemical method to achieve reversible cell cycle synchrony. Clowes (1965b), for example, used 5-amino uracil (5-AU) to induce synchrony in *Zea* roots. He suggested that 5-AU retarded the passage of cells through the entire cell cycle as well as stopping some cells in S. Scheuermann and Klaffke-Lobsien (1973) further examined 5-AU activity, but in *Vicia* root tip meristems. Their observations indicated that 5-AU reduced the rate of DNA synthesis, but that it did not block the S phase completely. They also pointed out that heterochromatic DNA seemed to be more reduced in 5-AU treated tissue.

Hydroxyurea (HU) has also been used as a cell cycle synchronization tool (Brulfert and Deysson 1971). Barlow (1969) examined HU effects on *Zea* root meristems. He observed that HU inhibited DNA synthesis but not protein or RNA synthesis. Cell division also stopped in HU treated roots, but only after a treatment time equivalent to the duration of G_2. Barlow further pointed out that in spite of DNA synthesis inhibition, cell growth in treated cells is still maintained. Other compounds such as barbital and ethidium bromide are known to inhibit DNA synthesis (Eriksson 1966; Bonaly and Deysson 1974; Lord 1974).

Perhaps the DNA synthesis inhibitor for which the best cell cycle kinetic data exists is 5-fluorodeoxyuridine (5-FUdR). Van't Hof and Kovacs (1970) and Kovacs and Van't Hof (1970, 1971) conducted extensive experiments using 5-FUdR to induce synchrony in cultured pea root tips. In order to do this, they first induced root cells into a parasynchronized condition (all cells in G_1 and G_2) by the sucrose starvation method already discussed. The roots were then cultured in medium with sucrose plus 5-FUdR. FUdR prevents cells from entering S but allows cells in G_2 to divide and then stop at the G_1/S boundary. Next, the cultured roots were returned to medium without sucrose in order to remove the FUdR; finally, the cells are induced to synchronously enter S by providing sucrose. Van't Hof and Kovacs (1970) determined that stationary phase meristems (cells in G_1 and G_2) and synchronized meristems (5-FUdR induced arrest at G_1/S) were different biochemically. To do this, they irradiated both types of meristems and then measured their respective irradiation induced DNA synthesis delay. Irradiated synchronized meristems did not require a delay time before S, indicating that possibly the proteins required to initiate DNA synthesis were already present in these meristems. This observation indicated, therefore, that 5-FUdR selectively affected DNA synthesis but not protein synthesis. Socher and Davidson (1971) observed that 5-FUdR also extended the G_2 period in *Vicia faba* roots.

Compounds Which Affect Other Cycle Stages

Table 5-3 gives a partial list of some compounds which also affect cell cycle progression. Many of the compounds at high concentrations cause mitotic aberrancies and exhibit the so called "colchicine effect." These same compounds (e.g., cycloheximide, chloramphenical, hydroquinone, maleic hydrazide) tend to arrest cells in G_2, thereby inhibiting passage into mitosis. Compounds, often considered protein synthesis inhibitors, may have a multiple effect, especially at high concentrations. Lytinskaya (1971) and Webster (1973) observed, for example, that some cells were actually arrested in mitosis when treated with cycloheximide, an effect usually considered to be caused by anaerobiosis or gross ATP inhibition. Other studies on cultured roots have reported the inhibition of cell progression into S and M in the presence of cycloheximide (Webster and Van't Hof 1969; Rost and Van't Hof 1973; Van't Hof et al. 1973) in root tips.

Antimitotic Agents

Hundreds of publications have been written on the subject of antimitotic agents (Eigsti and Dustin 1955; Kihlman 1966). There are basically two levels of direct mitotic interference: (1) Those compounds which interfere with spindle protein synthesis, such as colchicine (Pickett-Heaps 1967), or with spindle orientation, such as isopropyl carbamate, (Hepler and Jackson 1969), and (2) Compounds which inhibit cell plate formation, such as caffeine (Giménez-Martin et al. 1966).

Care must be taken, however, in categorizing chemical compounds as to the type of antimitotic behavior they exhibit. The difficulty in analyzing the antimitotic action of chemicals is further complicated by problems of concentration effects, tissue specificity, and duration of treatment (Mann et al. 1965a,b; D'Amato 1960).

Mann et al. (1965a,b) investigated the effects of carbamate and other herbicides on protein synthesis in a number of species and tissues. One of their most noteworthy observations is the striking difference in chemical-induced inhibition between species and between tissues in the same plant. Mann et al. (1965b) discussed the problem of the effects of chemical concentration and suggested that the following assumptions be made when considering chemical action: (1) at low concentrations herbicides (chemicals) influence only one or two reactions which are the herbicide target areas, and (2) for most herbicides the concentration at which these target areas are severely inhibited in at approximately 2 ppm. Herbicide or chemical effects generally caused by concentrations higher than this may, therefore, be due to secondary effects. It is important that workers first of all determine the lowest possible concentration at which a chemical compound causes any antimitotic effects.

Deysson (1968) and Rost and Bayer (1976) also pointed out the difficulty of interpreting the mechanism of action of chemical compounds because of the great differences in experimental conditions under which studies are conducted. There is little uniformity of technique and a lack of standardized criteria to measure chemical effects; this is particularly significant in herbicide research. Examination of a wide range of concentrations and durations of exposure to the chemical examined, plus consideration of the threshold levels which cause specific cytological reactions, will in most cases help clarify the mechanisms of antimitotic chemicals (D'Amato 1960).

Chemical inhibition of cell cycle progression, such as spindle inhibition, may not likewise inhibit cell enlargement or differentiation. Bayer et al. (1967) for instance reported that in cotton roots treated with the herbicide trifluralin the root tip increased in size to form a club shaped structure. Hess (1975) thoroughly investigated this phenomenon and determined that in *Chara* cells treated with trifluralin, the change from an elongated to a spherical form was due to a disorientation of cell wall microfibrils. D'Amato (1960) discussed the reduction of meristem size and the premature differentiation of vascular tissue which corresponds to chemical inhibition of the cell cycle. These observations are similar to those noted for other stress sources and strengthens the point that cell division and cell growth are separate events not under common control (Mitchison 1971).

Growth Regulators

The fact that plant growth regulators affect cell cycle stage has long been established. Fosket (in this volume) has discussed the possible cytokinin mode of action on the cell cycle, and Loy (in this volume) has described gibberellic acid effects in the shoot apex.

As previously discussed, root tips grown in sucrose deficient medium achieve a stationary phase where all cells become arrested in G_1 and G_2. Cells in mature tissues of plant organs also become arrested during normal development. Evans and Van't Hof (1974b, 1975) examined the cell population phase distribution in mature tissues of *Pisum, Vicia, Helianthus,* and *Triticum* and compared these distributions to that of cells in stationary phase meristems of the same species. *Pisum* and *Vicia* cells in mature tissues and meristems become arrested primarily in G_2 while in *Helianthus* and *Triticum* the majority stop in G_1. This indicates that cells may be preconditioned in the meristem to eventually become arrested according to a specific ratio. In *Pisum* and *Vicia* there is a naturally occurring substance, possibly a plant hormone, that determines whether some cells arrest in G_1 or G_2 (Evans and Van't Hof 1973, 1974a). Evans and Van't Hof (1973) showed that increased times in sucrose deficient medium caused a 39% decrease in the number of cells becoming stationary in G_2. Further, removal of the cotyledons before culture in a deficient medium likewise caused a decrease in number of G_2 stationary cells as compared to roots with cotyledons intact up to that time of culture. They considered this evidence that some type of compound comparable to the "chalones" found in animal cells was present in pea cotyledons which may promote G_2 arrest. Evans and Van't Hof (1974a) reported further investigations on these earlier observations. For these studies the growth substances were extracted by diffusion of the material from cotyledons into agar or liquid culture medium. Roots grown on medium showed a greater proportion of cells arrested in G_2. These experiments also demonstrated that a preset portion of cells could only stop in G_1 (40-50%) and in G_2 (20%) and that only the remaining fraction of cells could be manipulated by the G_2 factor. Pea cotyledons maintain enough of the substance to influence G_2 arrest in cultures for 8-10 days; after that time, cells became arrested primarily in G_1. The G_2 arrest material was not species specific in that the substance from *Pisum* cotyledons promoted arrest in *Vicia* roots; however, *Helianthus* and *Triticum* roots were unaffected.

RADIATION STRESS

Radiation-Induced Cell Cycle Responses

Evans (1965) surveyed the x-irradiation damage in *Vicia faba* primary root meristems. Three basic types of radiation-induced damage were reported: (1) mitotic cycle delay, (2) formation of chromosome aberrations, and (3) loss of proliferative capacity due either to premature differentiation or cell death. According to Evans, division delay and chromosome aberrations actually contribute little toward depression of root growth, while inhibition of cell progression followed by premature differentiation and root growth inhibition is considered to be the most significant factor induced by sublethal irradiation doses.

Van't Hof and Sparrow (1963a) tend to agree with this observation, based on their experiments with chronically γ-irradiated pea roots. In their experiments the size of the root meristem cell population decreased with increasing dose, while the mitotic cycle time remained constant regardless of dose. They concluded that growth inhibition and meristem size decrease were due to a radiation-induced decrease in the population of cells capable of entering division and not to a change in cycle time.

On the other hand, Gudkov and Drodzinskii (1972) examined the radiosensitivity of the mitotic cycle phases of pea roots after acute doses of γ-irradiation ranging from 50 to 800 rad. At increasing doses the mitotic index decreased as did the number of cells in the mitotic cycle (proliferative pool). They also noted that the duration of the mitotic cycle increased as the dose increased. A further analysis of each interphase stage revealed that G_2 and M remained constant regardless of dose, S increased from 7.7 hours in the control to 9.1 hours at 800 R, while G_1 increased quite dramatically with dose from 5.3 hours at 50 R to 9.0 hours at 800 R.

Hsu et al. (1962), Van't Hof (1963), Sparrow (1965) Leeper et al. (1972), Burholt and Van't Hof (1972), and others have established the generalization that among the stages of the mitotic cycle, G_2 cells are more radiosensitive in terms of cell cycle response than either S or G_1 cells. Dewey et al. (1972) and Dettor et al. (1972) have further clarified this phenomena of radiosensitivity in different stages of the cell cycle. By using several lines of natural and experimental evidence, these workers have established the hypothesis that increases in radiosensitivity may be due to increases in chromatin condensation. They cite, for instance, Natarajan and Ahnstrom's (1970) work which pointed out that more chromosomal aberrations were produced in condensed heterochromatin regions than in dispersed euchromatin regions in plant cells. They also cite their own work, where cells grown in hypertonic substances become more condensed and consequently exhibit more radiation damage.

Sparrow and his coworkers at Brookhaven National Laboratory have, over the years in a long series of papers, formulated a broad generalization relating certain cellular characteristics to radiosensitivity for a wide range of organisms. Sparrow et al. (1964) point out that tolerance of species to chronic exposure of ionizing radiation is related to the size and number of genetic targets, and also to the average length of the mitotic cycle. Organisms with long cycle times and long periods in G_2, for instance, would offer potential for greater radiation damage than organisms with short G_2 duration and short cycle times. This last part corresponds to the law of Bergonie and Tribondeau, which states, in part, that tissues which have most cells actively in the mitotic cycle are most radiosensitive.

The best correlation, however, from Sparrow's work, exists between radiation sensitivity and DNA content per chromosome. Yamakawa and Sparrow (1966) measured the radiosensitivity of pollen grains based upon dose dependent pollen abortion, and reported a linear relationship between interphase chromosome volume (ICV) and radiosensitivity. Sparrow et al. (1967), Underbrink et al. (1968), Price et al. (1973) have expanded this generalization to many other organisms. The relationship apparently exists, therefore, that as the interphase chromosome volume, or nuclear volume increases, the radiosensitivity of the organisms also increases. Other correlations have been reported by Van't Hof and Sparrow (1963b) and Van't Hof (1964, 1974) such as DNA content and mitotic cycle duration. The generalization expressed by Van't Hof (1964) that as DNA content increases, the duration of S increase is also true, with the exception of polyploid cells, which maintain a duration equal that in diploid nuclei of the same species (Troy and Wimber 1968). In Van't Hof (1975) the relationship was demonstrated to hold true for more than sixty plant species and varieties.

Bennett (1973) summarized many studies which show a similar relationship between the duration of meiosis, its stages and nuclear DNA content. Bennett (1974) also examined the role nuclear characters play in morphogenesis. He drew correlations from the literature showing the relationship of DNA content (nucleotype) and many plant metabolic, size, and developmental characters.

Cell Cycle Responses to Ultra Violet and Visible Light

Lemma pupusilla is a short day plant with a photoperiodic flowering response. Halaban (1972) examined mitotic cycle times and division rates to determine if division synchrony is correlated with photoperiodic treatment. She reported no essential change in cycle times, nor in cell cycle phases between long night and long day treated plants. Bernier et al. (1967) increased the day length period of *Sinapsis alba* from 8 to 20 hours. This duration increase was accompanied by the division of shoot apex cells previously held in G_2. Jacqmard and Miksche (1971) examined light induced flower induction in the same plant and showed that cycle progression occurred based on DNA value shifts. Crapo (1975) examined the effects of differential light intensity on growth and metabolism in barley, wheat, and tomato. In all three species the percentage of cells in mitosis decreased linearly with decreasing light intensity to the above ground shoot. The number of dividing cells and growth in general, as measured by biomass, were more sensitive to shoot light intensity change than were respiration and ion uptake. Of additional interest was the observation that in low light intensity treated wheat (100 ft-c) xylem differentiation extended down into the reduced root meristem. This is another example of inhibition of the cell cycle but not of the growth and differentiation cycle.

Klein (1967) demonstrated the reversible reduction of cell growth and cell division by ultraviolet light (254 μm). Brown and Klein (1973) studied pea roots grown in liquid culture and exposed directly to filtered light of various qualities. Near UV and visible radiation increased cycle time primarily by increasing G_1. Green radiation increased all cell cycle phases significantly, except G_2, which was reduced. All other wavelength radiation also decreased G_2 duration. Brown and Klein (1973) suggested a visible light influence on the protein and RNA synthetic mechanism controlling G_1 and G_2 progression.

Structural Responses

A massive amount of literature exists concerning radiation stress on plant organs, tissues, and cells. Reviews on structural and physiological effects of irradiation have demonstrated obvious teratological responses in a variety of plants (Gunckel and Sparrow 1953, 1961). These authors stated that both genetic and/or physiological radiation damage may be the direct cause of these effects, but that physiological damage is probably most significant. This is evidenced by their observation that plant parts removed from the irradiation source grow normally without any mutation effects. Mitotic delay, chromosome aberrations, and loss of proliferation capacity due to cell death or to premature differentiation have also been considered as causes of radiation-induced damage (Evans 1965).

Haber and Rothstein (1969) and Foard and Haber (1970) have pointed out that understanding radiosensitivity in complex tissues can not simply be related to cell division related responses. They also attempted to clarify the law of Bergonie and Tribondeau by showing that radiosensitivity, as measured by photodestruction in tobacco leaves after γ-irradiation, is unrelated to cell division rates. This observation strengthens the discussion of Mitchison (1971) concerning the organization of cell cycle metabolic systems in complex tissues. Mitchison described a DNA-division cycle (DD cycle) and a growth cycle in cells of higher organisms. The DD cycle consisted of events such as G_1, DNA synthesis, G_2, mitosis, and the accompanying required macromolecular syntheses. The growth cycle consisted of most other macromolecular synthesis which occurs in cells during the cycle. Presumably, the growth cycle continues at times when the DD cycle has been stopped, either naturally or artificially. Mitchison (1971) described situations in animal cells where radiation-induced mitotic inhibition resulted in, or rather did not likewise inhibit, cell enlargement. Somewhat similar observations have been reported in plant cells. Wangenheim (1970, 1971) irradiated *Oenothera* and *Hordeum* roots at doses high enough to inhibit or drastically delay cell division. During radiation-induced mitotic delay, he noted that cells continued to grow and to differentiate. Sometimes, cells were seen to complete differentiation before the cells were able to resume mitotic activity. Foard and Haber (1970) and Haber and Foard (1964) conducted interesting experiments where *Avena* seeds were irradiated at high doses (500–800 kR) and then germinated. In these instances, small plantlets were formed (gamma plantlets) solely by cell enlargement with no concomitant cell division. There is, therefore, a separation of the so-called DD cycle and the growth cycle in plants, as well as in animal cells. Two further examples of the separation of the DD and growth cycles by irradiation can be cited. Smith and Kersten (1941) irradiated dry seeds of *Vicia faba* with soft x-rays and examined root modifications after germination. The roots which emerged were shorter than the controls and formed no laterals. The meristem was reduced by premature differentiation of vascular cells close to the tip. The root tip was subsequently distorted by radial enlargement of cells from all root tissues. Decrease in meristem size was also reported by Van't Hof and Sparrow (1963a) in chronically irradiated pea roots. They also noted that root elongation was inhibited and dose dependent, and that the number of cells per root meristem decreased with time after irradiation. Measurements of mitotic cycle time (CT) indicated that CT remained the same regardless of protracted dose and that the decrease in cell number was due to the radiation-inhibited entry of cells into mitosis.

One further example of separation of the DD cycle and growth cycle due to a stress response, a magnetic field in this case, is appropriate here. Dunlop and Schmidt (1964, 1965) grew roots of *Allium cepa* and *Narcissus tozetta* in magnetic fields of less than 5000 gauss. This type of stress exhibited growth reaction patterns similar to particle irradiations. Among the effects were inhibition of mitosis and cytokinesis, retardation of growth in length, and induced radial swelling.

Quiescent Center

Several papers have examined the quiescent center (QC) and its response to irradiation (Barlow 1973; Clowes 1963; 1965a, 1967; 1970a, 1972b; Clowes and Hall 1962); consequently, I will not discuss this root region in detail. The QC cells in roots of *Zea mays* have a cycle time of approximately 174 hours as opposed to 14 hours for root tip initials and 22–23 hours for stele cells immediately above and 200 μm above the QC (Clowes 1965a). The QC cell cycle time consists of 151 hours in the G_1 phase while S, G_2, and M are of similar length to those in other regions. After x-irradiation (Clowes 1972a,b) the cells of the root meristems are inhibited from synthesizing DNA or dividing. The QC cells, on the contrary, show a decreased cycle time and began to enter S. A very interesting observation (Clowes 1970a) is that some QC cells previously in prolonged G_1 phase can reach prophase and even metaphase within 30 minutes. This shows that these cells are stimulated to divide by irradiation and that they somehow lose their requirement to progress through G_2 before initiating mitosis.

Clowes (1970b) cited suggestions that starvation may be the factor which induced and maintained the "quiescent" state of the root QC. He then disproved these suggestions by showing that labeled thymidine supplemented to the endosperm of germinating *Zea* seedlings pass through the root and are distributed throughout the apical regions. Further, labeled compounds, incorporated into root caps which were in turn grafted to decapitated roots, appeared in the cap initials, QC, and stele. These experiments demonstrate that soluble materials available in the root tissue are capable of passing into and through the QC cells. Macleod and Scadeng (1975) determined with pea roots that the use of various concentrations of IAA (10^{-6} M – 10^{-12} M) and sucrose (1-12%) had no effect on the size of the QC. The proposition can be made, therefore, that the semi-stationary condition of G_1 cells in the QC is not due to nutritional starvation but instead to some kind of internal regulatory mechanism. Webster and Langenauer (1973) proposed the type of mechanism this might be. In their experiments, corn roots were starved of sucrose so that cells became arrested in G_1 and G_2. When resupplied with sucrose all cells initiated cycle progression, including the QC, but the cells of the QC soon became arrested once again. They interpreted this behavior as indicating that the activity of the cells surrounding the QC control their progression. Feldman and Torrey (1976) reinforced this idea by experiments with decapitated *Zea mays* roots. Decapitation destroyed the root cap initial cells, but was followed by the activation of cells distal to the QC leading to its regeneration after 2-3 days. Feldman (1975) showed a correlation between extractible cytokinins and QC regeneration. In decapitated roots after 24 hours of regeneration the cytokinin level was lowest and the QC size was smallest, while after 36-72 hours the QC was restored to the intact root size and the cytokinin level was again at the level of the controls.

SUMMARY AND CONCLUSIONS

Nutritional stress. Cells of root tips grown in culture medium lacking sucrose will become arrested (stationary phase) in G_1 and in G_2 according to a species specific ratio. Starvation inhibits polyribosome formation to a high degree but does not completely inhibit protein synthesis. Replenishment of sucrose to stationary phase root tips allows cells to resume proliferation. This resumption requires O_2 and protein synthesis. Inhibition of RNA synthesis by actinomycin D does not immediately inhibit progression indicating the continued presence of some required RNAs.

Water stress. Reduction of cellular water content naturally in dry seeds or experimentally will inhibit RNA and protein synthesis. Cells in embryos of dry seeds become arrested in G_1 and G_2. During drying the S period is more sensitive to water stress than is M. The arrested ratios are both organ and species specific.

Temperature stress. Temperature extremes $< 5°C$ and $> 35°C$ inhibit mitosis and DNA synthesis reversibly. In intermediate temperatures (10-30°C) the cycle time duration decreases with increasing temperature. In addition, the total number of cells dividing decreases even though the mitotic index remains approximately constant. The durations of the G_1 and S phases are most plastic to temperature; G_1 tends to decrease with temperature increases while S increases in length. The results of studies on onion are contrary to this trend in that all cycle phases are reduced in duration by the same proportion with increasing temperature.

Oxygen stress. Oxygen is an absolute requirement for cell progression. G_2 cells are apparently most oxygen dependent followed in sensitivity by G_1, S, and M. Under hypoxia (0.0005% O_2) cells may be arrested in mitosis. Inhibition of respiration by such compounds as DNP and NaCN also impair cycle progression and may arrest cells in mitosis.

Biotic stress. Microorganisms and other pathogenic agents apparently act by inducing cell cycle progression. This may involve production or inhibition of plant regulatory hormones or possibly cAMP.

Chemical stress. Chemicals may act as gross inhibitors of cell metabolism and cell cycle progression, or they may, at specific concentrations, affect specific cycle stages. FUdR for example is considered to be a "preprophase" inhibitor in that it inhibits DNA synthesis. Some concentrations, however, may prolong G_2. Other compounds such as colchicine inhibit mitosis by inhibiting tubulin synthesis. A third category of chemicals such as caffiene inhibits cell plate formation. Care must be taken in considering chemical effects because different species and organs are affected by different concentrations. The precise mode of action of a chemical should be considered the measurable affect induced at the lowest concentration.

Radiation stress. Radiation-induced damage includes inhibition of mitosis and DNA synthesis, production of cell cycle delay, and formation of chromosome aberrations. A correlation apparently exists between DNA content (target size) and radiation sensitivity—as the amount of DNA increases per chromosome or per nucleus the sensitivity also increases. G_2 is apparently most sensitive to irradiation followed by S, G_1, and M.

Conclusions

Possibly three common criteria of stress expression become apparent after surveying the literature.

1. Inhibition of mitosis and the formation of chromosome and mitotic aberrations particularly as a result of chemical and radiation stress. The types of mitotic aberrancy are often related to direct interference with the spindle apparatus. This type of interference is often a result of spindle tubule inhibition or binding which causes spindle loss or disorientation.

2. Chemical and physical stresses have been reported to induce radial enlargement of stressed organs (Smith and Kersten 1941; Dunlop and Schmidt 1965; Bayer et al. 1967; Hess 1975). A similar phenomenon induced by irradiation has been reported by several workers (Van't Hof and Sparrow 1963a; Wagenheim, 1970, 1971), who observed a decrease in the size of meristems and the premature differentiation of cells. Differentiation implies the initiation of specific events involved in development of mature cell types and tissues, but it also implies cell enlargement in cells which would not be as enlarged without a stress factor. These observations have

also been made in animal cells by Mitchison (1971), who separated the animal cell life cycle into two interconnected cycles. The DNA/division cycle (DD cycle) and the growth cycle. Figure 5-5 shows this relationship diagrammatically for plant cells. The inference in this separation is that inhibition of the DD cycle by any means which causes cell arrest, but not death, will still "allow" cell enlargement.

3. The third common stress expression concerns cell cycle accumulation patterns in response to stresses. As pointed out by Evans and Van't Hof (1975), some cells cycle rapidly, some slowly, and other do not cycle at all. Under nutritional stress, Van't Hof and others in his laboratory have clearly demonstrated the predictable pattern of cell cycle arrest after sucrose starvation. This pattern of arrest has also been demonstrated in animal cells in culture (Gurley et al., this volume), in mature plant tissues, and in meristems due to natural and experimentally induced water, O_2, radiation, respiration, and chemical stresses. Ratios of cell arrest accumulation are species and tissue specific.

Van't Hof and Kovacs (1972) developed the Principal Control Point (PCP) hypothesis to explain this occurrence. The PCP hypothesis was developed primarily on the strength of sucrose starvation studies by Van't Hof and by the dry seed studies of D'Amato. This hypothesis is, however, more broadly applicable, as shown by the cell cycle arrest behavior induced by many other stress factors. The repeatability of the size of arrested G_1 and G_2 populations with other types of stress has not been well documented. But the fact that cell arrest does occur in both G_1 and G_2 is well accepted. It may be justified, therefore, to apply the PCP hypothesis to stress effects generally regardless of source.

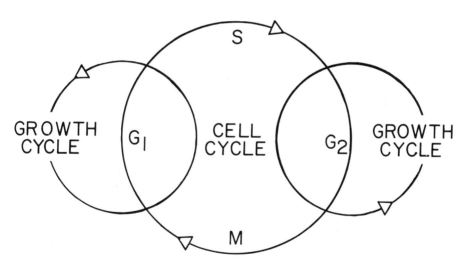

Figure 5-5.
Representation to show the separation of the cell cycle and growth cycles. During normal differentiation in plant tissues, cells leave the cycle in G_1 and G_2. After stress, cells often become arrested in G_1 and G_2. Inhibition of the cell cycle often leads to the premature enlargement and differentiation of these arrested cells.

In the investigations described, some were conducted with intact plants and others with isolated organs in tissue culture. The question of whether intact or cultured tissues are more appropriate for cell cycle stress studies should be considered. Physiological mechanisms exist in intact plants which enable them to self-adjust to accommodate for certain stress effects. Water stressed seedlings, for example, with intact cotyledons, may draw on cotyledon water to reverse high solute potentials in the roots. This internal adjustment would result in renewed cell cycle progression. Other sources of stress such as nutrient deficiency are also difficult to interpret in intact plants due to the possible transport of stored materials. In these two instances, perhaps organ cultures or free cell cultures would be more suitable.

Some chemical compounds, such as certain herbicides and plant growth regulators, are readily transported throughout a plant, whereas others are relatively immobile. Cell cycle responses in apical regions accompanying the action of mobile chemicals would be tempered by the secondary effects induced by them in other plant regions. Cultured plant parts treated with the same chemical would not show this difficulty.

No experimental system is clearly better than any other to study cell cycle stress responses. Regardless of the system selected, extreme care must be applied to insure that (1) secondary effects of the applied stress are considered, and (2) that the possible contribution of the experimental conditions themselves are examined. In addition, it is important to remember that species exhibit variable responses to different concentrations and doses of stress factors, and last, that not all tissues of the same organism respond in the same way.

LITERATURE CITED

Aitchison, P. A. and M. M. Yeoman. 1974. Control of periodic enzyme synthesis in dividing plant cells. In *Cell Cycle Controls,* eds. G. M. Padilla, I. L. Cameron and A. Zimmerman. Academic Press, N. Y., pp. 251–263.

Amoore, J. E. 1961a. Arrest of mitosis in roots by oxygen-lack or cyanide. *Proc. Royal Soc. Lond.,* Ser. B. **154**:95–108.

Amoore, J. E. 1961b. Dependence of mitosis and respiration in roots upon oxygen tension. *Proc. Royal Soc. Lond.,* Ser. B. **154**:109–129.

Amoore, J. E. 1962a. Oxygen tension and the rates of mitosis and interphase in roots. *J. Cell Biol.* **13**:365–371.

Amoore, J. E. 1962b. Participation of a nonrespiratory ferrous complex during mitosis in roots. *J. Cell Biol.* **13**:373–381.

Amoore, J. E. 1963a. Non-identical mechanisms of mitotic arrest by respiratory inhibitors in pea root tips and sea urchin eggs. *J. Cell Biol.* **18**:555–567.

Amoore, J. E. 1963b. Action spectrum of mitotic ferrous complex. *Nature* **199**:38–40.

Arnold, B. C. 1969. Gall mites and meristems. *Phytomorphology* **19**:92–95.

Ashton, F. M. and A. S. Crafts, 1973. *Mode of Action of Herbicides.* John Wiley and Sons, N.Y.

Avanzi, S., A. Brunori, and F. D'Amato. 1969. Sequential development of meristems in the embryo of *Triticum durum.* A DNA autoradiographic and cytophotometric analysis. *Dev. Biol.* **20**:368–377.

Badr, E. A. 1972. Kinetics of the cell cycle in *Nicotiana tabacum. Egyptian J. Gen. Cytol.* 1:18–27.

Badr, E. A., A. M. Mourad, and F. M. Naquib. 1972. Effects of nitrous acid on the cell cycle and mitotic apparatus in *Vicia faba. Egyptian J. Gen. Cytol.* 1:231–246.

Bal, A. K. 1970. Effect of actinomycin D on root meristem cells of *Allium cepa* L. *Z. Pflanzenphysiol.* 63:261–268.

Barlow, P. W. 1969. Cell growth in the absence of division in a root meristem. *Planta* 88:215–223.

Barlow, P. W. 1973. Mitotic cycles in root meristems. In *The Cell Cycle in Development and Differentiation,* eds. M. Balls and F. S. Billett. Cambridge U. Press., pp. 133–165.

Bayer, D. E., C. L. Foy, T. E. Mallory, and E. G. Cutter. 1967. Morphological and histological effects of trifluralin on root development. *Amer. J. Bot.* 54:945–952.

Benbadis, M. C. 1970. Proliferation et differenciation cellulaires dans le meristeme radiculaire d'*Allium sativum* L.: action du chloramphenicol. *Soc. Bot. France Mem.* 125–143.

Benbadis, M. C., F. Levy, C. Pareyre, and G. Deysson. 1971. Le cycle cellulaire dans les meristemes radicularies d'*Allium sativum* L. Sous L'influence d'inhibiteurs de la synthese proteique: action comparee du chloramphenicol, du cycloheximide et de la tubulosine. *C. R. Acad. Sci.* (Paris) Ser D 272:707–710.

Benbadis, M. C. and F. Levy. 1974. Influence of metabolic inhibitors of the growth and development of root meristems. A comparative study of the action of chloramphenicol and cyclohexamide. In *Proc. Symp. Structure and Function of Primary Root Tissues* ed. J. Kolek, pp. 137–146.

Bennett, M. D. 1973. The duration of meiosis. In *The Cell Cycle in Development and Differentiation,* eds. M. Balls and F. S. Billett. Cambridge U. Press, pp. 111–131.

Bennett, M. D. 1974. Nuclear characters in plants. *Brookhaven Symp, in Biol.* 25:344–366.

Bergman, H. F. 1959. Oxygen deficiency as a cause of disease in plants. *Bot. Rev.* 25:417–485.

Bernier, G., J. Kinet, and R. Bronchart. 1967. Cellular events at the meristem during floral induction in *Sinapis alba* L. *Physiol. Veg.* 5:311–324.

Bloch, R. 1953. Abnormal plant growth. *Brookhaven Symp. Biol.* 6:41–54.

Bodson, M. 1970. Influence d'un abaissement de la temperature et d'un retour a temperature ordinaire sur l'activite mitotique du meristeme caulinaire de *Sinapis alba. C. R. Acad. Sci.* (Paris) Ser. D 270:314–316.

Bonaly, J. 1971. Modifications du cycle cellulaire sous l'influence de l'hydrazide maleique dans les meristemes radiculaires d' *Allium sativum* L. *C. R. Acad. Sci.* (Paris), Ser. D. 273:150–153.

Bonaly, J. and G. Deysson. 1974. Modifications du cycle cellulaire sous l'influence du barbital dans les meristemes radiculaires d'*Allium sativum* L. *C. R. Seances Soc. Biol. Fil. Publ. Concours Cent. Natl. Rech. Sci.* 168:18–22.

Braun, A. C. 1969. Abnormal growth in plants. In *Plant Physiology,* Vol. VB, ed. F. C. Steward. Academic Press, N.Y. pp. 379–420.

Braun, A. C. 1975. The cell cycle and tumorigenesis in plants. In *Cell Cycle and Cell Differentiation,* eds. J. Reinert and H. Holtzer. Springer-Verlag, N.Y., pp. 177–196.

Brown, R. 1951. The effects of temperature on the duration of the different stages of cell division in the root tip. *J. Exp. Bot.* **2**:96-110.

Brown, R. and P. Rickless. 1949. A new method for the study of cell division and cell extension with some preliminary observations on the effect of temperature and of nutrients. *Proc. Royal Soc. Lond.*, Ser. B. **136**:110-125.

Brown, S. J. and R. M. Klein. 1973. Effects of near ultraviolet and visible radiations on cell cycle kinetics in excised root meristems of *Pisum sativum. Amer. J. Bot.* **60**:554-560.

Brulfert, A. and G. Deysson. 1971. Synchronisation des divisions cellulaires par action de l'hydroxyuree sur les meristemes radiculaires d'*Allium sativum* L. *C. R. Acad. Sci.* (Paris) Ser. D. **273**:146-149.

Brunori, A. 1967. Relationship between DNA synthesis and water content during ripening of *Vicia faba* seeds. *Caryologia* **20**:333-338.

Brunori, A., S. Avanzi, and F. D'Amato. 1966. Chromatid and chromosome aberrations in irradiated dry seeds of *Vicia faba. Mutation res.* **3**:305-313.

Brunori, A. and F. D'Amato. 1967. The DNA content of nuclei in the embryo of dry seeds of *Pinus pinea* and *Latuca sativa. Caryologia* **20**:153-160.

Bullough, W. S. 1952. The energy relations of mitotic activity. *Biol. Rev.* **27**:133-168.

Burholt, D. R. and J. Van't Hof. 1971. Quantitative thermal-induced changes in growth and cell population kinetics of *Helianthus* roots. *Amer. J. Bot.* **58**:386-393.

Burholt, D. R. and J. Van't Hof. 1972. Cell population kinetics of *Pisum* root meristem cells during and after a mitotic-inhibitory exposure to protracted gamma-irradiation. *Rad. Biol.* **21**:307-319.

Chen, D., S. Sarid, and E. Katchalski. 1968. The role of water stress in the inactivation of messenger RNA of germinating wheat embryos. *Proc. Nat. Acad. Sci.* **61**:1378-1383.

Clowes, F. A. L. 1963. The quiescent center in meristems and its behavior after irradiation. *Brookhaven Symp. Biol.* **16**:46-58.

Clowes. F. A. L. 1965a. The duration of the G_1 phase of the mitotic cycle and its relation to radiosensitivity. *New Phytol.* **64**:355-359.

Clowes, F. A. L. 1965b. Synchronization in meristem by 5-amino-uracil. *J. Exp. Bot.* **16**:581-586.

Clowes, F. A. L. 1967. The functioning of meristems. *Sci. Prog.* **55**:529-542.

Clowes, F. A. L. 1970a. The immediate response of the quiescent centre to X-rays. *New Phytol.* **69**:1-18.

Clowes, F. A. L. 1970b. Nutrition and quiescent centre of root meristems. *Planta* **90**:340-348.

Clowes, F. A. L. 1972a. Cell cycles in a complex meristem after X-irradiation. *New Phytol.* **71**:891-897.

Clowes, F. A. L. 1972b. The control of cell proliferation within root meristems. In *The Dynamics of Meristem Cell Populations,* eds. M. W. Miller and C. C. Kuehnert. Plenum Pub. Corp., N.Y., pp. 133-147.

Clowes, F. A. L. and E. J. Hall. 1962. The quiescent centre in root meristems of *Vicia faba* and its behavior after acute X-irradiation and chronic gamma irradiation. *Rad. Bot.* **3**:45-53.

Constantinescu, D. G. and D. Radu. 1970. Contribution a l'etude de l'action des ions Ni^{2+} sur la cellule vegetale. *Soc. Bot. France Mem.* 213-218.

Coulomb, P. and C. Coulomb. 1972. Restauration de meristemes radiculaires de courge anoxies. *C. R. Acad. Sci.* (Paris) Ser D **274**:856–858.

Crapo, N. 1975. Selected responses of root metabolism to decreased photosynthesis. Ph.D. Dissertation, University of California, Davis.

D'Amato, F. 1949. Prophase inhibition of mitosis in root meristems. *Caryologia* **1**:109–121.

D'Amato, F. 1960. Cyto-histological investigations of anti-mitotic substances and their effects of patterns of differentiation. *Caryologia* **13**:330–351.

D'Amato, F. 1972. Morphogenetic aspects of the development of meristems in seed embryos. In *The Dynamics of Meristem Cell Populations,* eds. M. W. Miller and C. C. Keuhnert. Plenum Pub. Corp., N.Y., pp. 149–163.

Dettor, C. M., W. C. Dewey, L. F. Winans, and J. S. Noel. 1972. Enhancement of X-ray damage in synchronous Chinese hamster cells by hypertonic treatments. *Rad. Res.* **52**:352–372.

Dewey, W. C., J. S. Noel, and C. M. Dettor. 1972. Changes in radiosensitivity and dispersion of chromatin during the cell cycle of synchronous Chinese hamster cells. *Rad. Res.* **52**:373–394.

Deysson, G. 1968. Antimitotic substances. *Int. Rev. Cytol.* **24**:99–148.

Deysson, G. 1970. Sur l'utilisation des meristemes radiculaires pour l'etude d'actions antimitotiques. *Soc. Bot. France Mem.* 95–110.

Dunlop, D. W. and B. L. Schmidt. 1964. Biomagnetics. I. Anomalous development of the root of *Narcissus tazetta* L. *Phytomorphology* **14**:333–342.

Dunlop, D. W. and B. L. Schmidt. 1965. Biomagnetics. II. Anomalies found in the root of *Allium cepa* L. *Phytomorphology* **15**:400–414.

Eigsti, D. H. and P. Dustin, Jr. 1955. *Colchicine in Agriculture, Medicine, Biology, and Chemistry.* Iowa State College Press, Ames, Iowa.

Eriksson, T. 1966. Partial synchronization of cell division in suspension cultures of *Haplopappus gracilis. Physiol. Plant.* **19**:900–910.

Evans, H. J. 1965. Effects of radiations on meristematic cells. *Rad. Bot.* **5**:171–182.

Evans, H. J., G. J. Neary, and S. M. Tonkinson. 1957. The use of colchicine as an indicator of mitotic rate in broad bean root meristems. *J. Genet.* **55**:487–502.

Evans, H. J. and J. R. K. Savage. 1959. The effect of temperature on mitosis and on the action of colchicine in root meristem cells of *Vicia faba. Exp. Cell Res.* **18**:51–61.

Evans, L. S. and J. Van't Hof. 1973. Cell arrest in G_2 in root meristems: a control factor from the cotyledons. *Exp. Cell Res.* **82**:471–473.

Evans, L. S. and J. Van't Hof. 1974a. Promotion of cell arrest in G_2 in root and shoot meristems in *Pisum* by a factor from the cotyledons. *Exp. Cell Res.* **87**:259–264.

Evans, L. S. and J. Van't Hof. 1974b. Is the nuclear DNA content of mature root cells prescribed in the root meristem? *Amer. J. Bot.* **61**:1104–1111.

Evans, L. S. and J. Van't Hof. 1975. The age-distribution of cell cycle populations in plant root meristems: complex tissues. *Exp. Cell Res.* **90**:401–410.

Feldman, L. J. and J. G. Torrey. 1976. The isolation and culture *in vitro* of the quiescent center of *Zea mays. Amer. J. Bot.* **63**:345–355.

Foard, D. E. and A. H. Haber. 1970. Physiologically normal senescence in seedlings grown without cell division after massive γ-irradiation of seeds. *Rad. Res.* **42**:372–380.

Gelfant, S. 1959. A study of mitosis in mouse ear epidermis *in vitro.* II. Effects of oxygen tension and glucose concentration. *Exp. Cell Res.* **18**:494–503.

Gelfant, S. 1962. Initiation of mitosis in relation to the cell division cycle. *Exp. Cell Res.* **26**:395–403.

Gelfant, S. 1963. Patterns of epidermal cell division. I. Genetic behavior of the G_1 cell population. *Exp. Cell Res.* **32**:521–528.

Gelfant, S. 1966. Patterns of cell division: The demonstration of discrete cell populations. In *Methods of Cell Physiology,* ed. D. M. Prescott, Vol. II. Academic Press, N.Y., pp. 359–395.

Giménez-Martín, G., A. González-Fernández, and J. F. López-Sáez. 1966. Duration of the division cycle in diploid, binucleate and tetraploid cells. *Exp. Cell Res.* **43**:293–300.

Giménez-Martín, G., A. González-Fernández, C. de la Torre, and M. E. Fernández-Gómez. 1971. Partial initiation of endomitosis by 3' -deoxyadenosine. *Chromosoma* **33**:361–371.

González-Fernández, A., G. Giménez-Martín, and C. de la Torre. 1971. The duration of the interphase periods at different temperatures in root tip cells. *Cytobiologie* **3**:367–371.

González-Fernández, A., G. Giménez-Martín, M. E. Fernández-Gómez, and C. de la Torre. 1974. Protein synthesis requirements at specific points in the interphase of meristematic cells. *Exp. Cell Res.* **88**:163–170.

Gudkov, I. N. and D. M. Drodzinskii. 1972. Radiosensitivity of mitotic cycle phases in plant meristems. *Radiobiologiya* **12**:151–161.

Gunckel, J. E. and A. H. Sparrow. 1953. Aberrant growth in plants induced by ionizing radiation. *Brookhaven Symp. Biol.* **6**:252–279.

Gunckel, J. E. and A. H. Sparrow. 1961. Ionizing radiations: biochemical, physiological and morphological aspects of their effects on plants. *Encycl. Plant Physiol.* **16**:555–611.

Haber, A. H. and D. E. Foard. 1964. Further studies of gamma-irradiated wheat and their relevance to use of mitotic inhibition for developmental studies. *Amer. J. Bot.* **51**:151–159.

Haber, A. H. and B. E. Rothstein. 1969. Radiosensitivity and rate of cell division: "Law of Bergonie and Tribondeau." *Science* **163**:1338–1339.

Hadder, J. C. and G. B. Wilson. 1958. Cytological assays of c-mitotic and prophase poison actions. *Chromosoma* **9**:91–104.

Halaban, R. 1972. The mitotic index and cell cycle of *Lemna perpusilla* under different photoperiods. *Plant Physiol.* **50**:308–310.

Hepler, P. K. and W. T. Jackson. 1969. Isopropyl N-phenylcarbamate affects spindle microtubule orientation in dividing endosperm cells of *Haemanthus katherinae. J. Cell Sci.* **5**:727–743.

Hess, F. D. 1973. The cytological effect of the herbicide trifluralin on root meristems of cotton (*Gossypium hirsutum* L. 'Acala 4–42'). M. S. Thesis, University of California, Davis.

Hess, F. D. 1975. The mechanism of action of the herbicide trifluralin: Influence on cell division, cell enlargement, and microtubular protein. Ph.D. Dissertation, University of California, Davis.

Howard, A. and S. R. Pelc. 1953. Synthesis of deoxyribonucleic acid in normal and irradiated cells and its relation to chromosome breakage. *Heredity* (suppl.) **6**:261–273.

Hsu, T. C., W. C. Dewey, and R. M. Humphrey. 1962. Radiosensitivity of cells of

Chinese hamster *in vitro* in relation to the cell cycle. *Exp. Cell Res.* 27:441–452.

Itai, C. and Y. Vaadia. 1968. The role of root cytokinins during water and salinity stress. *Isr. J. Bot.* 17:187–195.

Jacqmard, A. and J. P. Miksche. 1971. Cell population and quantitative changes of DNA in the shoot apex of *Sinapis alba* during floral induction. *Bot. Gaz.* 132:364–367.

Jakob, K. M. 1972. RNA synthesis during the DNA synthesis period of the first cell cycle in the root meristem of germinating *Vicia faba*. *Exp. Cell Res.* 72:370–376.

Jauffret, F., O. Rohfritsch, and C. Wasser. 1970. Observations sur l'action cecidogene du Taxomyia taxi Inch. sur le bourgeon de *Taxus baccata* L. et la transformation du meristeme apical en tissu nourricier. *C. R. Acad. Sci.* (Paris) Ser D 271:1767–1770.

Kehr, A. E. and H. H. Smith. 1953. Genetic tumors in *Nicotiana* hybrids. *Brookhaven Symp. Biol.* 6:55–76.

Kessler, B. 1961. Nucleic acids as factors in drought resistance of higher plants. *Recent Advances in Bot.* 1153–1159.

Kihlman, B. A. 1966. *Actions of Chemicals on Dividing Cells*. Prentice-Hall Inc., N.J.

Klein, R. M. 1967. Influence of ultraviolet radiation on auxin-controlled plant growth. *Amer. J. Bot.* 54:904–914.

Kovacs, C. J. and J. Van't Hof. 1970. Synchronization of a proliferative population in a cultured plant tissue. *J. Cell Biol.* 47:536–538.

Kovacs, C. J. and J. Van't Hof. 1971. Mitotic delay and the regulating events of plant cell proliferation: DNA replication by a G_1/S population. *Rad. Res.* 48:95–106.

Kunkel, L. O. 1953. Virus-induced abnormalities. *Brookhaven Symp. Biol.* 6:157–172.

Lea, D. E. 1947. *Actions of Radiations of Living Cells*. Cambridge University Press.

Leeper, D. R., M. H. Schneiderman, and W. C. Dewey. 1972. Radiation-induced division delay in synchronized Chinese hamster ovary cells in monolayer culture. *Rad. Res.* 50:401–417.

Levitt, J. 1972. *Responses of Plants to Environmental Stresses*. Academic Press, N.Y.

Levy, F. 1971. Activite antimitotique et inhibition de la proteogenese sous l'influence du cycloheximide dans les cellules meristematiques radiculaires d'*Allium sativium* L. *C. R. Acad. Sci.* (Paris) Ser D 272:553–556.

López-Sáez, J. F., G. Giménez-Martín, and A. González-Fernández. 1966. Duration of the cell division cycle and its dependence on temperature. *Z. Zellforschung und Mikroskopische Anatomie* 75:591–600.

Lord, A. 1974. Observations on *Raphanus sativus* meristematic cells treated with ethidium bromide. *J. Ultrastruct. Res.* 46:117–130.

Lytinskaya, T. K. 1971. Effect of cycloheximide on the root meristem. *Tsitoligya* 13:1358–1368.

Macleod, R. D. 1969. Some effects of 2,4,5-trichlorophenoxyacetic acid on the mitotic cycle of lateral root apical meristems of *Vicia faba*. *Chromosoma* 27:327–337.

Macleod, R. D. 1972. Cell progression through the mitotic cycle in lateral root apical meristems of *Vicia faba* L. following colchicine treatment. *Caryologia* 25:83-94.

Macleod, R. D. and D. W. F. Scadeng. 1975. The quiescent centre in excised roots of *Pisum sativum* L. *Protoplasma* 86:135-140.

Mann, J. D., L. S. Jordan, and B. E. Day. 1965a. The effects of carbamate herbicides on polymer synthesis. *Weeds* 13:63-66.

Mann, J. D., L. S. Jordan, and B. E. Day. 1965b. A survey of herbicides for their effect upon protein synthesis. *Plant Physiol.* 40:840-843.

Mazia, D. 1970. Regulatory mechanisms of cell division. *Fed. Proc.* 29:1245-1247.

Mazia, D. 1974. The cell cycle. *Sci. Amer.* 230:54-64.

Mitchison, J. M. 1971. *The Biology of the Cell Cycle.* Cambridge University Press, London.

Murin, A. 1966. The effects of temperature on the mitotic cycle and its time parameters in root tips of *Vicia faba. Naturwissenschaften* 53:312-313.

Nagl, W. 1970a. Differential inhibition by actinomycin D and histone F1 of mitosis and endomitosis *Allium carinatum. Z. Pflanzenphysiol.* 63:316-326.

Nagl, W. 1970b. On the mercaptoethanol-induced polyploidization in *Allium cepa. Protoplasma* 70:349-359.

Nagl, W. 1972. Selective inhibition of cell cycle stages in the *Allium* root meristem by colchicine and growth regulators. *Amer. J. Bot.* 59:346-351.

Natarajan, A. T. and G. Ahnstrom. 1970. The localization of radiation-induced chromosome aberrations in relation to the distribution of heterochromatin in *Secale cereale. Chromosoma* 30:250-257.

Pareyre, C. 1970. Inhibition de la multiplication cellulaire et de la proteogenese dans les cellules meristematiques de la racine d'*Allium sativum* L. sous l'influence de la tubulosine. *Soc. Bot. France Mem.* 167-174.

Pickett-Heaps, J. D. 1967. The effect of colchicine on the ultrastructure of dividing plant cells, xylem wall differentiation and distribution of cytoplasmic microtubules. *Dev. Biol.* 15:206-236.

Pilet, F., A. C. Hildebrandt, A. J. Riker, and F. Skoog. 1960. Growth *in vitro* of tissues isolated from normal stems and insect galls. *Amer. J. Bot.* 47:186-195.

Price, H. J., A. H. Sparrow, and A. F. Nauman. 1973. Correlations between nuclear volume, cell volume, and DNA content in meristematic cells of herbaceous angiosperms. *Experientia* 29:1028-1029.

Rose, R. J. 1970. The effect of cycloheximide on cell division in partially synchronized plant cells. *Aust. J. Biol. Sci.* 23:573-583.

Rost, T. L. and D. E. Bayer. 1976. Cell cycle population kinetics of propham inhibited root tips. *Weed Science* 24:81-87.

Rost, T. L. and J. Van't Hof. 1973. Radiosensitivity, RNA, and protein metabolism of "leaky" and arrested cells in sunflower root meristems *(Helianthus annuus). Amer. J. Bot.* 60:172-181.

Scheuermann, W. and G. Klaffke-Lobsien. 1973. On the influence of 5-amino uracil on the cell cycle of root tip meristems. *Exp. Cell. Res.* 76:428-436.

Smith, G. F. and H. Kersten. 1941. Root modifications induced in *Vicia faba* by irradiating dry seeds with soft x-rays. *Plant Physiol.* 16:159-170.

Socher, S. H. and D. Davidson. 1971. Evidence for an effect of 5-fluoro-deoxy-uridine on cells in the G2 phase of the mitotic cycle. *Can. J. Genet. Cytol.* 13:70-74.

Soifer, D (ed.) 1975. *The Biology of Cytoplasmic Microtubules.* Vol. 253 Ann. New York Acad. Sci.

Sparrow, A. H. 1965. Cellular radiation biology. In *Symp. on Fundamental Cancer Res.* Williams and Wilkins Co., Baltimore.

Sparrow, A. H., R. C. Sparrow, K. H. Thompson, and L. A. Schairer. 1964. The use of induced mutations in plant breeding. Report of the Meeting Organized by the FAO of the U.N. and the IAEA. Rome, Italy, pp. 101–132.

Sparrow, A. H., A. G. Underbrink, and R. C. Sparrow. 1967. Chromosomes and cellular radiosensitivity. I. The relationship of D_O to chromosome volume and complexity in seventy-nine different organisms. *Rad. Res.* 32:915–945.

Srivastava, S. and S. R. V. Rao. 1972. Effects of Vincaleukoblastine (VLB) on the mitotic cycle of *Vicia faba* lateral root meristem cells. *Nucleus* (Calcutta) 15: 48–51.

Troy, M. R. and D. E. Wimber. 1968. Evidence for a constancy of the DNA synthetic period between diploid-polyploid groups in plants. *Exp. Cell Res.* 53:145–154.

Underbrink, A. G., A. H. Sparrow, and V. Pond. 1968. Chromosomes and cellular radiosensitivity. II. Use of interrelationships among chromosome volume, nucleotide content and Do of 120 diverse organisms in predicting radiosensitivity. *Rad. Bot.* 8:205–238.

Valadaud-Barrieu, M. and M. C. Izard. 1973. Modifications du cycle cellulaire sous l'influence de l'hydroquinone dans les meristemes radiculaire de *Vicia faba. C. R. Acad. Sci.* (Paris) Ser D 276:33–35.

Van't Hof, J. 1963. Mitotic delay following x-radiation in the meristem cells of *Pisum sativum. Rad. Bot.* 3:311–314.

Van't Hof, J. 1964. Relationships between mitotic cycle duration, S-period duration and the average rate of DNA synthesis in the root meristem cells of several plants. *Exp. Cell Res.* 39:48–58.

Van't Hof, J. 1966a. Experimental control of DNA synthesizing and dividing cells in excised root tips of *Pisum. Amer. J. Bot.* 53:970–976.

Van't Hof, J. 1966b. Inhibition of mitosis in *Pisum* root meristems by continuous gamma radiation: The influence of temperature on the synthesis of DNA, RNA and protein during inhibition. *Amer. J. Bot.* 53:246–252.

Van't Hof, J. 1970. Postirradiation hypoxia and the recovery of G_1 and G_2 cells and DNA synthesis in cultured root meristems. *Rad. Res.* 41:538–551.

Van't Hof, J. 1973. The regulation of cell division in higher plants. *Brookhaven Symp. Biol.* 25:152–165.

Van't Hof, J. 1974. Control of the cell cycle in higher plants. In *Cell Cycle Controls,* eds. G. M. Padilla, I. L. Cameron, and A. Zimmerman. Academic Press, N.Y., pp. 77–86.

Van't Hof, J. 1975. The duration of chromosomal DNA synthesis, the mitotic cycle and meiosis in higher plants. In *Handbook of Genetics,* Vol. II, ed. R. C. King. Plenum Pub. Co., N.Y.

Van't Hof, J. and C. J. Kovacs. 1970. Mitotic delay in two biochemically different G1 cell populations in cultured roots of pea *(Pisum sativum). Rad. Res.* 44: 700–712.

Van't Hof, J. and C. J. Kovacs. 1972. Mitotic cycle regulation in the meristem of cultured roots: The principal control point hypothesis. In *The Dynamics of*

Meristem Cell Populations, eds. M. W. Miller and C. C. Keuhnert. Plenum Press, N.Y.

Van't Hof, J., D. P. Hoppin, and S. Yagi. 1973. Cell arrest in G1 and G2 of the mitotic cycle of *Vicia faba* root meristems. *Amer. J. Bot.* **60**:889–895.

Van't Hof, J. and T. L. Rost. 1972. Cell proliferation in complex tissues: The control of the mitotic cycle of cell populations in the cultured root meristem of sunflower *(Helianthus). Amer. J. Bot.* **59**:769–774.

Van't Hof, J. and A. H. Sparrow. 1963a. Growth inhibition, mitotic cycle time and cell number in chronically irradiated root meristems of *Pisum. Rad. Bot.* **3**:239–247.

Van't Hof, J. and A. H. Sparrow. 1963b. A relationship between DNA content, nuclear volume, and minimum mitotic cycle time. *Proc. Nat. Acad. Sci.* **49**: 897–902.

Van't Hof, J. and H. K. Ying. 1964. Relationship between the duration of the mitotic cycle, the rate of cell production, and the rate of growth of *Pisum* roots at different temperatures. *Cytologia* **29**:399–406.

Wangenheim, K. H. V. 1970. Genetische Defekte und Strahlenschädigung. *Rad. Bot.* **10**:469–490.

Wangenheim, K. H. V. 1971. Radiation damage, caused by interference with cellular control of differentiation and proliferation. *1st European Biophysics Congress, Proc.,* pp. 425–429.

Webster, P. L. 1973. Effects of cycloheximide on mitosis in *Vicia faba* root-meristem cells. *J. Exp. Bot.* **24**:239–244.

Webster, P. L. and H. D. Langenauer. 1973. Experimental control of the activity of the quiescent center in excised root tips of *Zea mays. Planta* **112**:91–100.

Webster, P. L. and J. Van't Hof. 1969. Dependence of energy and aerobic metabolism of initiation of DNA synthesis and mitosis by G_1 and G_2 cells. *Exp. Cell Res.* **55**:88–94.

Webster, P. L. and J. Van't Hof. 1970. DNA synthesis and mitosis in meristems: Requirements for RNA and protein synthesis. *Amer. J. Bot.* **57**:130–139.

Webster, P. L. and J. Van't Hof. 1973. Polyribosomes in proliferating and nonproliferating root meristem cells. *Amer. J. Bot.* **60**:117–121.

Wilson, G. B. 1966. *Cell Division and the Mitotic Cycle.* Reinhold Pub. Corp., N.Y.

Wimber, D. E. 1966. Duration of the nuclear cycle in *Tradescantia* root tips at three temperatures as measured with ^{3}H-thymidine. *Amer. J. Bot.* **53**:21–24.

Wood, H. N. and A. C. Braun. 1961. Studies on the regulation of certain essential biosynthetic systems in normal and crown-gall tissues or cells. *Proc. Nat. Acad. Sci.* **47**:1907–1913.

Wood, N. H. and A. C. Braun. 1973. 8-bromoadenosine 3':5'-cycle monophosphate as a promoter of cell division in excised tobacco pith parenchyma tissue. *Proc. Nat. Acad. Sci.* **70**:447–450.

Wood, H. N., M. C. Lin, and A. C. Braun. 1972. The inhibition of plant and animal adenosine 3':5'-cyclic monophosphate phosphodiesterases by a cell-division-promoting substance from tissues of higher plant species. *Proc. Nat. Acad. Sci.* **69**: 403–406.

Wood, H. N., M. E. Rennekamp, D. V. Bowen, F. H. Field, and A. C. Braun. 1974. A comparative study of cytokinesins I and II and zeatin riboside: A reply to Carlos Miller. *Proc. Nat. Acad. Sci.* **71**:4140–4143.

Yamakawa, K. and A. H. Sparrow. 1966. The correlation of interphase chromosome volume with pollen abortion induced by chronic gamma irradiation. *Rad. Bot.* **6**:21–38.

Yee, V. 1976. The effects of water stress on the mitotic cycle kinetics of *Vicia faba* root tip cells. M.S. Thesis, Department of Botany, University of California, Davis.

Yeoman, M. M. 1974. Division synchrony in cultured cells. In *Tissue Culture and Plant Science,* ed. H. E. Street. Academic Press, N.Y., pp. 1–17.

Part II. Nuclear Structure and Chromosome Movement

Nuclear Structures During Cell Cycles

Walter Nagl   The University of Kaiserslautern

INTRODUCTION

It is primarily the replication of chromosomal DNA and the segregation of the chromosomes which allows one to distinguish between several phases of the cell cycle (Howard and Pelc 1953). While it is easy to separate the mitotic stages from each other on grounds of striking morphological changes of the chromosomes, the stages of interphase are very difficult to be discerned, and in many species this is so far not possible by means of structural analysis.

In this essay I will summarize the available data on man, animal, and plant species, whose nuclei undergo structural changes during interphase and very early prophase. Some new findings will also be reported. In addition, some unique structural features of nuclei passing through endo-cycles will be described. Finally, possible reasons for the morphological changes of chromatin during the cell cycle, as well as their functional significance, will be discussed.

TECHNIQUES EMPLOYED

Optical Microscopy

In most cases, the nuclear structure is well preserved after fixation of material in acetic alcohol and staining with aceto-carmine. If the nuclear DNA content is to

147

be measured for classification, a standardized Feulgen method must be used (e.g., Stowell 1945). However, fine structural details may be lost by this procedure. After hydrolysis, staining with toluidine blue O also gives good results.

Permanent slides may be obtained by the solid CO_2 technique (Conger and Fairchild 1953), or by related methods.

A stage-specific differential staining technique has been developed by Alvarez and Valladares (1972). However, the specificity has been questioned by Dewse (1974).

The fluorochrome quinacrine dihydrochloride has been employed by Moser et al. (1975) to differentiate nuclei in various phases of the cell cycle.

Electron Microscopy

The fine structure of nuclei has been repeatedly investigated in ultrathin sections after fixation in glutaraldehyde and osmium tetroxide, and embedding in Epon or Araldite. If only the chromatin is to be studied, staining with uranyl acetate and lead citrate is usually not necessary. Good results have been obtained by staining with uranyl acetate during dehydration (Wohlfarth-Bottermann 1957). Specific staining of nucleic acids can be achieved with bismuth (Albersheim and Killias 1963) and indium (Watson and Aldridge 1961); RNA-containing structures can differentially be shown by the EDTA technique of Bernhard (1969). Fiber size and DNA packing ratios can be determined in ultrathin sections (de la Torre et al. 1975), or after spreading of the chromosome fibers and critical point drying (Golomb and Bahr 1974).

Recently, also, the scanning electron microscope has been applied to karyology; the material was fixed in ethanol:acetic acid (3:1), dried, and coated with gold and coal (Lima-de-Faria 1974a; Sato 1974), or spread on a water surface and dried by the critical point method (Golomb and Bahr 1974).

Image Analysis

Computer systems are now more and more used for automated structural analysis and morphometry (Müller 1973; Gray 1973). Some recent employment of image analyzers in the study of cell cycle-dependent changes in the nuclear structure will be reported in the following section.

Volumetry

The nuclear volume, which increases during interphase and prophase in a stepwise manner, can be used to classify nuclei with differing structure in terms of the stage in which they are at the moment of fixation (e.g., Patau and Swift 1953; Alfert 1955; Doležal and Tschermak-Woess 1955; Tschermak-Woess and Doležal-Janisch 1957; Woodard et al. 1961; Mäkinen 1963; Van't Hof and Sparrow 1963; Klinger et al. 1967; Mittwoch 1967; Rasch et al. 1967; Nagl 1968, 1969a). This is of importance, because nuclei with varying chromatin structures, but the same DNA content, can only be categorized by their volume increases throughout the cell cycle. The increase, however, is not linear, and varies from tissue to tissue, from species to species, and is, moreover, dependent on the metabolic activity of the nucleus and cell. The increase between G_1 and G_2 has been found to be between 1.5 and 1.8 (Nagl 1968). Only in preimplantation mouse embryos was it found that the chromatin area nearly triples between G_1 and G_2 (Sawicki et al. 1974). Duplication of the nuclear volume has been found between the G_1 stages of successive endomitotic cycles; on the other hand, the nuclear volume between G_1 and G_2 of the same endomitotic cycle increases only by a factor of 1.32 to 1.48 (Nagl 1968).

It is also important actually to compare the nuclear volumes with each other, and not the areas only, because the latter may lead to erroneous (Tschermak-Woess and Hasitschka 1953) or undistinguishable values (Comings 1967b). The volume of nuclei not too irregularly shaped can be calculated by the formula for the rotation ellipsoid, $V = 4/3\pi \times a \times b \times c$, where a and b can be measured by the aid of an ocular micrometer, and c by means of the scale on the micrometer screw of the microscope (Tschermak-Woess and Hasitschka 1953). In many instances, the relative values obtained by the multiplication of the three diameters may be sufficient.

In this connection it is worth noting that nuclear volume and DNA content are correlated, in some way, even when different species are compared. Baetcke et al. (1967) established the following conversion factor: log DNA = -1.497 + (1.22 × log VOLUME). Since the nuclear volume also displays, however, tissue- and function-specific changes, this conversion can be made only on the basis of a large number of measurements, and it may give rather a rough estimate.

DNA Measurement

The measurement of the nuclear DNA content is the most appropriate way to classify nuclei in G_1 and G_2, i.e., to prove that the structural differences found among nuclei are really related to different stages of the cell cycle. The tissue is hydrolyzed in 1N HCl at 60°C for 10 minutes, Feulgen-stained under standardized conditions, and squashed. This ensures measurements of complete nuclei (not only sections of them), and enough separation from each other to obtain reference areas between them. Using a conservative cytophotometer, the two-wavelength method should be employed to minimize the distributional error due to inhomogeneities in the chromatin structure (e.g., chromocenters, unstained nucleoli). The method has been published independently by Ornstein (1952) and Patau (1952); tables to facilitate the calculations were given by Mendelsohn (1958). By use of a scanning densitometer, the distributional error can be neglected. For several scanning photometers, computer programs are commercially available that are adjusted to investigations on both basic cytological research and clinical routine check of nuclei and chromosomes.

The relative extinction values can be converted into absolute DNA amounts by including a reference species with known DNA content (e.g., *Allium cepa*, 2C = 35.5 pg; Van't Hof 1965); the nuclear DNA content of a number of animal and plant species is listed in Rees and Jones (1972), that of various plants in Bennett (1972).

Autoradiography

Nuclei exhibiting a structure characteristic for DNA replication can be most accurately identified by ^{3}H-thymidine autoradiography, although the nuclear DNA content, intermediate between that of G_1 and G_2 nuclei, is an indication also. The nuclear volume cannot be used to distinguish nuclei in S period from those in other stages, because the volume often is the same in G_1 and S.

To elucidate the actual relationship between structure and DNA synthesis, it is necessary to take very short pulse incubation times, otherwise the label will be found over a structure not identical with that during the incubation period. Good results have been obtained with a 5 to 10 minute pulse with ^{3}H-thymidine of a specific activity between 5 and 25 Ci/mM and a final concentration of 2 to 10 μCi/ml. Exposure time should not be more than a few days to avoid too dense labeling, which hinders a view of the structure under the silver grains.

For electron microscopic autoradiography, the grids carrying the ultrathin sections have to be covered with a homogeneously monolayer of the photographic emulsion; proved methods have been given by Caro (1962), Hay and Revel (1963); Haase and Jung (1964); for a review see Jacob (1971). Exposure time normally is much longer than in light autoradiography, some 4 to 8 weeks. To obtain small silver grains instead of net-like traces, Lettré and Paweletz (1966) composed a special developer containing ascorbic acid and phenidon (= 1-phenyl-3-pyrazolidon).

Biochemical Techniques

There are only a few biochemical attempts to elucidate changes of the nuclear structure in relation to the cell cycle, because synchronized material of species with nuclei suited for structural analysis are difficult to obtain. The findings obtained so far are of a more indirect nature, and particularly are concerned with the role of nuclear proteins in the control of chromosome condensation at prophase. Details will be found in other chapters of this book.

Pederson (1972) introduced the accessibility of chromatin to DNase, and Pederson and Robbins (1972) the capacity of binding actinomycin D - [3]H to distinguish between nuclei in various phases of the cell cycle (see the Discussion and Conclusions section).

Terminology

The inexact use of cytological terms has led to some confusion and misunderstanding in the literature. Therefore, some terms will be defined as they will be used in this essay.

Chromocenters are heterochromatic portions as seen in an interphase or working nucleus. They can have their origin from a single heterochromatic segment of a certain chromosome, or originate by fusion of several segments of nonhomologous chromosomes (collective chromocenters); in endopolyploid nuclei they may be the product of the fused corresponding heterochromatic sections of the sister chromosomes, which arose through repeated endo-cycles; then they are called endochromocenters.

Chromomeres are knob-like thickenings of the chromonema because of a local coiling of the chromatin fibril. In pachytene, the chromomeres are visible like beads on a string, in polytene nuclei the corresponding chromomeres fuse to bands, and in many interphase and working nuclei they are more or less homogeneously distributed through the nucleus. Their visibility and their number depends solely on the degree of condensation of the chromatin fibril. In genetical and molecular terms, one chromomere apparently corresponds to one gene and a number of repetitive DNA sequences.

Chromonema is the term used in light microscopy to denote the chromosomes as seen in interphase and working nuclei, i.e., the decondensed chromosomes. The visibility of a chromonema depends on a certain degree of condensation that must be less than in mitosis, but more than in cases where chromomeres become distinguishable. In the electron microscope, nuclei of the chromonematic type display a reticulum of threads, whose electron-density does not differ from that of the heterochromatin (chromocenters). It can also be seen that the threads themselves are composed of much thinner fibrils, the so-called elementary fiber of the chromosome. Thus, the chromonema represents a state of the chromosome, in that the elementary fiber is coiled or otherwise condensed to a high extent.

Euchromatin is that portion of the chromatin which undergoes a characteristic condensation-decondensation (coiling-uncoiling) cycle during the mitotic cell cycle. In the light microscope, euchromatin appears less stained than heterochromatin; in the electron microscope this is not always the case (see above under chromonema). In genetical and molecular terms, the euchromatin contains the single genes and unique DNA sequences as well as the intermediate-repetitive sequences. It is active in RNA synthesis.

Heterochromatin is the portion of chromatin which remains condensed throughout the cell cycle, except for a short period in early prophase and endomitosis of plants. If not stated otherwise, the term denotes *constitutive* (or *karyotypical*) *heterochromatin*. This contains among others the highly repetitive sequences which may appear as a satellite in a CsCl gradient, if their base composition is different enough from that of the bulk of the DNA. It is inactive in RNA synthesis, often replicates asynchronously (normally later than the euchromatin), and can be stained differentially by the Giemsa and fluorescence banding techniques. The term *facultative heterochromatin* refers to the second X chromosome in mammalian females that becomes condensed due to gene dosage compensation. (The chromocenter formed by this X chromosome at interphase is called Barr body or sex chromatin.) Facultative heterochromatin contains, of course, the same DNA sequences as the active homologous X, but is functionally inactive and late replicating like the constitutive heterochromatin.

Although widely used, it is absolutely wrong to call all *condensed chromatin* heterochromatin. In many cells, particularly of higher animals, euchromatin becomes functionally condensed (inactivated), e.g., in lymphocytes. It must lead to contradictions if this is named heterochromatin.

Throughout biology there is no rule without exception. The nucleolus-associated heterochromatin (nucleolus organizers) is not genetically inert and inactive, but contains the redundant ribosomal genes together with non-coding spacer sequences between them. It forms a puff-like structure, the intranucleolar chromatin, on which high transcriptional activity can be found.

Prochromosome is the term for a chromocenter, when the number of chromocenters equals the number of mitotic chromosomes, i.e., when each chromosome forms a single individually visible chromocenter at interphase.

Species-specific nuclear structures are known from many species, particularly of plants and lower animals. The main types known from light microscopy are the *chromomeric* type, the *chromonematic* type, and the *chromocenter* type. Often, however, chromocenters also occur in nuclei of the chromomeric and chromonematic type, and the nuclear structure may change during cell differentiation in relation to functional changes.

In electron microscopy, nuclei of the *chromocenter* type are those which display a diffuse euchromatin and distinct heterochromatic patches. The *reticular* type is characterized by an electron-dense reticulum of euchromatin threads; in addition, heterochromatin patches may be visible. Nuclei of the *diffuse* type are predominately found in active cells with highly decondensed chromatin; in some cases the chromatin subunits (nucleosomes or *ν*-bodies) may be visible.

It must be emphasized that it is not often possible to correlate a certain light microscopic structural type with an electron microscopic image of the same nucleus. The main reason for this inability may be seen in the lack of enough studies in that field.

Detailed definitions with examples and references can be found in Nagl (1976a, b, and in press).

NUCLEAR STRUCTURE AND ULTRASTRUCTURE DURING THE MITOTIC CYCLE

Mazia (1963) suggested that chromosomes undergo a condensation-decondensation sequence extending throughout the entire cell cycle with the two extreme conformational states existing in mitosis (condensed) and S phase (decondensed). Schor et al. (1975) found evidence for this suggestion on the basis of stage-specific response of the chromatin to ultraviolet light irradiation. But what are the structural changes seen in the light and electron microscopes?

Altmann (1966) and Müller (1966) first reported structural differences of mouse nuclei in various stages of the cell cycle. Although nuclei in G_1 display a tissue-specific and function-dependent structure, they generally contain a fine euchromatic background, in which several compact chromocenters (heterochromatic masses) are located. During early S period, the euchromatin becomes more granular, while the chromocenters show a more diffuse surface than in G_1. At late S, the heterochromatin also becomes granular, so that no distinct chromocenters can be distinguished. During G_2 all chromatin becomes roughly granular, and later fibrillar, thus proceeding to the prophasic appearance of the chromosomes (Figure 6-1). Altmann (1966) and Müller (1966) based their interpretations on autoradiographic and cytophotometric data, as well as on observations made in living cells.

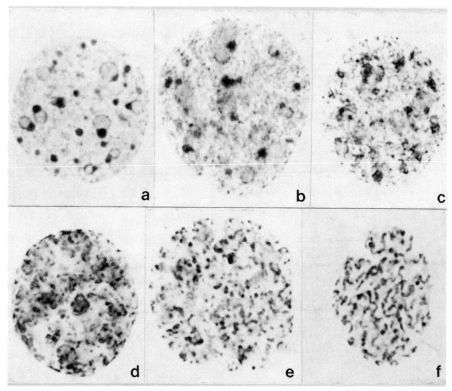

Figure 6-1.
Changes of nuclear structure in liver epithelium cells of the mouse in the course of the mitotic cycle, *a*, G_1 (compact chromocenters, diffuse chromatin); *b*, early S ("corroded" chromocenters, granular euchromatin); *c*, late S (granular heterochromatin); *d*, early G_2 (formation of granular chains); *e*, late G_2 (short chromosomal threads become visible); *f*, prophase. Orcein-stained. x 1500. Reproduced, with permission, from Altmann (1966).

Recently, Moser et al. (1975) described differential nuclear fluorescence during the cell cycle, evidently without knowledge of the former results. Human cells stained with quinacrine dihydrochloride could be classified by their fluorescence pattern into early and late G_1, S, and G_2.

Most findings of changes in nuclear structure during the cell cycle have been obtained in plants. For instance, striking differences between G_1 and G_2 nuclei can be observed in root tips of *Allium carinatum* and *A. flavum* (Nagl 1968, 1970a). In the light microscope, the euchromatin of G_1 nuclei appears in the form of granules (chromomeres) and fibrils (chromonemata) of an average diameter of 0.25 μm; the euchromatic elements of G_2 nuclei exhibit diameters between 0.4 and 0.5 μm (Figures 6-2-6-7; Table 6-1). The heterochromatin appears in the form of chromocenters, the number of which is reduced during interphase due to fusion of neighboring masses. Also the nucleoli show a tendency to fuse during interphase, so that

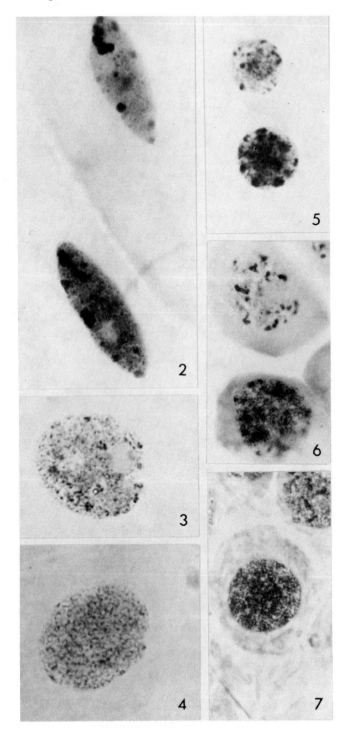

only a single large nucleolus appears at the beginning of mitosis. The nuclei, in G_1 and G_2, differ also in their average volume and, of course, in their DNA content (Table 6-1).

At the electron microscopic level, the structural changes are also evident (Figures 6-8-6-11). The nuclei in G_1 display a nonhomogeneous pattern of chromatin lumps and thin irregular threads, and also show heterochromatic masses. The G_2 nuclei exhibit a more regular pattern of thicker threads (see also Table 6-1). S phase nuclei have been detected in both electron microscopic autoradiography, and for routine studies, by light autoradiography of semithin sections of the same blocks from which the ultrathin sections for electron micrographs have been taken. Such semithin sections are not only much easier to dip into the emulsion, but also re-

Table 6-1
Nuclear characteristics at G_1, G_2, and dispersion (Z) stages of the mitotic cycle in *Allium carinatum* root tips

Parameter	G_1	G_2	Z
DNA content	32.94 pg	65.88 pg	65.88 pg
DNA content, ratio	1.00	2.00	2.00
Volume	450.24 μm^3	739.68 μm^3	879.04 μm^3
Volume, ratio	1.00	1.49	1.66
ϕ euchromatic elements (LM)[1]	0.25 μm	0.45 μm	0.52 μm
N chromocenters (LM)[1]	17	12	0
ϕ chromocenters (LM)[1]	15.0 μm	20.0 μm	—
ϕ chromatin threads (EM)	0.212 μm	0.392 μm	0.472 μm
Extinction (IPM 2)	0.25	0.50	0.45
N micropuffs + RNP clusters	7.5	11.0	13.3
N grains (^{3}H-uridine, LM)[2]	375±5	490±88	717±142

Abbreviations used: ϕ diameter, (LM) measured in the light microscope, (EM) measured in electronmicrographs, N number.

All data given are means. The DNA content may show some variation from population to population because of frequently observed karyotype polymorphism and generative polyploidy.

Sources: 1. Adapted from Nagl (1968).
          2. Adapted from Nagl (1973a).

Figures 6-2-6-7.
Changes of nuclear structure in root tip cells of *Allium flavum* (Figures 6-2-6-4) and *Allium carinatum* (Figures 6-5-6-7) in the course of the mitotic cell cycle. Figure 6-2. Nucleus in G_1 (top) and nucleus in G_2 (bottom). Figure 6-3. Nucleus in early dispersion stage (Z phase); notice granulation of chromocenter areas. Figure 6-4, Nucleus in Z phase; notice homogeneous euchromatic appearance of chromatin. Figure 6-5, Nuclei in G_1 (top) and G_2 (bottom). Figure 6-6. Nucleus in G_1 (top) and nucleus in early Z (bottom). Figure 6-7. Nucleus in Z phase. Figures 6-2-6-4, Feulgen. Figures 6-5-6-7. Acetocarmine. x 1000.

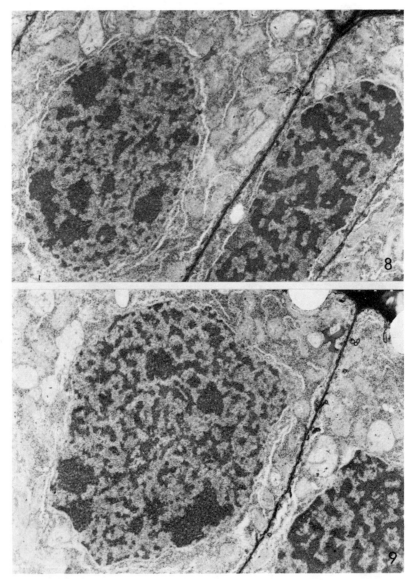

Figures 6-8 and 6-9.
Changes of nuclear ultrastructure in root tip cells of *Allium carinatum* from early G_1 to early S phases. Figure 6-8. Nucleus in late telophase to early G_1 (right) and nucleus in early G_1. Figure 6-9. Nucleus in late G_1 to early S. x 6000.

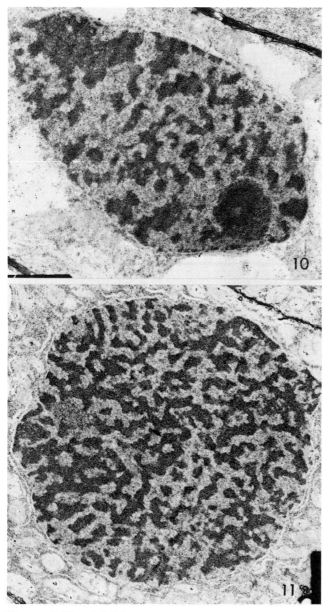

Figures 6-10 and 6-11.
Changes of nuclear ultrastructure in root tip cells of *Allium carinatum* from late S to early prophase. Figure 6-10. Nucleus in late S to early G_2. Figure 6-11. Nucleus in Z phase. x 6000.

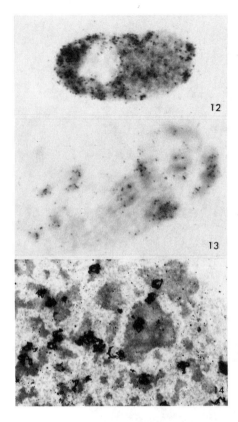

Figures 6-12-6-14.
Nuclear structure and ultrastructure of *Allium carinatum* nuclei at late S period as revealed by light and electron microscopic autoradiography. Notice incorporation of ^{3}H-thymidine during a 5 min pulse into the chromocenters. Figure 6-12. Toluidine blue (stained after development). x 1400. Figure 6-13. Feulgen. x 2000. Figure 6-14. Electron micrograph of an uranyl acetate-stained ultrathin section. x 20,000.

quire an exposure time of only two days. The findings indicate a stepwise transition from the structure typical for G_1 to that typical for G_2, although no dramatic changes have been found (Figures 6-12-6-14).

Structural differences between plant nuclei in G_1 and G_2 have also been found in other species. They are most clear in species with nuclei of the "reticulate type" of structure, as found in many monocots (Lafontaine and Lord 1969; Chouinard 1970; Speta 1972). Since the reticulum is only the appearance of chromatin in fixed nuclei, the terms chromomeric and chromonematic nuclei according to Tschermak-Woess (1963) should be preferred, and the expression "reticulate nuclei" used only with care.

Lafontaine and Lord (1969) found differences in the diameter of the chromonema in G_1 and G_2 nuclei of *Triticum aestivum*: in G_1 it was between 0.1 and 0.15 μm, in G_2 about 0.3 μm. Similar differences were reported by Speta (1972) for *Allium triquetrum*, by Klueva et al. (1974) for *Allium fistulosum*, and by Lafontaine (1974) and Lafontaine and Lord (1974) for *Allium porrum*. In the latter species, electron micrographs reveal two threads of about 0.3 μm in late telophase nuclei and a coarse reticulum of 0.1 to 0.15 μm threads in G_1 nuclei. During early S phase, portions of these strands unravelled extensively, while a chromatin reticulum was reported to appear throughout the nuclear cavity during middle S, which consisted of tortuous strands of about 0.25 μm in diameter. At late S, strands of approximately 0.3 μm in diameter were found to form a loose reticulum. At the light microscopic level, the authors found that the nuclear cavity is occupied by convoluted strands during G_1. This strand-like organization of the chromatin is partly lost during early S, because chromocenters appear which are connected with each other with finely twisted filaments. At mid-S, densely stained strands reappear which give rise to an elaborate reticulum; there are fewer and smaller chromocenters observed at this stage. At late S, the strands are said to be slightly thicker, but no chromocenters appear at all. Finally at G_2, the chromatin reticulum has assumed its most complex organization: paired chromatids forming two tortuous strands and large chromocenters are present.

However, in my opinion, *Allium porrum* is not very well suited for investigations of this kind, because its chromocenters are very small, and, therefore, might be easily missed in ultrathin sections. Because there is some controversy about heterochromatin decondensation during DNA replication, this point will be discussed in more detail in a following paragraph.

Structural differences among nuclei staying at different stages of the mitotic, endomitotic, and amplification cycles have been found in aseptically cultured protocorms of the orchid *Cymbidium* (Nagl 1972a, and in press). The nuclear structure in this genus can deal as a marker for several cell cycle changes occurring during differentiation, and will, therefore, be reviewed in more detail in a consequent section.

Finally, the diameter of the chromatin fibrils appears to increase throughout the interphase. De la Torre et al. (1975) found a mean thickness of 13.5 nm in G_1 nuclei of *Allium cepa*, of 15.0 nm during S, and of 17.4 nm during G_2. In frequency distribution diagrams, a peak occurred at 12.5 nm for the diameter of fibers in G_1 nuclei, while a bimodal distribution was found in G_2 nuclei, with peaks at 15.0 and 22.5 nm. Golomb and Bahr (1974), on the other hand, stated that there is a constant 20 nm fiber in human chromatin and chromosomes, in spite of their diagram showing a bimodal distribution of interphase fibers with peaks at approximately 6 and 18.2 nm. Whether this may be related to certain stages of the cell cycle, or to various degrees of chromatin template activity, has not been discussed.

S Period

Hay and Revel (1963) reported that in S phase nuclei of *Amblystoma* the meshwork-like structure of the chromatin is looser in G_1 than in G_2. Blondel (1968) found decondensation during DNA replication of heterochromatin in synchronous cell cultures at the ultrastructural level, but his electron micrographs are not conclusive, and may also be interpreted in an opposite manner. Milner (1969a) again interpreted electron microscopic autoradiographs of PHA-stimulated lymphocytes in terms of decondensation of heterochromatin during DNA replication. Nagl (1970b), however, pointed out that decondensation did not occur in constitutive or facultative heterochromatin, but in functionally condensed euchromatin, i.e., chromatin inactivated by changes in the chromosomal proteins (compare also Hsu 1962; Littau et al. 1965; Billett and Barry 1974; Fellenberg 1974; Cowden and Curtis 1975). Milner (1969b) did not find a change in the activation-dependent decondensation when DNA replication was inhibited.

On the other hand, Citoler and Gropp (1969) emphasized the condensed state of autosomal heterochromatin during the S phase of *Microtus agrestis*. The large chromocenters of this species, formed by autosomal and sex chromosomal constitutive heterochromatin and also facultative sex chromatin, were subjected to electron microscopic autoradiography by Comings and Okada (1973). After a pulse label of 10 minutes, the heterochromatin as seen in ultrathin sections was clearly labeled in its condensed state (Figure 6-15).

Wide evidence for incorporation of ^{3}H-thymidine into condensed chromatin comes from studies in dipteran salivary gland chromosomes. Our present knowledge indicates that DNA replication always takes place in condensed bands (Rudkin 1963; Gabrusewycz-Garcia 1964; Rodman 1968; Arcos-Terán and Beermann 1968; Plaut 1969; Berendes 1973). Decondensation of bands is apparently a prerequisite for RNA synthesis (puffing!), but not for DNA synthesis.

In spite of these findings, Sanchez and Oberti (1974) again suggested that DNA replication in the heterochromatin of HeLa cells requires transformation of the condensed chromatin to a diffuse state. The micrographs of quinacrine dihydrochloride-stained HeLa cells and fibroblasts as given by Moser et al. (1975) do, however, show distinct heterochromatin masses during early and late S period.

In plants, some Japanese workers (e.g., Tanaka 1965; Masubuchi 1973, 1974) described decondensation of heterochromatin during DNA replication in orchids and liverworts; moreover, this DNA synthesis in heterochromatin was said to take place at *early* S period. These findings have not as yet been confirmed in other laboratories, but early replication fractions of heterochromatin have been also found in the mouse (Hsu and Markvong 1975) and in *Tetrahymena* (Andersen and Engberg 1975). Kuroiwa (1973, 1974) interpreted electron microscopic autoradiographs of early S phase nuclei of *Physarum polycephalum* and *Crepis capillaris* in terms of decondensation of the replicating heterochromatin. The author stated that condensation of the late replicating chromatin occurs within 15 minutes after S, so that decondensation of heterochromatin could not be observed if too long pulse periods were taken.

However, the latter statement does not fit the experiments in *Allium carinatum* (Nagl 1970b), where a ^{3}H-thymidine pulse of 5 to 10 minutes has been applied. In these studies it has been clearly shown that the chromocenters remain in the condensed state during the replication of their DNA (Figures 6-12-6-14). Similar results have already been obtained in several other plants (Lima-de-Faria and Jaworska 1968; Żuk 1969), and were substantiated for *Euglena gracilis* recently (Bertaux et al. 1976). Lafontaine and Lord (1969) reported that S phase nuclei of *Raphanus sativus* and *Tropaeolum majus*, two species with distinct prochromosomes, replicate heterochromatin in the condensed state. The authors, however, suggested that this image may be caused by the movement of the nuclei from a

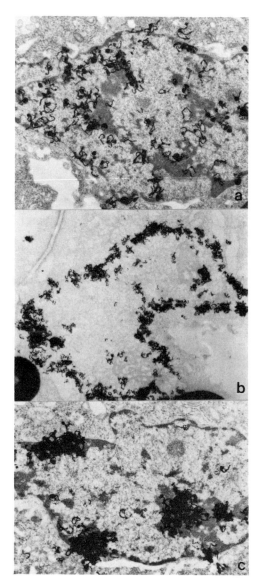

Figure 6-15.
Electron microscope autoradiography of *Microtus agrestis* nuclei labeled with ^{3}H-thymidine for 10 min. *a*, Diffuse pattern (replication of euchromatin); *b*, peripheral pattern (presumably replication of intercalary heterochromatin); *c*, label localized to the centromeric type of constitutive heterochromatin forming large chromocenters. *a*, *b*: x 8000. *c*: 4,500. Reproduced, with permission, from D.E. Comings, and T.A. Okada, *J.Mol.Biol.* 75:609–618 (1973).

diffuse stage, during which the label was incorporated, to the compact state, at which time the cells were fixed.

The problem has been independently reinvestigated by J. Greilhuber in Vienna, P. Barlow in Cambridge, and myself in Kaiserslautern. The—so far unpublished—results indicate the following. Nuclei showing decondensed chromatin often have a DNA content lower than expected. Although this could indicate that replication is not finished in these nuclei, i.e., they are in late S, and the difference to G_2

nuclei with the complete 4C DNA content has never been significant and might just be a systematic measurement error due to the changed nuclear structure. Light and electron microscopic autoradiographs also do not allow an unequivocal interpretation. In addition to chromocenters which were labeled in the condensed state (Figure 6-14), structures have been observed that evidently represent somewhat decondensed chromocenters (Figures 6-16–6-18). However, the degree of decondensation in these late S phase nuclei is apparently different from that in mitotic or endomitotic Z phase (see p. 164). Thus, heterochromatin may replicate in both the condensed and decondensed state. The point needs, however, further experiments to come to a clear statement.

Sex Chromatin

The heterochromatic X chromosome of mammalian and human females also is replicated in its tightly coiled, dense state. This has been shown in various studies: The area of the Barr body increases during the cell cycle by about 38% (Comings 1967b), and does not disappear during any stage (Comings 1967b; Klinger et al. 1967; Mittwoch 1967; Schwarzacher and Schnedl 1967; Schnedl 1969). Comings (1967a) also found incorporation of ^{3}H-thymidine into condensed sex chromatin bodies, and Pera and Wolf (1967) into the condensed facultative (and constitutive) heterochromatin of the sex chromosomes of *Microtus agrestis*.

Recently, Back (1974) investigated the euchromatic and heterochromatic behavior of the human X_2 chromosome in the cell cycle. He found that the inactivated X chromosome decondensed in retardation after mitosis and appeared only for a short time as sex chromatin. In cells with a sufficiently long cell cycle, a heterochromatization occurred again during interphase; this behavior was also retained during a G_0 period. In cells with a short cycle, the X_2 chromosome condensed before mitosis and became heterochromatic. The figures given in this essay show that the condensation behavior and DNA replication are independent from each other: both sex chromatin-positive and sex chromatin-negative nuclei can be labeled over the visible condensed sex chromatin and an uncondensed area of corresponding size, respectively.

Wyandt and Hecht (1973a,b) reported that the human Y chromosome shows "dispersion" during DNA replication. However, the microphotos published suggest slight uncoiling or unfolding of the chromocenter *without* loss of the heterochromatic state.

Although there is, in my opinion, good evidence for the replication of DNA in the condensed state of heterochromatin, it is too early to generalize this idea. There may actually be differences between species and tissues, dependent upon the type of heterochromatin present. In this connection I should mention a hypothesis by Lafontaine (1974): He suggests that the gross organization of interphase nuclei (reticulate or chromomeric *versus* chromocentric) is a function of their genome size, or in detail, of the DNA content per chromosome. A high DNA content due to much repetitive DNA may allow the development of a reticulate organization, while nuclei with low DNA content may tend to concentrate their DNA in chromo-

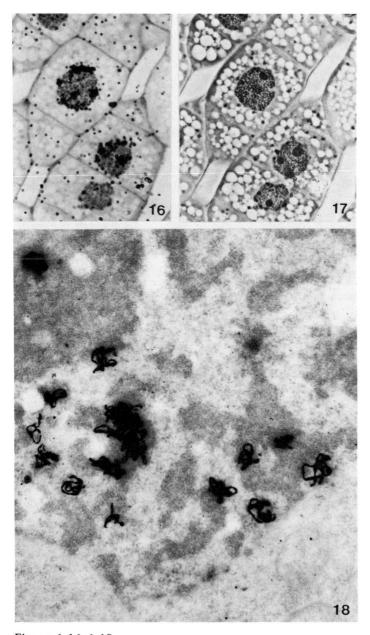

Figures 6-16-6-18.
Examples of ³H-thymidine incorporation into decondensed chromocenters of *Allium carinatum* (30 min pulse). Figures 6-16 and 6-17. Semithin section with and without photographic emulsion (Figure 6-17, phase contrast). x 900. Figure 6-18. Ultrathin section. x 35,000.

164   Walter Nagl

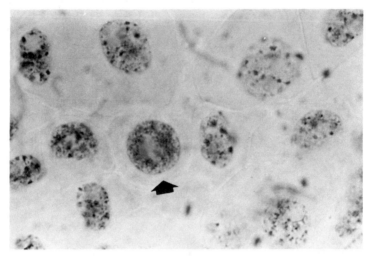

Figure 6-19.
Meristematic cells of *Cymbidium* protocorms cultured *in vitro*.
Notice the chromocenters in interphase nuclei, and nucleus at Z
phase (arrow). Acetocarmine. x 1200.

centers. However, this hypothesis should be reflected in the light of other findings,
such as (1) the location of highly repetitive DNA sequences in heterochromatin,
i.e., in chromocenters (Yunis and Yasmineh 1971), (2) the high content of intermediate-repetitive DNA in the monocots (Flavell et al. 1974), the appearance of
satellite DNA in species with low DNA content (Ingle et al. 1975), and (3) the
general interspersion of repetitious and single-copy DNA in the genome of eukaryotes
(Davidson et al. 1973). I believe it is worthwhile to compare the behavior of euchromatin and heterochromatin in species with different amounts of both intermediate
and highly repetitive DNAs.

Dispersion Stage (Z Phase)

This review would not be complete without discussing the changes in nuclear
structure occurring during the onset of prophase in many plants. Although those
changes have been known since 1929, when Heitz termed the earliest stage of
prophase "Zerstäubungsstadium," i.e., dispersion stage, these findings have been
widely ignored. In the dispersion stage euchromatin and heterochromatin regions
form the same diffuse structure which results in the disappearance of the chromocenters (Figures 6-4, 6-7, 6-11, 6-19; Doležal and Tschermak-Woess 1955; Tschermak-Woess and Doležal 1956; Hasitschka-Jenschke 1961; Taschermak-Woess 1967;
Nagl 1968, 1969a, 1970b, 1972b,c, 1973a). The dispersion stage (or Z phase is the
prophasic stage of the longest duration, because it is most frequently seen among all
stages (Nagl 1968). It is hard to understand why the Z phase could be neglected by
many authors; the most plausible explanation of this curious fact is that many

workers misinterpreted the Z phase as late S phase, during which the heterochromatin underwent DNA replication. The incorrectness of this interpretation has been shown by both cytophotometric DNA measurement and autoradiography of Z phase nuclei (Tschermak-Woess 1959; Nagl 1968, 1972b), and in earlier studies by volumetry.

The dispersion stage starts with the desintegration of the chromocenters into larger chromomeres (heterochromomeres), reaches its maximum under nearly complete dissappearance of the heterochromatin, and is finished when the chromocenters reappear. The latter stage was also called "stage of new organization" (Nagl 1968), because the euchromatic elements appear to have the same size as in G_1 nuclei, and the chromocenters mostly have the form of twin-bodies (Figure 6-31). Now the chromosomes begin to spiralize. In several species, a differential rate of recondensation of heterochromatin occurs before the chromosomes reach their homogeneously condensed state approximately at mid-prophase (Doležal and Tschermak-Woess 1955; Tschermak-Woess and Doležal 1956; Nagl 1969a). As will be shown later, the dispersion stage is thought to be the only structural expression of endomitosis in angiosperms. This means that in the endomitotic cell cycle of plants mitosis is curtailed at that very early stage of prophase. Recently, however, the Z phase has again been interpreted as expression of DNA replication in the chromocenters rather than as expression of prophasic or endomitotic stages (see discussion on p. 169). As there is only the paper by Lafontaine and Lord (1974), in which the Z phase is interpreted in terms of DNA replication, and all other arguments are still in the stage of unpublished discussion and speculation, I have given the conservative interpretation, but will draw the reader's attention to the possibility that Z phase is identical with late S phase. If the latter is right, many of the structures described here would be the matter of new questions.

The structural changes occurring in nuclei during the cell cycle may be much more complex and differentiated than described above. Nescović (1968), for example, divided the stages of interphase in several substages, based on observations made in mammalian cell cultures. Klueva et al. (1974) found six ultrastructural types of interphase nuclei in *Allium fistulosum, Nigella sativa*, and *N. damascena*.

Nuclear Density

Automated image analyzers, which allow one to differentiate gray levels of particles or area portions, can be used to distinguish between cell cycle stages by the distribution of gray levels within nuclei. Up to this point no uniform results have been obtained. De la Torre and Navarrete (1974) found that in *Allium cepa* the chromatin of G_2 nuclei is denser than that of G_1 nuclei. One would expect this also on grounds of the non-doubled volume of G_2 versus G_1 nuclei, which increases the optical density. These findings correspond with older results obtained by Swift and Kleinfeld (1953) and Das and Alfert (1968), who described a higher stain-binding capacity in G_2 than in G_1 nuclei. In contrast, Sawicki et al. (1974) reported that the mean density of chromatin decreases from G_1 to G_2 in preimplantation mouse embryos (Figure 6-20).

Recent attempts of Zeiss (unpublished) to compare the average density of electron microscopic negatives of ultrathin-sectioned nuclei of *Allium carinatum* in various stages of the cell cycle with the integrating microdensitometer IPM 2 yielded the following results (Table 6-1): The extinction value doubles between G_1 and G_2, but is somewhat lower in Z phase than in G_2, probably due to the decondensation of heterochromatin.

Micropuffing

In interphase nuclei of plants, decondensed regions of the chromatin and clusters of granules can be detected, referred to as micropuffs and their synthetic product. The granules evidently contain ribonucleoprotein as shown by the EDTA technique (Lafontaine 1968; Colman and Stockert 1970; Esponda and Giménez-Martin 1971; Colman et al. 1972). The ultrastructure of these micropuffs resembles that occurring in puffs of polytene chromosomes (compare, e.g., Sorsa 1969; Gabrusewycz-Garcia and Garcia 1974).

A preliminary count of the number of micropuffs and clusters of obviously RNP granules has been made during the cell cycle of *Allium carinatum* (Nagl, unpublished). The results (Table 6-1) indicate a significant increase in the number of these functional structures from G_1 to Z phase and correspond well with previously reported findings on an increase of RNA synthesis during the cell cycle of this species (Nagl 1973a).

Changes in the functional ultrastructure of nuclei may also occur during the cell cycle of animal cells. Sanchez and Oberti (1974) described nuclear inclusions containing ribonucleoprotein in G_1 and S phase nuclei of HeLa cells that could not be detected in G_2 nuclei.

Mitotic Condensation of Chromosomes

The striking transformation of chromosomes during prophase into their transport form are well known, but not understood in detail. Whereas DuPraw (1970) put forward the folded fiber model, there is still more evidence for a higher order coiling imposed on a coiled (or folded) chromosome (e.g., Ohnuki 1968; Chentsov et al. 1973; Greilhuber 1973). Problems of this kind and of the possible control of chromosome condensation are not the subject of this essay.

STRUCTURE OF ENDO-CYCLING CELLS

Definition and Occurrence

Once a cell has left the mitotic cycle, its chromosome complement may be doubled and further multiplied in the course of endo-cycles, i.e., the endomitotic cycle and its variations. In these cell cycles, the nuclear envelope is conserved throughout all stages, and no spindle apparatus is formed (Geitler 1939). The endomitotic cycle is characterized by structural changes comparable with some of the mitotic transformations of the nucleus. In insects, chromosome condensation takes place comparable

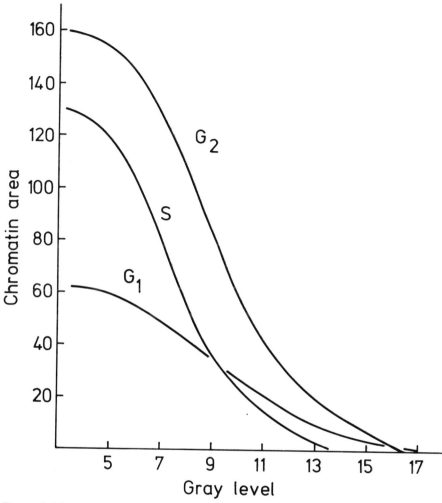

Figure 6-20.
Gray levels of nuclei from two-celled mouse embryos plotted against chromatin area by "Quantimet." Notice changes from G_1 to S and G_2. Redrawn from Sawicki et al. (1974).

to that in late prophase or metaphase of mitosis, but then the chromatids separate and uncoil to give rise to a polyploid nucleus (Geitler 1939). In angiosperms, endomitosis parallels mitosis only through the dispersion stage, then the nucleus enters G_1 of the consequent endomitotic cycle (Geitler 1953; Tschermak-Woess and Hasitschka 1953, 1954; Tschermak-Woess 1956, 1959, 1971; Nagl 1962, 1968, 1970b, c, 1972b, c, and in press).

Endomitotic polyploidization is particularly common in angiosperms and insects (reviews by Geitler 1953; D'Amato 1964; Torrey 1965; Tschermak-Woess 1971;

Nagl 1976a, b, and in press), but has also been detected in a few algae, mosses, ciliates, snails, and mammals. It seems to be rare in the Apiaceae (Umbelliferae) and Asteraceae (Compositae), and no endopolyploidy at all has so far been found in gymnosperms. The occurrence of endopolyploidy seems to be restricted to species with low basic nuclear DNA content (Nagl 1976c). It must be emphasized that endopolyploidization is not a pathological, but a regular event occurring during differentiation in about 75% to 85% of all cells of angiosperms (e.g., Butterfass, 1966), and in 99% of the cells of an insect showing endo-cycles. The highest degrees of endopolyploidy reached are 24,756 n in the endosperm haustorium of *Arum maculatum* (Erbrich 1965), 4,096 n to 8,192 n in the suspensor of *Phaseolus* species (Nagl 1962, 1969b, 1974a; Brady 1973) and in the elaiosomes of *Scilla bifolia* (Speta 1972), and 3,072 n in the endosperm of *Echinocystis lobata* (Turala 1966). In animals, a maximum degree of 200,000 C has been reported for the giant neurons of the gastropod *Aplysia californica* (Lasek and Dower 1971), and 500,000 C for the silk gland of the lepidopter *Bombyx mori* (Gage 1974); up to 32,000 n have been found in the salivary gland nuclei of *Chironomus* (Beermann 1952) and 4,096 n in the giant cells of the rat trophoblast (Nagl 1972d). Thus, a cell may undergo a number of endo-cycles, until it has reached the cell- or tissue-specific degree of endopolyploidy, i.e., DNA content (for recent, extensive reviews see Nagl 1976a, and in press).

The endoreduplication cycle represents a type of endo-cycle without structural changes in the nucleus during either DNA replication (endo-S) or the intervening period (endo-G). Here all stages of mitosis have completely been omitted during evolution (Levan and Hauschka 1953). The best examples of cells engaged in endo-reduplication cycles are the salivary gland nuclei of dipteran flies. Most other reports have to be taken with some reservation, because the term is widely used *without* knowledge of the cell cycle which leads to multiple DNA amounts. The occurrence of diplochromosomes in polyploid mitoses is not—despite its frequent reference as evidence for endoreduplication—restricted to postendoreduplication mitoses (Geitler 1953; Nagl 1970d; Poroshenko et al. 1970; Nagl, in press).

Finally, it must be stressed that the restitution cycle is not a type of endo-cycle. Polyploid restitution nuclei frequently originate by blockage of mitosis after spindle formation and breakdown of the nuclear envelope. The chromosomes separate slightly as in colchicine-induced C-mitosis and decondense. Therefore, nuclear structure after restitution cycles does not differ from the structure in the diploid ancestor nucleus (Figure 6-21).

Significance of Endo-cycles for Chromatin Structure

In contrast to polyploid restitution nuclei, manifold changes are frequent in endopolyploid nuclei, i.e., nuclei that arise through endo-cycles (reviewed by Geitler 1953; Tschermak-Woess 1963, 1971). These changes occur because the sister (or "endo"-) chromosomes which originate by endo-cyles, often do not separate from each other. At least their heterochromatic portions normally join to endochromo-centers, from which the euchromatic portions of the chromosomes radiate in all directions (Figure 6-28). In other cells, the endochromosomes may join along their whole length, thus forming polytene or giant chromosomes. Plant polytene chromosomes normally do not exhibit the banded appearance so characteristic for dipteran salivary gland chromosomes. The reason for this may be the differences in the endo-cycles by which the polytene chromosomes are formed: In dipterans, endoreduplication cycles allow a complete "pairing" of homologous loci so that the sister chromomeres can join to form a band; in plants, the endomitotic cycles cause a slight separation throughout the dispersion stage, so that the "pairing" is not as correct as in the dipteran larvae.

It takes some experience to distinguish structures in giant chromosomes that are (1) the expression of RNA synthesis (called functional structures, e.g., puffs), and (2) the consequence of the life history in terms of cell cycle variants. On the one hand, the polytene chromosomes of the *Phaseolus* suspensor may assume various structures due to changes in functional activity (Nagl 1969c, 1970e, 1973b, 1974a). On the other hand, they may show differences in gross morphology because of endomitosis (Nagl 1965) or interruption of polytenization by a polyploid mitosis (Nagl 1970c). Figures 6-21-6-27 illustrate some of the structural changes of *Phaseolus* polytene chromosomes as they occur in connection with changes of functional activity and changes in the cell cycles.

Brady and Clutter (1974) described the disintegration of giant chromosomes in *Phaseolus coccineus* during DNA replication, and, in addition, decondensation of the heterochromatin. However, their figures suggest artifacts introduced by the preparation technique. Therefore, their interpretation should be questioned. Moreover, there is multiple evidence that dipteran polytene chromosomes undergo no structural changes during DNA replication (see the previous section).

In contrast, Nagl (1965, 1967) observed the characteristic dispersion stage in *Phaseolus* giant chromosomes as an indication of endomitosis. Barlow (unpublished) also found the Z phase in polytene nuclei of *Bryonia* anther hair cells. Whether these observations are so rare because of (1) the short duration of endomitosis in those highly polytenic nuclei or (2) the suppression of endomitosis to endoreduplication during the formation of giant chromosomes, is not yet known.

The characteristic changes in nuclear structure occurring during the polytenization cycle have been reviewed by Beermann (1962) for salivary gland cells, and Nagl (in press) for plant and animal cells.

Nuclei passing through an endomitotic cycle may show structural differences between G_1 and G_2 as in the mitotic cycle. This is the case in *Sauromatum guttatum* and *Allium carinatum* and *flavum*, where mitosis is changed to endomitosis in the course of cell differentiation (Tschermak-Woess 1954; Nagl 1968, 1976a). Whereas no characteristic chromatin pattern could be found in S phase, the typical dispersion of heterochromatin was displayed by nuclei in endomitosis (Figures 6-28-6-32). This has been demonstrated by volumetry, DNA measurement, and autoradiography.

The structural differences occurring in G_1 and G_2 nuclei passing through the endomitotic cycle allow a distinction to be made between diploid nuclei at G_2 and tetraploid nuclei at G_1, both of which have the same DNA content. With more

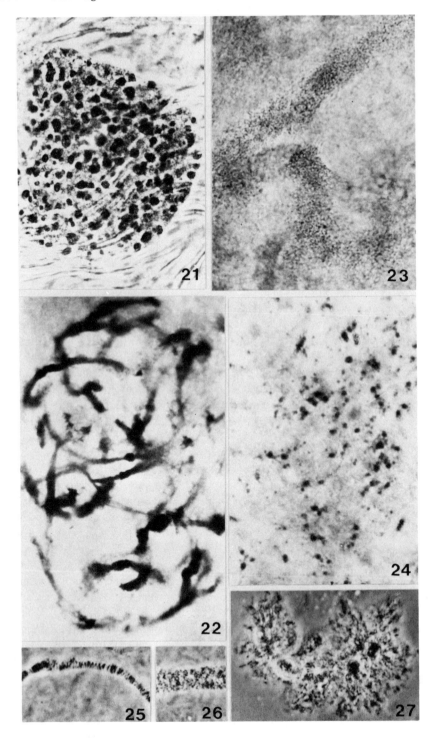

uncertainty this is also possible in the protocorms of the orchid *Cymbidium* (Figure 6-33). The higher the degree of endopolyploidy, the more difficult the stages are to distinguish, because of the formation of structures specific for polytenization and functional activity. Moreover, it cannot yet be ruled out that the "dispersion endomitosis" is a misinterpreted late S phase (see p. 160).

Endo-cyles with DNA Underreplication

A common aspect of endopolyploidization and polytenization in insects is the occurrence of an endo-cycle variant with selective underreplication of portions of the genome; so far as we know, only portions of the heterochromatin are affected. Most cases of underreplication were reported from the centromeric heterochromatin which contains highly reinterated DNA (and appearing sometimes as satellite DNA in CsCl gradients). Also heterochromatin containing redundant genes, as the ribosomal cistrons, can be involved. In such instances the euchromatin alone becomes highly endopolyploid, while the heterochromatin remains diploid, or at a lower stage of ploidy than the euchromatin. Such partially endopolyploid nuclei can be detected by the size of their chromocenters; they are too small in relation to the nuclear size. The situation is more accurately revealed by differential scanning densitometry of euchromatin and heterochromatin.

Fox (1970, 1971) analyzed microphotos of nuclei from locusts and beetles and found underrepresentation of heterochromatin in the polyploid nuclei. Also the α-heterochromatin in the salivary gland chromosomes of *Drosophila* is not, or is only a few times replicated, while the euchromatin of these chromosomes is highly polytenic; thus, the chromocenter mass is relatively too small in the polytene nuclei as compared with prophase chromosomes (Hennig 1972b; Laird 1973). In the mealy bug, *Planococcus citri*, the paternally derived chromosome set becomes heterochromatic in most somatic tissues; within some organs, the heterochromatin participates in DNA replication during endopolyploidization, but fails to do so in other organs, which then display nearly euchromatic endopolyploid nuclei (Nur 1966, 1968).

Figures 6-21-6-27.
Chromatin structures in polyploid nuclei after various types of cell cycles. Figure 6-21. 64-ploid restitution nucleus in a root tip of *Allium carinatum* treated with colchicine. The nuclear structure resembles that of diploid nuclei. Figure 6-22. Highly endopolyploid nucleus of the suspensor of *Phaseolus coccineus* displaying giant chromosomes. Figure 6-23. Giant chromosomes of Phaseolus during endomitosis (Z phase). Figure 6-24. Portion of a highly endopolyploid nucleus from the *Phaseolus* suspensor; the formation of polytene chromosomes in this nucleus was inhibited as a result of passing through a polyploid mitotic cycle during early differentiation (Nagl 1970c). Figures 6-25-6-27. Functional structures of polytene chromosomes in *Phaseolus*. Figure 6-25. Inactive, banded state (see Nagl 1969c). Figure 6-26. Active, granular state. Figure 6-27. Hyperactive, "lampbrush state" (Nagl 1970e). Figures 6-21 and 6-23. Acetocarmine. Figures 6-22 and 6-24. Toluidine blue. Figures 6-25-6-27. Unstained, phase contrast. Figures 6-21, 6-22, 6-24, x 400. Figures 6-23, 6-25-6-27. x 1,200.

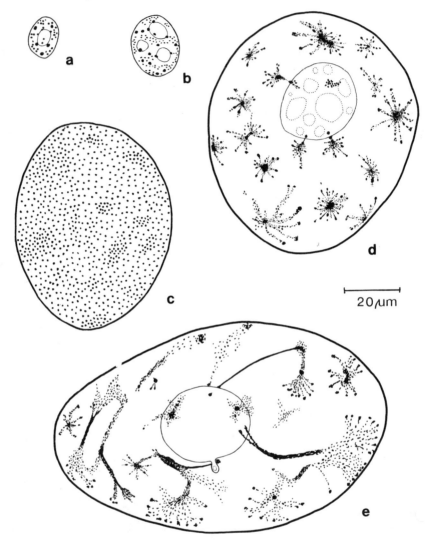

Figure 6-28.
Camera lucida drawings of diploid and polyploid nuclei in the suspensor and endosperm of *Phaseolus coccineus*. *a*, diploid nucleus from the suspensor; *b*, triploid nucleus from the endosperm; *c*, nucleus from the suspensor in endomitosis (Z phase) from the 64n level to the 128n level; *d*, 96-ploid nucleus from the endosperm: the sister chromosomes synthesized during the five endocycles are held together in their centrometric heterochromatin forming "endochromocenters" (Geitler 1953), while their euchromatic arms radiate out from them; *e*, 96-ploid nucleus from the endosperm: portions of the sister chromosomes synthesized during the 5 endo-cycles join to form polytene sections, while other sections, particularly the euchromatin, form diffuse areas. This is the beginning of polytenization observable in several nuclei of the chalazal part of the endosperm. According to Nagl (1962).

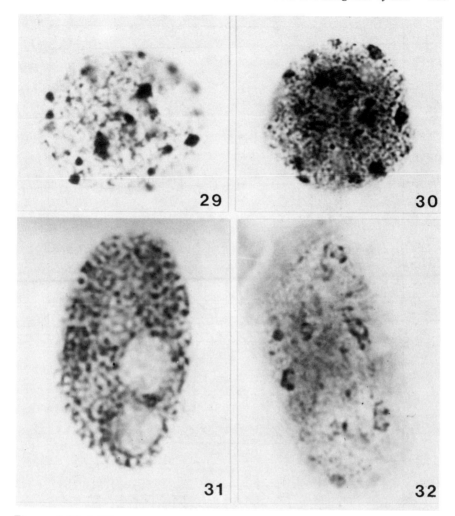

Figures 6-29-6-32.
Changes in nuclear structure during the tetraploid endomitotic cell cycle in root tips of *Allium carinatum.* Figure 6-29. Nucleus in endo-G_1. Figure 6-30. Nucleus in endo-G_2. Figure 6-31. Nucleus in endomitosis (Z phase). Figure 6-32. Nucleus in very early endo-G_1 ("stage of new organization"); notice the duplicate appearance to the chromocenters, and the G_1-sized euchromatic elements. Figures 6-29-6-31. Feulgen. Figure 6-32. acetocarmine. x 1,900.

Underreplication of heterochromatin has also been found in nuclei isolated from various tissues of fruits of Cucurbitaceae (Pearson et al. 1974; Ingle and Timmis 1975). Using microdensitometer scans of photographs of Feulgen-stained nuclei, the authors found that in approximately 95% of the fruit cells the ratio of euchromatin to heterochromatin was 1.00 : 0.05. In meristematic cells of root tips the corresponding ratio was 1.00 : 0.25. Based on the absolute DNA content it was

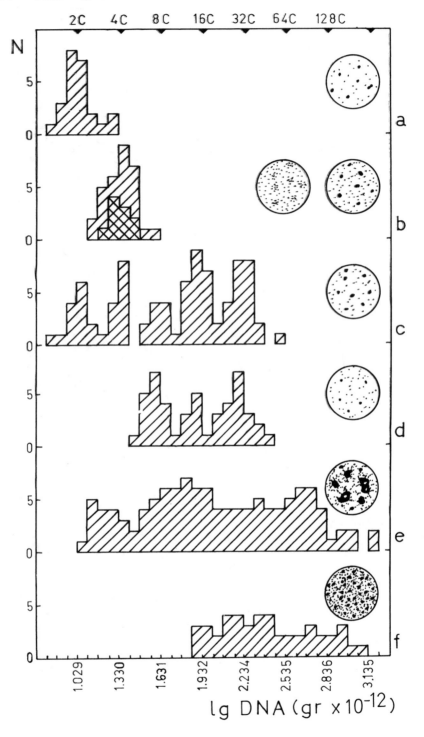

concluded that the euchromatin had undergone three rounds of replication more than the heterochromatin during the endopolyploidization process. The differential replication of euchromatin and heterochromatin in this and other plant families could be confirmed with CsCl gradient studies in which a reduction in the proportion of satellite DNA in maturing tissues could be detected (reviewed by Nagl 1974b, 1975, 1976a, and in press).

Lima-de-Faria et al. (1975) analyzed the nuclear DNA from suspensor cells and other tissues of *Phaseolus coccineus* by biochemical techniques, and found evidence for differential replication of the genome during the polytenization process in the suspensor cells. While the ribosomal genes were apparently underreplicated, a tissue-specific DNA satellite originated by extra replication. Structural analysis of the polytene chromosomes also indicated an amplification process (i.e., extra replication of small portions of the genome; Avanzi et al. 1970). Recently, diploid mitotic and polytenic interphase chromosomes were compared by the differential Giemsa staining technique. The results also revealed evidence for differential replication of certain portions of heterochromatin during the endo-cycles occurring in the suspensor cells (D. Schweizer, pers. comm.).

Recently, a case of drastic heterochromatin underreplication has been detected in early embryogenesis of the nasturtium, *Tropaeolum majus* (Figures 6-34-6-37). The proembryo of this species develops a very large suspensor (total length about 5 mm) which is composed of highly endopolyploid nuclei (Nagl 1962, 1976d). Certain suspensor cells, however, contain nuclei that are deficient in heterochromatin (Figures 6-36 and 6-37). The DNA content of these nuclei is always relatively low and does not fit a geometric order which is starting with the 2C DNA content (Nagl et al. 1976). Biochemical analysis of the DNA extracted from the suspensor cells did not yet yield clear results upon the question, what DNA fraction is underreplicated, because it has not yet been possible to separate the closely intermingled cells exhibiting underreplication and those which do not.

Figure 6-33.
Frequency diagram of nuclei with different DNA contents and chromatin structures from an aseptically cultured *Cymbidium* protocorm: *a*, diploid nuclei in G_1; *b*, diploid nuclei in G_2 (hatched diagram, right drawing.) and Z phase (cross-hatched diagram, left drawing); *c*, endopolyploid nuclei which have passed through endomitotic cycles; *d*, endopolyploid nuclei which probably have passed through underreplication cycles (see the discussion in Nagl, 1974b); *e*, heterochromatin-rich nuclei which have passed through endo-amplification cycles; *f*, "dense" nuclei which release the extra DNA from the chromocenters. While diploid nuclei show distinct peaks corresponding to a geometrical order of DNA increase, the peaks of the nuclei arisen through underreplication cycles are all at the lower side of a duplication series, and nuclei showing DNA amplification and release of amplified DNA do not fall into distinct classes. The different nuclear volumes have not been represented in the schematic drawings of the nuclear structures.

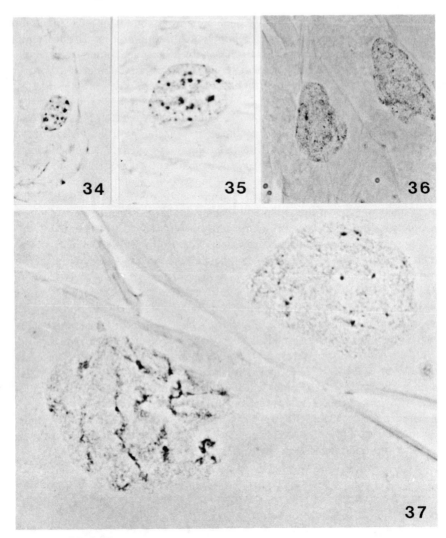

Figures 6-34-6-37.
The nuclear structure in *Tropaeolum majus* as a marker of endo-cycles without and with heterochromatin underreplication. Figure 6-34. Low-endopolyploid nucleus with distinct chromocenters. Figure 6-35. Higher-endopolyploid nucleus with proportionally enlarged chromocenters. Figure 6-36. Polyploid nuclei with very little chromocenters (sized as in diploid nuclei). Cytophotometric DNA measurements showed that the heterochromatin is not decondensed, but actually has not been replicated. Figure 6-37. Highly endopolyploid suspensor cell nuclei with different proportions of heterochromatin. Feulgen, phase contrast. Figures 6-34-6-36. x 400. Figure 6-37. x 900.

DNA Amplification

This term denotes the reduction of DNA replication to a very small portion of the genome—often only a few genes such as those for ribosomal RNA. Although this phenomenon commonly is interpreted as extrareplication of the DNA sequences involved, I prefer to interpret this event as a further and consequently an extreme reduction of DNA replication in an endo-cycle of those genes or DNA stretches which are actually required by a cell. Thus, the amplification cycle represents, in evolutionary terms, the latest and most economic type of cell cycle (Nagl 1974b, 1975).

Amplification of ribosomal genes in oocytes is very common and leads to various structural peculiarities in the nuclei involved. In oocytes of amphibians, for example, loops from the nucleolus organizers of the lampbrush chromosomes become replicated thousands of times; the detached replicas then form additional nucleoli (e.g., Callan 1966; Macgregor 1972); in the salivary gland nuclei of sciarid flies, DNA puffs are the morphological expression of DNA amplification (e.g., Crouse and Keyl 1968; Brito da Cunha et al. 1969, 1973; Stocker and Pavan 1974); in oocytes of *Acheta* certain chromomeres enlarge like a puff (Lima-de-Faria 1974b); in many other insects, DNA amplification takes place during oogenesis and is visible by various morphological features such as DNA bodies etc. (Gall et al. 1969; Allen and Cave 1972; Cave and Allen 1974; for a review see Nagl 1976a). Amplified DNA becomes detached from the chromosomes and represents a type of "metabolic DNA" (reviewed by Buiatti, in press). An excellent analysis of release of extra DNA from the chromomeres of *Acheta* chromosomes has been performed by Lima-de-Faria (1974b). Figure 6–38 shows a computer display of this release.

In plants, structural and cytochemical evidence of DNA amplification has not only been obtained from the suspensor cells of *Phaseolus coccineus* (see above), but also from differentiating metaxylem cells of *Allium cepa* (Innocenti and Avanzi 1971; Avanzi et al. 1973; Innocenti 1973). Further known possible cases are listed by Nagl (1975).

Aseptically cultured protocorms of the orchid *Cymbidium* are some of the best examples to show how the nuclear structure can serve as a marker of the cell cycle stage, particularly because of the clear chromocenters (Figure 6–44) which undergo dramatic changes during the various cycles. Therefore, this system will be reviewed in more detail.

Many cells of *Cymbidium* protocorms and roots become endopolyploid; *in vitro*, the occurrence and degree of endopolyploidy can be experimentally controlled by omitting or adding certain phytohormones to the culture medium (Nagl and Rücker 1972, 1974). However, a portion of the cells within various organs display nuclei extremely rich in heterochromatin as expressed by the size and DNA content of several chromocenters (Figure 6–39). These nuclei contain DNA amounts which do not fall into multiples of the 2C value, i.e., the DNA increase during endopolyploidization does not follow a geometric order as in the "standard" nuclei. Since the heterochromatin-rich nuclei incorporate ^{3}H-thymidine mainly into the growing chromocenters, but lose their extra-large masses of heterochromatin after blocking

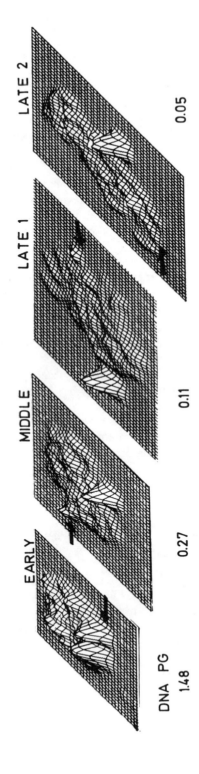

Figure 6-38.
Computer display of the release of extra DNA from chromomere 11 in the amplification cycle in oocytes of the house cricket, *Acheta domestica* during the pachytene. From Lima-de-Faria (1974b), changed.

DNA synthesis by hydroxyurea, we suggested that those nuclei underwent DNA amplification (Nagl 1972a; Nagl et al. 1972). Recent studies showed that the amplified DNA may also be lost during normal development of the cultures, because it is metabolically labile and not incorporated into the genome (Nagl 1974b). During the active stage of cytodifferentiation, the heterochromatic DNA has completed up to as many as five rounds of replication more than the euchromatin (Nagl 1972a).

Figure 6-33 shows, diagrammatically, the changes of nuclear structure in *Cymbidium* protocorms, as they occur during the mitotic, the endomitotic, the amplification, and probably the underreplication cycle. The classification of nuclei is based on DNA measurements, volumetry, autoradiography, physiological experiments, and DNA measurement of isolated heterochromatin. The main results are the following:

1. There is a slight structural difference between G_1 and G_2 nuclei.
2. There is some tendency to polyteny in the endopolyploidization cycles.
3. There are some nuclei with small chromocenters and the DNA content falls into the lower shoulder of the frequency peak, indicating the occurrence of underreplication; however, loss of chromosomes, or of heterochromatic portions of chromosomes during the culture period cannot yet be excluded.
4. In nuclei undergoing the endo-amplification cycle, several chromocenters grow more than we would expect from the nuclear volume. Later, the chromocenters break down and give rise to many small heterochromatic spots; these types of nuclei are called "dense nuclei" (Figure 6-45), and are also characterized by loss of incorporated ^{3}H-thymidine. Finally, the nuclear structure and DNA content become the same as in the other endopolyploid nuclei. Figures 6-40 to 6-43 show some nuclei stained with the fluorochrome quinacrine, which is specific for AT-rich heterochromatin (Borisova et al. 1974; Latt et al. 1974). It can clearly be seen that in the standard nuclei, endopolyploidization takes place under parallel increase of total nuclear size and heterochromatin content; in the heterochromatin-rich nuclei, the heterochromatin-content is very much higher in all stages of polyploidy, i.e., the amplification process occurs in addition to the endopolyploidization. The system has also been studied in terms of DNA ultracentrifugation and denaturation profiles (see Capesius et al. 1975; Schweizer and Nagl 1976).

DISCUSSION AND CONCLUSIONS

In this section, the mechanisms responsible for the cell cycle-dependent structural changes, and their significance for the nuclear function will be discussed briefly. In addition, I will try to clarify my own point of view of controversial interpretations of structural changes.

Nuclear structural changes may either be the expression and consequence of changes in chromosomal function, and cell metabolism in general, or the prerequisite and basis of functional changes. Studying the cell cycle-related changes, one must consider both views. The thickening of chromatin threads between G_1 and G_2 is, without any doubt, caused by the reduplication of the chromosomal mass during the

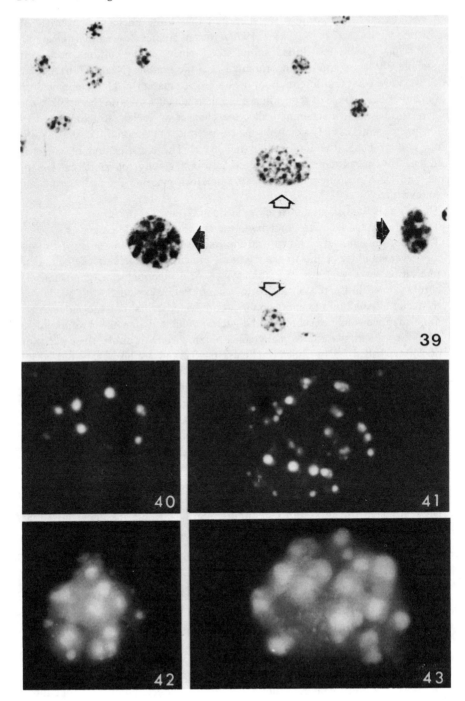

S period, or, in other words, by the replication of chromosomal DNA and the synthesis and binding to the DNA of chromosomal proteins. It remains to be explained, however, why the reduplication of the chromosomes becomes structurally visible only in the nuclei of a few species. The basis might be seen in the different contents of repetitive DNA between species: Chromatin with repetitive DNA shows the same electron density as heterochromatin and might behave in a similar way, e.g., show increased tendency to stickiness and fusion. As a consequence, the newly reduplicated chromosomes do not separate at S phase as in species with less repetitive DNA, but remain fused during G_2, giving rise to the thicker chromatin elements.

The suggestions given above can also lead to a hypothesis with regard to the meaning of the dispersion stage (Z phase). Since the heterochromatin of all species studied so far, and the euchromatin of species rich in repetitive DNA, do not separate into the new chromatids after reduplication during S, the mechanical separation and reorganization of the chromatin must occur at a later point, but before the onset of chromosome coiling. The cleaved appearance of the chromocenters, and the G_1-like size of the euchromatic elements of *Allium carinatum* immediately after the dispersion stage, and immediately before onset of chromosome coiling, is consistent with the hypothesis put forward. A confirmation of the interpretation of Z phase as a period of chromatid separation comes from the study of polytene chromosomes. The failure of plant giant chromosomes to display a banding pattern under normal conditions is evidently caused by the endomitotic dispersion stage occurring in angiosperms. Passing through this stage, the endochromosomes separate slightly so that the fusion of corresponding sister chromomeres to bands is inhibited. A later fusion is, however, possible in spite of this little separation. If, in some species, future experiments will show that the Z phase is identical with late S period, then the endo-cycle occurring in these species would be an endoreduplication cycle, and the mechanisms described here for endomitosis would be active during DNA replication.

The molecular basis for the control of a cell through the cell cycle is apparently very complex. Changes in the conformation of the chromatin (e.g., Nicolini et al.

Figures 6-39-6-43.
Structures of nuclei in aseptically cultured protocorms of the orchid *Cymbidium* (hybrid "In memoriam Cyrill Strauss) which had passed through either the endomitotic cycle or the endo-amplification cycle. Figure 6-39. Diploid, endopolyploid (open arrows) and heterochromatin-rich endopolyploid (black arrows) nuclei. Feulgen-stained. Figures 6-40-6-43. Fluorescense patterns of *Cymbidium* nuclei after quinacrine staining. Figure 6-40. Diploid nucleus with a few chromocenters. Figure 6-41. Endopolyploid nucleus with proportionate increase in heterochromatin. Figures 6-42, 6-43. Heterochromatin-rich endopolyploid nuclei; the amplified chromocenters show more clearly than the regular chromocenters the differentiation into very brightly fluorescing areas, probably due to differences in AT-content (Schweizer and Nagl 1976). Figure 6-39. x 160. Figures 6-40-6-43. x 1,000.

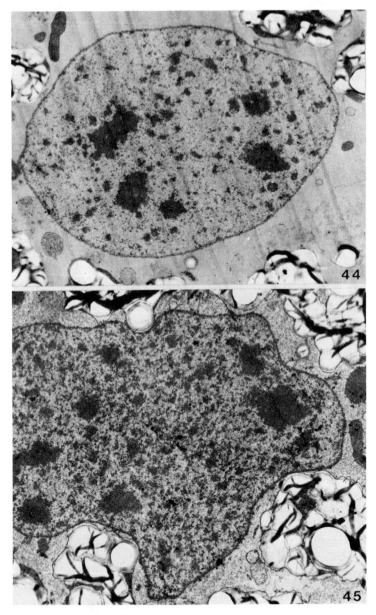

Figures 6-44 and 6-45.
Ultrastructure of *Cymbidium* nuclei in the endomitotic and amplification cycles, respectively. Figure 6-44. Endopolyploid nucleus with distinct chromocenters. Figure 6-45. "Dense" nucleus releasing the amplified DNA from the chromocenters. x 5,000.

1975) may be caused by histone modification, in particular, by phosphorylation (e.g., Bradbury et al. 1974). Non-histone chromosomal proteins evidently play a key role in both changes in template activity and structure of nuclei (e.g., Chiu and Baserga 1975; McClure 1974). This again demonstrates that structure and function cannot be examined independently from each other. As already stressed, changes in nuclear structure may also be interpreted in terms of their functional significance. As best demonstrated in puffs of polytene chromosomes, decondensation of chromatin is a general prerequisite for increased RNA synthesis. The increasing capability of RNA synthesis during the cell cycle is also accompanied by an increasing number of micropuffs of ribonucleoprotein granules. This, in turn, results in doubling the mean density of ultrathin sections of nuclei between G_1 and G_2, although we would expect from enlargement of the nuclear volume that the duplication of chromatin during S is not reflected by such a high increase in mean density.

The generalization of the concept of the activity:decondensation relationship evidently has inspired some workers to over-interpret autoradiographs in terms of decondensation of heterochromatin as a prerequisite for DNA replication. The facts obtained from plants, animals, and man, in diploid, polyploid, and polytene nuclei, exclude, however, the necessity of a gross change in structure and ultrastructure of nuclei during the S period. Nevertheless, it will be one of the most important cytological studies to be done in the coming years to elucidate the actual relationship between late S period and the dispersion (Z) stage.

The following conclusion would seem to be true at this time: The nuclear structure may well be suited to serve as a *marker* of certain cell cycle stages, at least in a number of organisms. Moreover, the chromatin structure can indicate quite accurately the life history of a nucleus in terms of the type of cell cycle from which the nucleus has originated. This is particularly important for the distinction of polyploid nuclei arisen either through restitution cycles, endomitotic cycles, endoreduplication cycles, underreplication cycles, or amplification cycles. During the last ten years the enthusiasm about the progress in eukaryotic genetics and molecular karyology, substantiated by the introduction of biochemical techniques, led to the restriction of scientific interest to DNA synthesis. Recently the significance of the various types of cell cycles, and their stages, for differentiation and development of eukaryotes have been rediscovered (Cameron et al. 1971; Reinert and Holtzer 1975; Nagl 1976a,b). This essay was intended to stimulate interest not only in molecules and chromatin fractions during the cell cycle, but also in the nucleus as a structural and functional entity.

ACKNOWLEDGMENTS

I thank Drs. C. de la Torre and M. H. Navarrete (Institute of Cell Biology, Department of Cytology, Madrid, Spain) for providing unpublished material; Professors Dr. H. -W. Altmann (Pathological Institute, University of Würzberg, German Federal Republic) and Dr. C. D. Comings (City of Hope Medical Center, Duarte, California) for original microphotos and micrographs; and Drs. Shelly and Christian Puff (Institute of Botany, University of Vienna, Austria) for correcting the manuscript. I am also grateful to Mrs. E. Mossmann for preparing some of the electron micrographs published for the first time in this essay, and to Mrs. S. Kühner for assistance in writing the manuscript.

LITERATURE CITED

Albersheim, P. and U. Killias. 1963. The use of bismuth as an electron stain for nucleic acid. *J. Cell Biol.* **17**:93–103.

Alfert, M. 1955. Quantitative cytochemical studies on patterns of nuclear growth. *Leiden 1954 Symp. on Fine Structure of Cells.* Groningen.

Allen, E. R. and M. D. Cave. 1972. Nucleolar organization in oocytes of gryllid crickets: subfamilies Gryllinae and Nemobiinae. *J. Morph.* **137**:433–448.

Altmann, H. -W. 1966. Der Zellersatz, insbesondere an den parenchymatösen Organen. *Verhdl. Deutsch. Ges. Pathol.* **50**:15–51.

Alvarez, Y. and Y. Valladares. 1972. Differential staining of the cell cyle. *Nature New Biol.* **238**:279–280.

Andersen, H. A. and J. Engberg. 1975. Timing of the ribosomal gene replication in *Tetrahymena pyriformis. Exptl. Cell Res.* **92**:159–163.

Arcos-Terán, L. and W. Beermann. 1968. Changes of DNA replication behavior associated with intragenic change of the white region in *Drosophila melanogaster. Chromosoma* **25**:347–391.

Avanzi, S., P. G. Cionini, and F. D'Amato. 1970. Cytochemical and autoradiographic analyses on the embryo suspensor cells of *Phaseolus coccineus. Caryologia* **23**:605–638.

Avanzi, S., F. Maggini, and A. M. Innocenti. 1973. Amplification of ribosomal cistrons during the maturation of metaxylem in the root of *Allium cepa. Protoplasma* **76**:197–210.

Back, F. 1974. Euchromatic and heterochromatic behavior of the human X2 chromosome in cell cycle and cell development. *Humangenetik* **25**:315–329.

Baetcke, K. P., A. H. Sparrow, C. H. Nauman, and S. S. Schwemmer. 1967. The relationship of DNA content to nuclear and chromosome volumes and to radiosensitivity (LD_{50}). *Proc. Natl. Acad. Sci.* **58**:533–540.

Beermann, W. 1952. Chromomerenkonstanz und spezifische Modifikationen der Chromosomenstruktur in der Entwicklung und Organdifferenzierung von *Chironomus tentans. Chromosoma* **5**:139–198.

Beermann, W. 1962. Riesenchromosomen. *Protoplasmatologia VI/D.* Springer, Vienna & New York.

Bennett, M. D. 1972. Nuclear DNA content and minimum generation time in herbaceous plants. *Proc. Roy. Soc. London* B **181**:109–135.

Berendes, H. D. 1973. Synthetic activity of polytene chromosomes. *Int. Rev. Cytol.* **35**:61–116.

Bernhard, W. 1969. A new staining procedure for electron microscopical cytology. *J. Ultrastr. Res.* **27**:250–265.

Bertaux, O., C. Frayssinet, and R. Valencia. 1976. Ultrastructural modifications of the nucleus and nucleolus of *Euglena gracilis* Z. during the cell cycle. *C. R. Acad. Sci. Paris* D **282**:1293–1296.

Billett, M. A. and I. M. Barry. 1974. Role of histones in chromatin condensation. *Europ. J. Biochem.* **49**:477–484.

Blondel, B. 1968. Relation between nuclear fine structure and ^{3}H-thymidine incorporation in a synchronous cell culture. *Exptl. Cell Res.* **53**:348–356.

Borisova, O. F., A. P. Razjivin, and V. I. Zaregorodzew. 1974. Evidence for the quinacrine fluorescence on three AT pairs of DNA. *FEBS Letters* **46**:239–242.

Bradbury, E. M., R. J. Inglis, and H. R. Matthews. 1974. Control of cell division by very lysine-rich histone (F1) phosphorylation. *Nature* **247**:257–261.

Brady, T. 1973. Cytological studies on the suspensor polytene chromosomes of *Phaseolus*: DNA content and synthesis, and the ribosomal cistrons. *Caryologia* **25**:Suppl. 233–259.

Brady, T. and M. E. Clutter. 1974. Structure and replication of *Phaseolus* polytene chromosomes. *Chromosoma* **45**:63–79.

Brito da Cunha, A. C. Pavan, I. S. Morgante, and M. C. Garrido. 1969. Study on cytology and differentiation in Sciardae. II. DNA redundancy in salivary gland cells of *Hybosciara fragilis* (Diptera, Sciaridae). *Genetics* **61**:Suppl. 335–349.

Brito da Cunha, I. S. Morgante, C. Pavan, M. C. Garrido, and J. Marques. 1973. Studies on cytology and differentiation in Sciaridae. IV. Nuclear and cytoplasmic differentiation in salivary glands of *Bradysia elegans* (Diptera, Sciaridae). *Caryologia* **26**:83–100.

Buiatti, M. In press. Gene amplification and tissue cultures. In *Applied and Fundamental Aspects of Plant Cell and Tissue Culture*, eds. J. Reinert, and Y. P. S. Bajaj. Springer, Berlin & Heidelberg & New York.

Butterfass, T. 1966. Neue Aspekte der Polyploidieforschung und-züchtung. *Mitteilungen der Max-Planck-Gesellschaft* **1**:47–58.

Callan, H. G. 1966. Chromosomes and nucleoli of the axolotl, *Ambystoma mexicanum*. *J. Cell Sci.* **1**:85–108.

Cameron, I. L., G. M. Padilla, and A. M. Zimmermann, eds. 1971. *Developmental Aspects of the Cell Cycle*. Academic Press, New York & London.

Capesius, I., B. Bierweiler, K. Bachmann, W. Rücker, and W. Nagl. 1975. An A+T-rich satellite DNA in a monocotyledonous plant, *Cymbidium*. *Biochim. Biophys. Acta* **395**:67–73.

Caro, L. G. 1962. High resolution autoradiography II. The problem of resolution. *J. Cell Biol.* **15**:189–199.

Cave, M. D. and E. R. Allen. 1974. Nucleolar DNA in oocytes of crickets: representatives of the subfamilies Oecanthinae and Gryllotalpinae (Orthoptera: Gryllidae). *J. Morph.* **142**:379–394.

Chentsov, Yu. S., V. Yu. Polyakov, and V. I. Vasin. 1973. Ultrastructure of mitotic chromosomes in whole preparations of tulip chromosomes. *Doklady Biol. Sci.* **209**:engl. 94–96.

Chiu, N. and R. Baserga. 1975. Changes in template activity and structure of nuclei from WI-38 cells in the prereplicative phase. *Biochemistry* **14**:3126–3132.

Chouinard, L. A. 1970. Localization of intranucleolar DNA in root meristematic cells of *Allium cepa*. *J. Cell Sci.* **6**:73–85.

Citoler, P. and A. Gropp. 1969. DNS- Replikation von autosomalem Heterochromatin. *Exptl. Cell Res.* **54**:337–346.

Colman, O. D. and J. C. Stockert. 1970. Observación de Diferenciaciones Morfológicas Características "Micropuffs") en la Cromatina de Células vegetales. *Cienc. e Investig.* **26**:509–511.

Colman, O. D., I. L. Stockert, P. Esponda, and M. d. C. Risueño. 1972. Intranuclear differentiation in *Allium cepa* meristematic cells. *Cytobiologie* **6**:104–108.

Comings, D. E. 1967a. The duration of replication of the inactive X-chromosome in humans based on the persistence of the heterochromatic sex chromatin body during DNA synthesis. *Cytogenetics* **6**:20–37.

Comings, D. E., 1967b. Sex chromatin, nuclear size, and the cell cycle. *Cytogenetics* **6**:120–144.

Comings, D. E. and T. A. Okada. 1973. DNA replication and the nuclear membrane. *J. Mol. Biol.* **75**:609–618.

Conger, A. D. and L. M. Fairchild. 1953. A quick freeze method for making smear slides permanent. *Stain Technol.* **28**:281–283.

Cowden, R. R. and S. K. Curtis. 1975. Microspectrophotometric estimates of non-histone proteins in cell nuclei displaying differing degrees of chromatin condensation. *J. Morphol.* **145**:1–12.

Crouse, H. V. and H. G. Keyl. 1968. Extra replication in the "DNA puffs" of *Sciara coprophila. Chromosoma* **25**:357–364.

D'Amato, F. 1964. Endopolyploidy as a factor in plant tissue development. *Caryologia* **17**:41–52.

Das, N. K. and M. Alfert. 1968. Cytochemical studies on the concurrent synthesis of DNA and histone in primary spermatocytes of *Urechis caupo. Exptl. Cell Res.* **49**:51–58.

Davidson, E. H., B. R. Hough, C. S. Amenson, and R. J. Britten. 1973. General interspersion of repetitive with nonrepetitive sequence elements in the DNA of *Xenopus. J. Mol. Biol.* **77**:1–23.

De la Torre, C. and M. H. Navarrete. 1974. Estimation of chromatin patterns at G_1, S, and G_2 of the cell cycle. *Exptl. Cell Res.* **88**:171–174.

De la Torre, C., A. Sacristàn-Gàrante, and M. H. Navarrete. 1975. Structural changes in chromatin during interphase. *Chromosoma* **51**:183–198.

Dewse, C. D. 1974. Observations on the technic for differential staining of the cell cycle using safranin and indigo-picro-carmine. *Stain Technol.* **49**:57–60.

Dolezal, R. and E. Tschermak-Woess. 1955. Verhalten von Eu- und Heterochromatin, und interphasisches Kernwachstum bei *Rhoeo discolor.* Vergleich von Mitose und Endomitose. *Österr. Bot. Z.* **102**:158–185.

Du Praw, E. J. 1970. *DNA and Chromosomes.* Holt, Rinehart and Winston, New York.

Erbrich, P. 1965. Über Endopolyploidie und Kernstrukturen im Endospermhaustorien. *Österr. Bot. Z.* **112**:197–262.

Esponda, P., and G. Giménez-Martín. 1971. Chromatin differentiations in microspores. *Experientia* **27**:855–856.

Fellenberg, G. 1974. Developmental physiology. *Progr. Bot.* **36**:147–166.

Flavell, R. B., M. D. Bennett, J. B. Smith, and D. B. Smith. 1974. Genome size and the proportion of repeated nucleotide sequence DNA in plants. *Biochem. Genet.* **12**:257–279.

Fox, D. P. 1970. A non-doubling DNA series in somatic tissues of the locusts *Schistocerca gregaria* (Forskal) and *Locusta migratoria* (Linn). *Chromosoma* **29**:446–461.

Fox, D. P., 1971. The replicative status of heterochromatic and euchromatic DNA in two somatic tissues of *Dermestes maculatus* (Dermestidae, Coleoptera). *Chromosoma* **33**:183–195.

Gabrusewycz-Garcia, N. 1964. Cytological and autoradiographic studies in *Sciara coprophila* salivary gland chromosomes. *Chromosoma* **15**:312–344.

Gabrusewycz-Garcia, N. and A. M. Garcia. 1974. Studies on the fine structure of puffs in *Sciara coprophila*. *Chromosoma* 47:385–401.

Gage, L. P. 1974. Polyploidization of the silk gland of *Bombyx mori*. *J. Mol. Biol.* 86:97–108.

Gall, J. G., H. C. MacGregor, and M. E. Kidston. 1969. Gene amplification in the oocytes of Dytiscid water beetles. *Chromosoma* 26:169–187.

Geitler, L. 1939. Die Entstehung der polyploiden Somakerne der Heteropteren durch Chromosomenteilung ohne Kernteilung. *Chromosoma* 1:1–22.

Geitler, L. 1953. Endomitose und endomitotische Polyploidisierung. *Protoplasmatologia* VI/C. Springer, Vienna & New York.

Golomb, H. M. and G. F. Bahr. 1974. Human chromatin condensation from interphase to metaphase: a scanning electron microscope study. *Exptl. Cell Res.* 84:79–87.

Gray, P., ed. 1973. *The Encyclopedia of Microscopy and Microtechnique*. Van Nostrand Reinhold Comp. New York.

Greilhuber, J. 1973. Differentielle Heterochromatinfärbung und Darstellung von Schraubenbau sowie Subchromatiden an pflanzlichen somatischen Chromosomen in der Meta- und Ana- phase. *Österr. Bot. Z.* **121**:1–11.

Haase, G. and G. Jung. 1964. Herstellung von Einkornschichten aus photographischen Emulstionen. *Naturwissenschaften* 51:404–405.

Hasitschka-Jenschke, G. 1961. Das Längenverhältnis der eu- und hetero-chromatischen Abschnitte riesenchromosomenartiger Bildungen verglichen mit dem der Prophase- Chromosomen bei *Bryonia dioica*. *Chromosoma* 12:466–483.

Hay, E. D. and J. P. Revel. 1963. The fine structure of the DNP component of the nucleus. An Electron Microscopic study utilizing autoradiography to localize DNA-synthesis. *J. Cell Biol.* 16:29–51.

Hennig, W. 1972a. Highly repetitive DNA sequences in the genome of *Drosophila hydei*. I. Preferential localization in the X-chromosome heterochromatin. *J. Mol. Biol.* **71**:407–417.

Howard, A. and S. R. Pelc. 1953. Synthesis of DNA in normal and irradiated cells and its relation to chromosome breakage. *Heredity* 6:Suppl. 261–273.

Hsu, T. C. 1962. Differential rate in RNA synthesis between euchromatin and heterochromatin. *Exptl. Cell Res.* 27:332–334.

Hsu, T. C. and A. Markvong. 1975. Chromosomes and DNA of *Mus*: Terminal DNA synthetic sequence in three species. *Chromosoma* 51:311–322.

Ingle, J. and J. N. Timmis. 1975. A role for differential replication of DNA in development. In *Modifications of the Information Content of Plant Cells* eds. R. Markham et al. North Holland/American Elsevier, Amsterdam & Oxford & New York, pp. 37–52.

Ingle, J., J. N. Timmis and J. Sinclair. 1975. The relationship between sat DNA, rib RNA gene redundancy and genome size. *Plant Physiol.* **55**:496–501.

Innocenti, A. M. 1973. Aspects ultrastructuraux des premiers stades de la différenciation du métaxylème chez les racines de l'*Allium cepa* L. cultivar fiorentina. *C. R. Acad. Sci. Paris* D 277:2153–2156.

Innocenti, A. M. and S. Avanzi. 1971. Some cytological aspects of the differentiation of metaxylem in the root of *Allium cepa*. *Caryologia* 24:283–292.

Klinger, M. P., H. G. Schwarzacher, and J. Weiss. 1967. DNA content and size of sex chromatin positive female nuclei during the cell cycle. *Cytogenetics* 6:1–19.

Klueva, T. S., G. E. Onishchenko, V. Yu. Poliakov, and Yu. S. Chentsov. 1974. Ultrastructure of the cell interphase nuclei of some plants in different periods of mitotic cycle. *Tsitologiya* **16**:1465-1469.

Kuroiwa, T. 1973. Fine structure of interphase nuclei. VI. Initiation and replication sites of DNA synthesis in the nuclei of *Physarum polycephalum* as revealed by electron microscopic autoradiography. *Chromosoma* **44**:291-299.

Kuroiwa, T. 1974. Fine structure of interphase nuclei. III. Replication site analysis of DNA during the S period of *Crepis capillaris. Exptl. Cell Res.* **83**:387-398.

Lafontaine, J. -G. 1968. Structural components of the nucleus in mitotic plant cells. In *The Nucleus*, eds. A. I. Dalton and F. Haguenau. Acad. Press, New York, pp. 151-196.

Lafontaine, J. -G. 1974. Ultrastructural organization of plant cell nuclei. In *The Cell Nucleus*, Vol. 1, ed. H. Busch. Acad. Press, New York & London, pp. 149-185.

Lafontaine, J. -G. and A. Lord. 1969. Organization of nuclear structures in mitotic cells. In *Handbook of Molecular Cytology*, ed. A. Lima-de-Faria. North Holland, Amsterdam, pp. 381-411.

Lafontaine, J. -G. and A. Lord. 1974. An ultrastructural and radioautographic study of the evolution of the interphase nucleus in plant meristematic cells. (*Allium porrum*). *J. Cell Sci.* **14**:263-287.

Laird, C. D. 1973. DNA of *Drosophila* chromosomes. *Ann. Rev. Genet.* **7**:177-204.

Lasek, R. J. and W. J. Dower. 1971. *Aplysia californica*: analysis of nuclear DNA in individual nuclei of giant neurons. *Science* **172**:278-280.

Latt, S. A., R. L. Davidson, M. S. Lin, and P. S. Gerald. 1974. Lateral asymmetry in the fluorescence of human Y chromosomes with 33 258 Hoechst. *Exptl. Cell Res.* **87**:425-429.

Lettré, H. and N. Paweletz. 1966. Probleme der elektronenmikroskoskopischen Autoradiographie. *Naturwissenschaften* **53**:268-271.

Levan, A. and T. S. Hauschka. 1953. Endomitotic reduplication mechanisms in ascites tumors of the mouse. *J. Natl. Cancer Inst.* **14**:1-43.

Lima-de-Faria, A. 1974a. Amplification of rib DNA in *Acheta*. IX. The isolated rib DNA-RNA complex in the scanning electron microscope. *Hereditas* **78**:255-263.

Lima-de-Faria, A. 1974b. The molecular organization of the chromosomes of *Acheta* involved in ribosomal DNA amplification. *Cold Spring Harbor Symp. Quant. Biol.* **38**:559-571.

Lima-de-Faria, A. and H. Jaworska. 1968. Late DNA synthesis in heterochromatin. *Nature* **217**:138-141.

Lima-de-Faria, A., R. Pero, S. Avanzi, M. Durante, U. Stahle, F. D'Amato, and H. Granström. 1975. Relation between ribosomal RNA genes and the DNA satellites of *Phaseolus coccineus. Hereditas* **79**:5-20.

Littau, V. C., C. J. Burdick, V. G. Allfrey, and A. E. Mirsky. 1965. The role of histones in the maintenance of chromatin structure. *Proc. Sci.* **54**:1204-1212.

MacGregor, H. C. 1972. The nucleolus and its genes in amphibian oogenesis. *Biol. Rev. Cambr. Phil. Soc.* **47**:177-210.

Mäkinen, V. 1963. The mitotic cycle in *Allium cepa*, with special reference to the diurnal periodicity and to the seedling aberrations. *Annal. Bot. Soc. Zool. Bot. Fennica "Vanamo"* **34**:1-60.

Masubuchi, M. 1973. Evidence of early replicating DNA in heterochromatin. *Bot. Mag.* (Tokyo) **86**:319-322.

Masubuchi, M. 1974. Early replicating DNA in heterochromatin of *Plagiochila ovalifolia* (liverworts). *Bot. Mag.* (Tokyo) **87**:229-235.

Mazia, D. 1963. Synthetic activities leading to mitosis. *J. Cell. Comp. Physiol.* **62**, Suppl. **1**:123-140.

McClure, M. E. 1974. Fluctuations in chromatin function and composition during cell proliferation. In *Cell Cycle Controls*, eds. G. M. Padilla, I. L. Cameron, and A. Zimmermann. Academic Press, New York & London, pp. 337-359.

Mendelsohn, M. L. 1958. The two-wavelength method of microspectrophotometry. II. A set of tables to facilitate the calculations. *J. Biophys. Biochem. Cytol.* **4**:415-416.

Milner, G. R. 1969a. Nuclear morphology and the ultrastructural localization of DNA synthesis during S-phase. *J. Cell Sci.* **4**:569-582.

Milner, G. R. 1969b. Nuclear ultrastructure of the transforming lymphocyte during inhibition of DNA synthesis with hydroxyurea. *J. Cell Sci.* **4**:583-591.

Mittwoch, U. 1967. Barr bodies in relation to DNA values and nuclear size in cultured human cells. *Cytogenetics* **6**:38-50.

Moser, G. C., H. Müller, and E. Robbins. 1975. Differential nuclear fluorescence during the cell cycle. *Exptl. Cell Res.* **91**:73-78.

Müller, H. -A. 1966. Die Chromozentren in den Leberzellkernen der Maus unter normalen und pathologischen Bedingungen. *Ergebn. Allgem. Pathol.* **47**:143-185.

Müller, W. 1973. Das Leitz-Textur-Analyse-System. Leitz-Mitt. *Wiss. Techn. Suppl.* **1**:101-116.

Nagl, W. 1962. Über Endopolyploidie, Restitutionskernbildung und Kernstrukturen im Suspensor von Angiospermen und einer Gymnosperme. *Österr. Bot. Z.* **109**: 431-494.

Nagl, W. 1965. Die SAT-Riesenchromosomen der Kerne des Suspensors von *Phaseolus coccineus* und ihr Verhalten während der Endomitose. *Chromosoma* **16**: 511-520.

Nagl, W. 1967. Die Riesenchromosomen von *Phaseolus coccineus* L.: Baueigentümlichkeiten, Strukturmodifikationen, zusätzliche Nukleolen und Vergleich mit den mitotischen Chromosomen. *Österr. Bot. Z.* **114**:171-182.

Nagl, W. 1968. Der mitotische und endomitotische Kernzyklus bei *Allium carinatum*. I. Struktur, Volumen und DNS-Gehalt der Kerne. *Österr. Bot. Z.* **115**:322-353.

Nagl, W. 1969a. Über den unterschiedlichen Formwechsel des Heterochromatins in der mitotischen Prophase von *Lithospermum* und *Lilium*. *Österr. Bot. Z.* **116**: 295-306.

Nagl, W. 1969b. Banded polytene chromosomes in the legume *Phaseolus vulgaris*. *Nature* **221**:70-71.

Nagl, W. 1969c. Correlation of structure and RNA synthesis in the nucleolus-organizing polytene chromosomes of *Phaseolus vulgaris*. *Chromosoma* **28**:85-91.

Nagl, W. 1970a. Correlation of chromatin structure and interphase stage in nuclei of *Allium flavum*. *Cytobiologie* **1**:395-398.

Nagl, W. 1970b. The mitotic and endomitotic nuclear cycle in *Allium carinatum*. II. Relations between DNA replication and chromatin structure. *Caryologia* **23**: 71-78.

Nagl, W. 1970c. Inhibition of polytene chromosome formation in *Phaseolus* by polyploid mitoses. *Cytologia* **35**:252-258.

Nagl, W. 1970d. On the mercaptoethanol-induced polyploidization in *Allium cepa*. *Protoplasma* **70**:349-360.

Nagl, W. 1970e. Temperature-dependent functional structures in the polytene chromosomes of *Phaseolus*, with special reference to the nucleolus organizers. *J. Cell. Sci.* **6**:87-107.

Nagl, W. 1972a. Evidence of DNA amplification in the orchid *Cymbidium in vitro*. *Cytobios* **5**:145–154.

Nagl, W. 1972b. Selective inhibition of cell cycle stages in the *Allium* root meristem by colchicine and growth regulators. *Amer. J. Bot.* **59**:346–351.

Nagl, W. 1972c. Molecular and structural aspects of the endomitotic chromosome cycle in angiosperms. *Chrom. Today* **3**:17–23.

Nagl, W. 1972d. Giant sex chromatin in endopolyploid trophoblast nuclei of the rat. *Experientia* **28**:217–218.

Nagl, W. 1973a. The mitotic and endomitotic nuclear cycle in *Allium carinatum*. IV. ^{3}H-Uridine incorporation. *Chromosoma* **44**:203–212.

Nagl, W. 1973b. Photoperiodic control of activity of the suspensor polytene chromosomes in *Phaseolus vulgaris*. *Z. Pflanzenphysiol.* **70**:350–357.

Nagl, W. 1974a. The *Phaseolus* suspensor and its polytene chromosomes. *Z. Pflanzenphysiol.* **73**:1–44.

Nagl, W. 1974b. DNA synthesis in tissue and cell cultures. In *Tissue Culture and Plant Science*, ed. H. E. Street. Acad. Press, London-New York, pp. 19–42.

Nagl, W. 1975. Organization and replication of the eukaryotic chromosome. *Progr. Bot.* **37**:186–210.

Nagl, W. 1976a. *Zellkern und Zellzyklen*. Ulmer, Stuttgart, GFR.

Nagl, W. 1976b. Nuclear organization. *Ann. Rev. Plant Physiol.* **27**:39–69.

Nagl, W. 1976c. Endopolyploidy and polyteny understood as evolutionary strategies. *Nature* **261**:614–615.

Nagl, W. 1976d. Early enbryogenesis in *Tropaeolum majus* L.: Evolution of DNA content and polyteny in the suspensor. *Plant Sci. Lett.* **7**:1–8.

Nagl, W. In press. *Endopolyploidy and Polyteny in Differentiation*. Springer, Berlin & Heidelberg & New York.

Nagl, W. and W. Rücker. 1972. Beziehungen zwischen Morphogenese und nuklearem DNS-Gehalt bei aseptischen Kulturen von *Cymbidium* nach Wuchstoffbehandlung. *Z. Pflanzenphysiol.* **67**:120–134.

Nagl, W. and W. Rücker. 1974. Shift of DNA replication from diploid to polyploid cells in cytokinin-controlled differentiation. *Cytobios* **10**:137–144.

Nagl. W. J. Hendon, and W. Rücher. 1972. DNA amplification in *Cymbidium* protocorms *in vitro*, as it relates to cytodifferentiation and hormone treatment. *Cell Diff.* **1**:229–237.

Nagl, W., C. Peschke, and R. Van Gyseghem. 1976. Heterochromatin underreplication in *Tropaeolum* embryogenesis. *Naturwissenschaften* **63**:198–199.

Nešković, B. A. 1968. Developmental phases in intermitosis and the preparation for mitosis of mammalian cells *in vitro*. *Intern. Rev. Cytol.* **24**:71–97.

Nicolini, C., K. Ajiro, T. W. Borun, and R. Baserga. 1975. Chromatin changes during cell cycle of HeLa cells. *J. Biol. Chem.* **250**:3381–3385.

Nur, U. 1966. Nonreplication of heterocrhomatic chromosomes in a mealy bug, *Planococcus citri* (Coccidea: Homoptera) *Chromosoma* **19**:439–448.

Nur, U. 1968. Endomitosis in the mealy bug, *Planococcus citri* (Homoptera: Coccidea). *Chromosoma* **24**:202–209.

Ohnuki, Y. 1968. Structure of chromosomes. I. Morphological studies of the spiral structure of human somatic chromosomes. *Chromosoma* **25**:402–428.

Ornstein, L. 1952. The distribution error in microspectrophotometry. *Lab. Investig.* **1**:250–262.

Patau, K. 1952. Absorption microphotometry of irregularly shaped objects. *Chromosoma* 5:341-362.

Patau, K. and H. Swift. 1953. The DNA content (Feulgen) of nuclei during mitosis in a root tip of onion. *Chromosoma* 6:149-169.

Pearson, G. G., J. N. Timmis, and J. Ingle. 1974. The differential replication of DNA during plant development. *Chromosoma* 45:281-294.

Pederson, T. 1972. Chromatin structure and the cell cycle. *Proc. Natl. Acad. Sci.* 69:2224-2228.

Pederson, T. and E. Robbins. 1972. Chromatin structure and the cell division cycle. *J. Cell Biol.* 55:322-327.

Pera, F. and U. Wolf. 1967. DNS-Replikation und morphologie der X-chromosomen wahrend der syntheseperiode bei *Microtus agrestis*. *Chromosoma* 22:378-389.

Plaut, W. 1969. On ordered DNA replication in polytene chromsomes. *Genetics* 61:Suppl. 239-244.

Poroschenko, G. G., M. M. Fomina, and T. A. Nikol'skaya. 1970. On diplochromosomes in endoreduplication. *Tsitologiya* 12:1575-1578.

Rasch, R. W., E. W. Rasch, and J. W. Woodard. 1967. Heterogeneity of nuclear population in root meristems. *Caryologia* 20:87-100.

Rees, H. and R. N. Jones. 1972. The origin of the wide species variation in nuclear DNA content. *Intern. Rev. Cytol.* 32:53-92.

Reinert, J. and H. Holtzer, eds. 1975. *Cell Cycle and Cell Differentiation.* Springer, Berlin & Heidelberg & New York.

Rodman, T. 1968. Relationship of developmental stage to initiation of replication in polytene nuclei. *Chromosoma* 23:271-287.

Rudkin, G. T. 1963. The structure and function of heterochromatin. *Genetics Today* 2:359-374.

Sanchez, S. and J. Oberti. 1974. Organisation nucláire de la cellule HeLa au cours du cycle cellulaire. Étude ultrastructurale et autoradiographique. *J. Microscopie* 19:25-36.

Sato, S. 1974. Scanning electron microscopy on chromosome aberrations in root-tip cells of *Vicia faba* treated with gallic acid. *Cytologia* 39:747-751.

Sawicki, W., J. Rowinski, and J. Abramczuk. 1974. Image analysis of chromatin in cells of preimplantation mouse embryos. *J. Cell Biol.* 63:227-233.

Schnedl, W. 1969. Die DNS-Synthese im Sexchromatin. *Cytobiologie* 1:115-120.

Schor, S. L., R. T. Johnson, and C. A. Waldren. 1975. Changes in the organization of chromosomes during the cell cycle: Response to ultraviolet light. *J. Cell Sci.* 17:539-565.

Schwarzacher, H. G. and W. Schnedl. 1967. Elektronenmikroskopische Untersuchungen menschlicher Metaphase-Chromosomen. *Humangenetik* 4:153-165.

Schweizer, D. and W. Nagl. 1976. Heterochromatin diversity in *Cymbidium*, and its relationship to differential DNA replication. *Exptl. Cell Res.* 98:411-423.

Sorsa, M. 1969. Ultrastructure of puffs in the proximal part of chromosome 3R in *Drosophila melanogaster*. *Annal. Acad. Sci. Fenn. A. IV.* 150:1-21.

Speta, F. 1972. Entwicklungsgeschichte und Karyologie von Elaiosomen an Samen und Früchten. *Naturkdl. Jahrb. Linz* 1972:9-65.

Stocker, A. J. and C. Pavan. 1974. The influence of ecdysterone on gene amplification, DNA synthesis, and puff formation in the salivary gland chromosomes of *Rhynchosciara hollaenderi*. *Chromosoma* 45:295-319.

Stowell, R. E. 1945. Feulgen reaction for thymonucleic acid. *Stain Technol.* **20**: 45-58.

Swift, H. and R. Kleinfeld. 1953. DNA in grasshopper spermatogenesis, oogenesis and cleavage. *Physiol. Zool.* **26**:301-311.

Tanaka, R. 1965. ^{3}H-thymidine autoradiographic studies on the heteropycnosis, heterochromatin and euchromatin in *Spiranthes sinensis. Bot. Mag.* (Tokyo) **78**:50-62.

Torrey, J. G. 1965. Cytological evidence of cell selection by plant tissue culture media. In *Proc. Int. Conf. Pl. Tissue Culture*, eds. P. R. White and A. R. Grove. McCutchan Publ. Corp., Berkeley, pp. 473-484.

Tschermak-Woess, E. 1954. Über die Phasen der Endomitose, Herkunft und Verhalten der "nuklealen Körper" und Beobachtungen zur karyologischen Anatomie von *Sauromatum guttatum. Planta* **44**:509-531.

Tschermak-Woess, E. 1956. Karyologische Pflanzenanatomie. *Protoplasma* **46**:798-834.

Tschermak-Woess, E. 1959. Die DNS-Reproduktion in ihrer Beziehung zum endomitotischen Strukturwechsel. *Chromosoma* **10**:497-503.

Tschermak-Woess, E. 1963. Strukturtypen der Ruhekerne von Pflanzen und Tieren. *Protoplasmatologia VI/1*. Springer, Wien.

Tschermak-Woess, E. 1967. Der eigenartige Verlauf der I. meiotischen Prophase von *Rhinanthus,* die Riesenchromosomen und das besondere Verhalten der kurzen Chromosomen in Mitose, Meiose und hoch-endopolyploiden Kernen. *Caryologia* **20**:135-152.

Tschermak-Woess, E. 1971. Endomitose. In *Handbuch der allgemeinen Pathologie* II/1/2, ed. H. -W. Altmann. Springer Berlin, Heidelberg, New York. pp. 569-625.

Tschermak-Woess, E. and R. Doležal. 1956. Der Formwechsel des Heterochromatins im Verlauf der Mitose von *Vicia faba. Österr. Bot. Z.* **103**:457-468.

Tschermak-Woess, E. and R. Doležal-Janisch. 1957. Über das Chromosomen-und Kernwachstum in der Wurzel von *Haemanthus. Chromosoma* **9**:81-90.

Tschermak-Woess, E. and G. Hasitschka. 1953. Veränderungen der Kernstruktur während der Endomitose, rhythmisches Kernwachstum und verschiedenes Heterochromatin bei Angiospermen. *Chromosoma* **5**:574-614.

Tschermak-Woess, E. and G. Hasitschka. 1954. Über endomitotische Polyploidisierung im Zuge der Differenzierung von Trichomen und Trichocyten bei Angiospermen. *Öster. Bot. Z.* **101**:79-117.

Turala, K. 1966. Endopolyploidie im Endosperm von *Echinocystis lobata. Österr. Bot. Z.* **113**:235-244.

Van't Hof, J. 1965. Relationship between mitotic cycle duration, S-period duration and the average rate of DNA synthesis in root meristems of several plants. *Exp. Cell Res.* **39**:48-58.

Van't Hof, J. and A. H. Sparrow. 1963. A relationship between DNA content, nuclear volume, and minimum mitotic cycle time. *Proc. Natl. Acad. Sci.* **49**: 897-902.

Watson, M. C., and W. G. Aldridge. 1961. Methods for the use of indium as an electron stain for nucleic acids. *J. Biophys. Biochem. Cytol.* **11**:257-272.

Wohlfarth-Bottermann, K. E. 1957. Cytologische Studien IV. Die Entstenhung, Vermehrung und Sekretabgabe der Mitochondrien von *Paramecium. Zeitschrift für Naturforschung.* Teil b. **12**:164-167.

Woodard, J., E. Rasch, and H. Swift. 1961. Nucleic acid and protein metabolism during the mitotic cycle in *Vicia faba. J. Biophys. Biochem. Cytol.* **9**:445–462.

Wyandt, H. E. and F. Hecht. 1973a. Human Y-chromatin. I. Dispersion and condensation. *Exptl. Cell Res.* **81**:453–461.

Wyandt, H. E. and F. Hecht. 1973b. Human Y-chromatin. II. DNA replication. *Exptl. Cell Res.* **81**:462–467.

Yunis, J. J. and W. G. Yasmineh. 1971. Heterochromatin, satellite DNA, and cell function. *Science* **174**:1200–1209.

Żuk, J. 1969. Analysis of Y-chromosome heterochromatin in *Rumex thyrsiflorus. Chromosoma* **27**:338–353.

The Nucleolus During Mitosis in Plants: Ultrastructure of Persistent Nucleoli in Mung Beans

James P. Braselton   Ohio University

INTRODUCTION

The nucleolus is the most conspicuous morphological component of the interphase, eukaryotic nucleus; biochemical, cytochemical, and ultrastructural studies have established the essential role of the nucleolus in ribosome biogenesis (for reviews see Birnstiel 1967; Perry 1969). Nucleolar behavior during mitosis, however, is not consistent throughout either animals, plants, or fungi, and the variations in behavior have hindered our understanding of the role(s) of nucleoli in mitotic cells.

As was brought out by Pickett-Heaps (1970), within the Plant Kingdom there seems to be at least four major degrees of nucleolar behavior during mitosis: (a) "dispersive behavior," in which the nucleolus disappears during prophase; (b) "semi-persistent behavior," in which portions of the nucleolus remain as discrete units during mitosis, but appear to become excluded passively from the spindle; (c) "persistent behavior," in which the nucleolus loosens, and a large portion of the loosened material coats the chromosomes; and (d) "autonomous behavior," in which the nucleolus remains as a discrete entity, which goes through an apparently orderly division with the halves becoming incorporated into daughter nuclei. Care must be taken when using the available terminology to describe nucleolar behavior during mitosis. Earlier light microscopists and electron microscopists used the terms "persistent" and "persisting" in reference to nucleoli, or nucleolar remnants, that were observed in metaphase, anaphase, or telophase. And despite Pickett-Heaps' effort to distinguish between degrees of nucleolar behavior, "persistent" and "persisting" are still used by most workers to denote the type of nucleolar behavior which was labeled "semi-persistent" by Pickett-Heaps. In the remainder of this article, I will adhere to the terminology used by most investigators and use the term "persistent" in reference to nucleoli remaining during mitosis.

In higher plants dispersive nucleolar behavior is the most prominent type, and there have been several excellent ultrastructural studies (Lafontaine and Chouinard 1963; Lafontaine and Lord 1974) and reviews (Lafontaine 1968, 1974; Lafontaine and Lord 1969) of dispersion of nucleoli during prophase and subsequent nucleolar reformation in plants. Although there have been many light microscopic observations of persistent nucleoli in mitosis of higher plants, only cursory reports have been made regarding ultrastructural aspects of such nucleoli. Most recent research of persistent nucleoli has been directed towards mammalian cell lines (Brinkley 1965; Heenan and Nichols 1966; Hsu et al. 1965; Noel et al. 1971), in which some persistent nucleoli apparently are naturally occurring, whereas others may be due in part to various experimental conditions. Further investigation into persistent nucleolar behavior in mitosis of higher plants seems warranted because the phenomenon occurs more often than generally believed. The phenomenon is a natural one (i.e., it is not the result of experimental conditions), and our total understanding of mitosis will remain incomplete until we fully understand variations in nucleolar behavior as well as typical, dispersive nucleolar behavior.

This study was initiated, therefore, to better characterize persistent nucleoli in mitosis of a vascular plant, and to compare and contrast persistent behavior to dispersive behavior.

MATERIALS AND METHODS

Seeds of mung bean (*Phaseolus mungo* L.) were purchased from Carolina Biological Supply Company (Burlington, N.C.) and were germinated in moist vermiculite. The first 1–2 mm of root tips of primary roots 2–4 cm long were prepared for either light or electron microscopy.

Light microscopy. Root tips were fixed in ethanol-formalin-glacial acetic acid (30:15:1) for 12–24 hours at room temperature, hydrolyzed in 1 N HCl at 60° for 10 minutes, and squashed in aceto-carmine (Rattenbury 1952).

Electron microscopy. Root tips were fixed in 2%–2% glutaraldehyde-formaldehyde in 0.1 M phosphate buffer at pH 6.7 for 1–2 hours at room temperature; rinsed in buffer; post-fixed for 1 hour in 1% OsO_4 in 0.1 M phosphate buffer at pH 6.7 at 4°C; dehydrated in an ethanol-propylene oxide series; and embedded in Araldite-Epon. Thin-sections were stained with aqueous uranyl acetate and lead citrate, and examined with a Hitachi HS-8 electron microscope.

OBSERVATIONS

Light Microscopy

Interphase nucleoli (Figure 7-4) appear spherical, are 3–5μm in diameter, and stain deep, brownish-red against a lightly stained, brownish-pink nucleus and cytoplasm. Metaphase mitotic figures stain deep, brownish-red, and individual chromosomes cannot be detected easily. Nucleoli in metaphase cells stain much less dense than either interphase nucleoli or metaphase chromosomes, and have been observed in the following positions: (a) at metaphase, at the equatorial plate or slightly positioned towards one of the poles (Figure 7-1); (b) located between the equatorial plate and one of the poles (Figure 7-2); (c) elongated perpendicularly to, and extending through, the equatorial plate (Figure 7-3); or (d) located at one of the poles (Figure 7-4). Nucleoli on or near the metaphase plate generally appear irregular in shape, whereas nucleoli at or near the poles (Figure 7-4) appear spherical and more compact than nucleoli at the other locations.

Nucleoli frequently are not observed in either anaphase or telophase, but when present (Figures 7-5 and 7-6) they are located at either pole, are not associated with the chromosomes, and are spherical (1-2 μm in diameter).

Electron Microscopy

Interphase nucleoli (Figure 7-7) are circular in profile, are 3-5 μm in diameter, and have components similar to those typically observed in nucleoli of meristematic cells in various higher plants (Chouinard 1966; Hyde 1966, 1967; Hyde et al. 1965; Lafontaine 1968, 1974; Lafontaine and Chouinard 1963; Lafontaine and Lord 1969, 1974). Electron-dense granules are contained within regions throughout the nucleolus; fibrils are tightly packed into 0.5 μm diameter, thread-like skeins, which are often termed "nucleolenema"—an amorphous substance which pervades the nucleolenema often obscures the fibrous components; and less electron-dense "vacuoles" are interdispersed among fibrillar and granular regions.

In early prophase the nucleolus becomes less regular in shape and the granular material becomes localized towards the periphery (Figure 7-8). By late prophase when the nuclear envelope is beginning to disrupt (Figure 7-9), the majority of the granules are dispersed into the nucleoplasm. The fibrillar regions appear more loosened than in early prophase, but the amorphous background material continues to reduce the clarity of individual fibrils as was the case in interphase nucleoli. Less electron-dense nucleolar "vacuoles" can still be seen in the fibrillar region.

Metaphase nucleoli located near the metaphase plate (Figure 7-10) are irregular in profile and are much less electron-dense and less compact than nucleoli in interphase or prophase. At low magnification, metaphase nucleoli (Figure 7-10) appear fairly homogeneous. At higher magnification (Figures 7-11 and 7-12), each metaphase nucleolus can be seen to consist predominantly of loosely arranged fibrils and granules distributed among the fibrils. An amorphous background substance which is less electron-dense than the fibrils and granules is irregularly dispersed throughout the nucleolus. Metaphase nucleoli at or near the poles (Figure 7-12) are more circular in profile than nucleoli located on or near the equatorial plate.

Although close associations of nucleoli with metaphase chromosomes were observed (Figure 7-11), no specific regions interpreted as nucleolus organizers were seen. A microtubule was observed within one metaphase nucleolus (Figure 7-11), but the microtubule was not oriented parallel with the spindle. Despite serial sec-

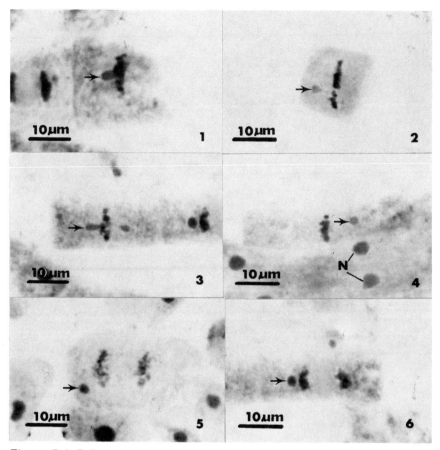

Figures 7-1-7-6
Light micrographs of aceto-carmine squash preparations showing nucleoli in various stages of mitosis. Figure 7-1. Metaphase with persistent nucleolus (arrow) near the metaphase plate. Figure 7-2. Persistent nucleolus (arrow) located between the metaphase plate and one pole. Figure 7-3. Persistent nucleolus (arrow) elongated perpendicular to and extending through the metaphase plate. Figure 7-4. Metaphase with persistent nucleolus (arrow) at one pole. Also note appearance of interphase nucleoli (N). Figure 7-5. Late anaphase with persistent nucleolus (arrow) located near one of the anaphase complements. Figure 7-6. Telophase with persistent nucleolus (arrow) located near one of the daughter nuclei.

tions of other metaphase nucleoli, no additional microtubules other than this one were detected.

Persistent nucleoli in anaphase (Figure 7-13) and telophase (Figures 7-14 and 7-15) appear circular in profile, have diameters 1-2 μm, and consist essentially of tightly packed, electron-dense fibrils within a less electron-dense background sub-

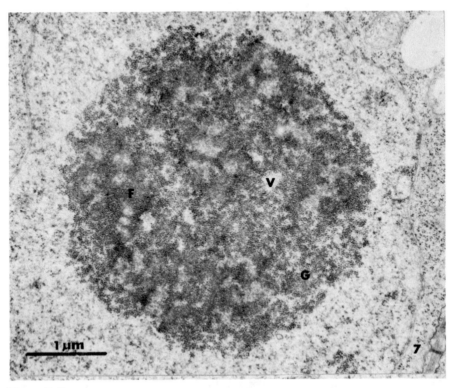

Figure 7-7.
Electron micrograph of interphase nucleolus exhibiting fibrillar (F), granular (G), and "vacuolar" (V) regions.

stance. Granular appearing profiles are occasionally observed dispersed among the fibrils, but it is difficult to determine if these are indeed granules or cross-sections of fibrils. As seen in Figure 7-14, nucleoli persisting into telophase remain outside the nuclear envelope of the reformed nucleus, and the reforming nucleolus of the daughter nucleus appears similar in ultrastructure to reforming nucleoli as documented for other higher plants (Lafontaine and Chouinard 1963; Lafontaine and Lord 1966, 1969).

DISCUSSION

Although there have been many light microscopic (Avanzi 1950; Brown and Emery 1957; D'Amato-Avanzi 1952; Ehrenberg 1946; Fabbri 1960; Frew and Bowen 1929; Gori 1951, 1956; Håkansson and Levan 1942; Iyengar 1939; Jacob 1942; Levan 1944; Mensinkai 1939; Pathak 1940; Ramanujam 1938; Tjio 1948; Underbrink et al. 1967; Wager 1904; Zirkle 1928) and several electron microscopic (Allen and Bowen 1966; Hanzely 1974; Lafontaine and Lord 1966; Underbrink et al. 1967) reports of nucleoli persisting during mitosis of vascular plants, the extent

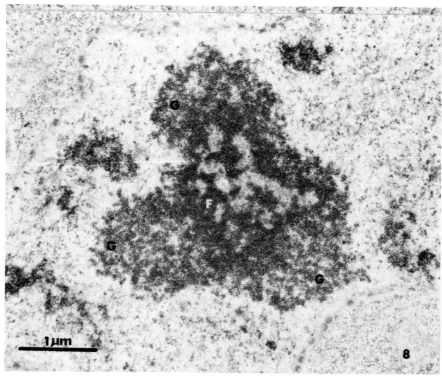

Figure 7-8.
Electron micrograph of early prophase nucleolus in which the granular material (G) is localized on the periphery of the centralized fibrillar material (F).

to which persistent nucleoli occur throughout vascular plants is not known. Within the angiosperms, legumes (Avanzi 1950; D'Amato-Avanzi 1952; Håkansson and Levan 1942; Iyengar 1939; Jacob 1942; Wager 1904) and grasses (Brown and Emery 1957; Ramanujam 1938; Zirkle 1928) are the groups cited most often in reports of persistent nucleoli. In lower vascular plants, the psilophyte *Psilotum nudum* (Allen and Bowen 1966; Fabbri 1960), *Equistem* sp. (Frew and Bowen 1929), and the fern *Ophioglossum petiolatum* (Underbrink et al. 1967) are the only representatives reported to exhibit nucleoli that persist during mitosis. I am not aware of persistent nucleoli in gymnosperms.

Unfortunately, popular cytological techniques such as Feulgen, aceto-carmine, and aceto-orcein that clearly reveal chromosome morphology generally employ acetic-alcohol fixation and nucleoli are neither preserved nor stained. Also, medium and large chromosomes (i.e., greater than 5μm in length) may obscure nucleoli on the metaphase plate, even in preparations designed to stain nucleoli. This could explain why most light microscopic reports of persistent nucleoli of vascular plants have been made on organisms possessing small (0.5-2 μm) chromosomes (e.g., grasses, *Gladiolus, Phaseolus,* and *Cucurbita*).

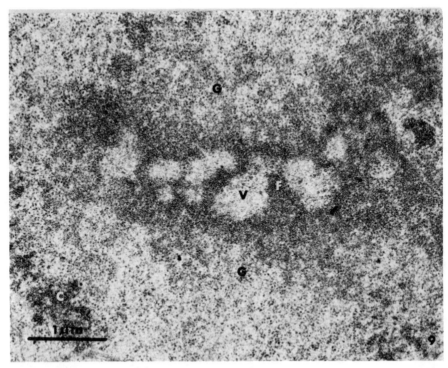

Figure 7-9.
Portion of late prophase nucleolus and adjacent nucleoplasm, which contains a portion of a condensing chromosome (C). The fibrillar region (F) of the nucleolus is still intact and "vacuoles" (V) are located throughout it. Granules (G) have dispersed into the nucleoplasm.

Another problem concerning observations of organisms expressing persistent nucleolar behavior is that nucleoli do not persist in all mitotic divisions within organisms possessing the phenomenon. For example, in the grasses that possess persistent nucleoli (Brown and Emery 1957), depending on the genus, 6% through 96% of metaphases contain nucleoli. In animal cell lines (Hsu et al. 1965), depending on the cell line, persistent nucleoli may occur in 4% (mouse LM) through 98% (Chinese hamster Dede) of metaphases. In addition, Ehrenberg (1946) showed that in *Salix* hybrids, roots grown at relatively low temperatures (11.5-16.5°C) have persistent nucleoli in mitosis, whereas roots grown at higher temperatures (23.5-29°C) have no or few mitotic divisions with nucleoli persisting during metaphase. Mung beans appear to be ideal for the study of persistent nucleoli because the small chromosomes do not obscure nucleoli during metaphase, and because in root tips over 90% of the metaphases have nucleoli present. Also, a preliminary study (unpublished) indicates that temperature does not affect the frequency of persistent nucleoli in mung bean mitosis.

Ultrastructurally, persistent nucleolar behavior in mitosis differs from dispersive

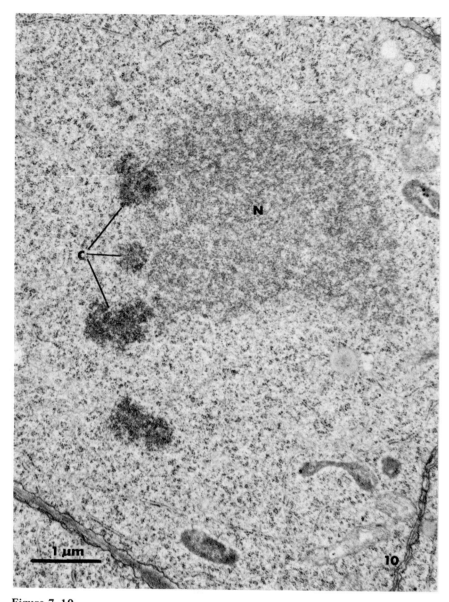

Figure 7-10.
Survey electron micrograph of metaphase cell with persistent nucleolus (N) located near the metaphase plate and in close association with several chromosomes (C).

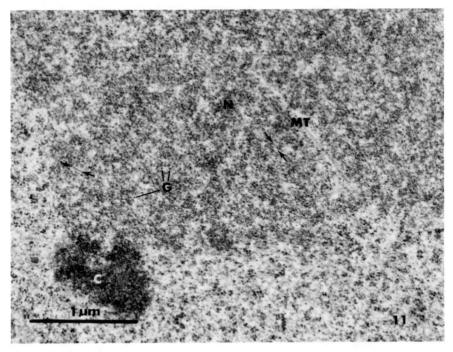

Figure 7-11.
Portion of a metaphase, persistent nucleolus (N) in close association with a chromosome(C). The nucleolus consists predominantly of fibrous material (arrows), granules (G) dispersed throughout, and an amorphous background substance. Also note the microtubule (MT).

behavior of vascular plants as was reported for *Vicia faba* (Lafontaine and Chouinard 1963) and *Allium porrum* (Lafontaine and Lord 1974). In *V. faba,* after breakdown of the nuclear envelope at the end of prophase, both fibrillar and granular components of nucleoli disperse into the spindle regions and cannot be recognized from similar structures already in the spindle. In *A. porrum* the granular portion behaves similarly to the granular material in *V. faba.* However, although the nucleolenema of *A. porrum* loses a large amount of material during prophase, a portion of the nucleolenema, presumably consisting in part of nucleolar organizing DNA, condenses with the chromosomes and is interpreted as forming the secondary constriction recognized in metaphase and anaphase chromosomes.

The predominant structural components of persistent mitotic nucleoli, or nucleolar remnants, have been reported as either fibrillar as seen in Chinese hamster cell line Dede (Brinkley 1965), *Cyperus alternifolius* (Papsidero and Braselton 1973), *Allium sativum* (Hanzely 1974), *Ophioglossum petiolatum* (Underbrink et al. 1967), and as observed here in mung beans; or as granular as reported for Chinese hamster cell line Don (Noel et al. 1971) and *Psilotum nudum* (Allen and Bowen 1966). It is not known what determines the fibrillar or granular composition of per-

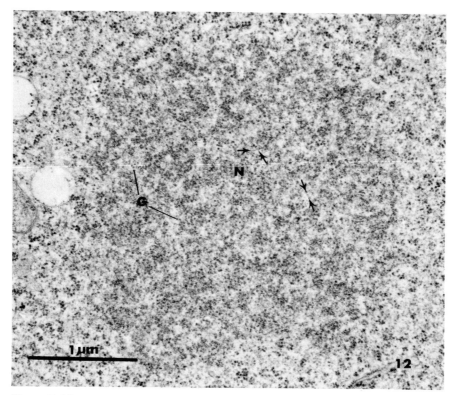

Figure 7–12.
Metaphase nucleolus (N) located in a polar region. The same structural components
(fibrils—arrows; granules—G) that are seen in metaphase nucleoli at the metaphase
plate are present.

sistent nucleoli. But further investigation into the two opposing situations reported
between the two Chinese hamster cell lines may provide insight into mechanisms
controlling nucleolar activity during cell division.

The most thorough study to date of persistent nucleoli was by Brinkley (1965)
on nucleoli of mitosis in Chinese hamster cell line Dede. Brinkley observed that
nucleolar remnants, which consisted predominantly of fibrillar material with some
granules dispersed throughout, existed freely in the cytoplasm or associated with
chromosomes. Remnants associated with chromosomes became incorporated into
daughter, telophase nuclei. Hanzely (1974) similarly reported that fibrous nucleolar
remnants were carried with (or by ?) the chromosomes to daughter nuclei during
mitosis in *Allium sativum*. In mung beans, however, although nucleoli were closely
associated with chromosomes at metaphase (Figures 7–10 and 7–11), the nucleoli
were not carried by the chromosomes to telophase nuclei; instead, nucleoli
preceded the chromosomes to the poles (often just to one or the other pole) and
remained outside the reconstructed nuclear envelope (Figure 7–14). Also, between

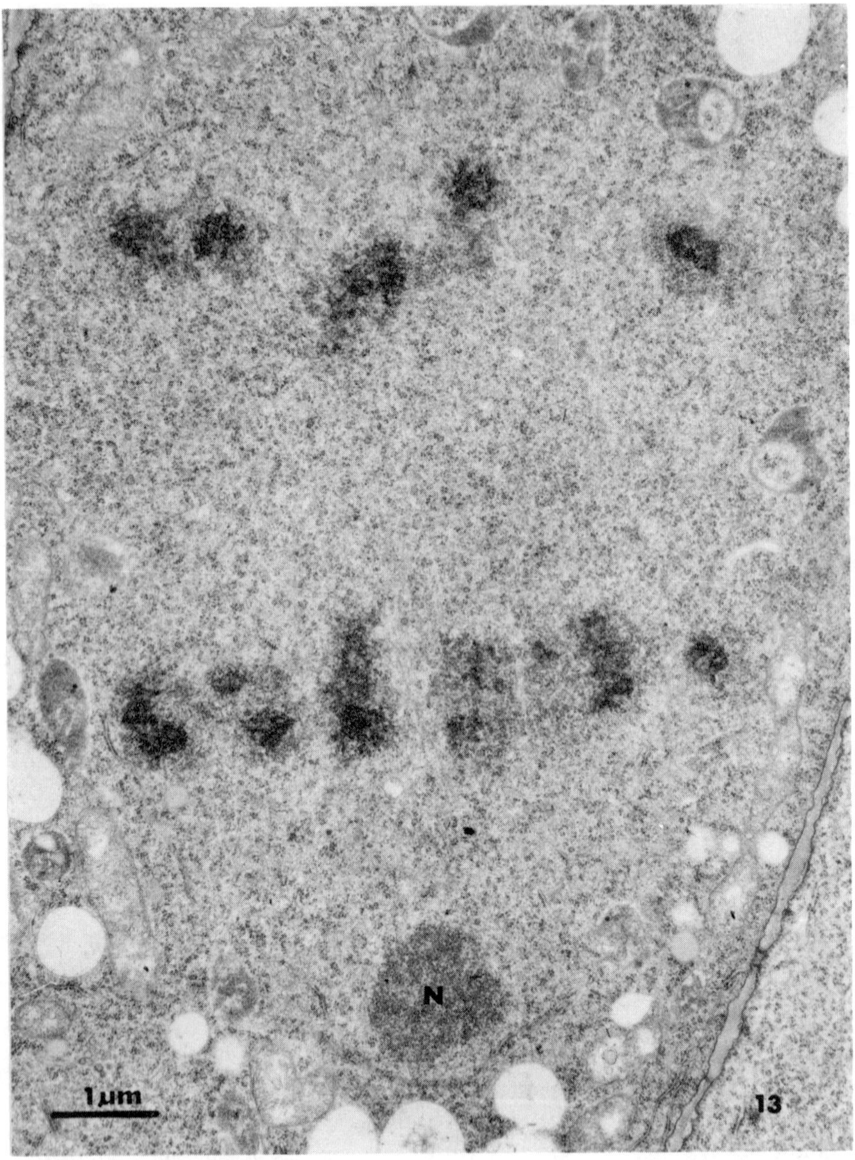

Figure 7–13.
Survey electron micrograph of an anaphase cell with persistent nucleolus (N) located near one of the poles. The nucleolus is more compact than nucleoli in metaphase.

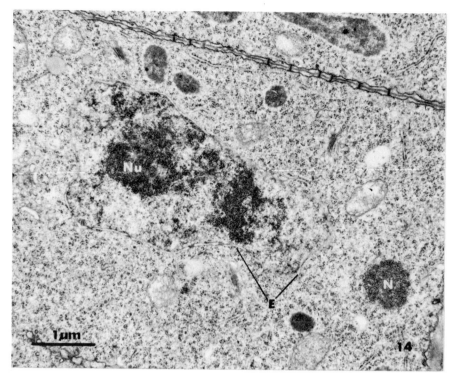

Figure 7-14.
Portion of a cell in telophase showing a nucleus with a typical reforming nucleolus
(Nu) and a persistent nucleolus (N) which remains to the exterior of the nuclear
envelope (E).

metaphase and anaphase, mung bean nucleoli became more compact and by telophase, at low magnification, resembled "cytoplasmic nucleoloids" observed in
microsporogenesis of *Lilium* (Dickinson and Heslop-Harrison 1970).

Cytoplasmic nucleoloids in *Lilium,* however, were described as being predominantly granular, as opposed to the fibrous nucleolar remnants reported here for
mung beans. Also, cytoplasmic nucleoloids appear to be involved in ribosome
biogenesis during late stages of microsporogenesis (Williams et al. 1973). At this
time, there is neither experimental evidence for the role(s) of persistent nucleoli in
mung beans, nor an indication of what determines the morphological changes of the
nucleoli throughout cell division.

Of primary concern in studies of nucleolar behavior during mitosis is the origin of
prenucleolar material in telophase and how the prenucleolar material organizes into
typical, interphase nucleoli. Several excellent articles have reviewed this subject
with regards to higher plants (Lafontaine 1968, 1974; Lafontaine and Lord 1966,
1969). From early light microscopic observations arose the concept of the
"chromosomal RNA cycle" (Mazia 1961), which is an interpretation of nucleolus-
chromosomal association based largely on changes in staining properties of mitotic

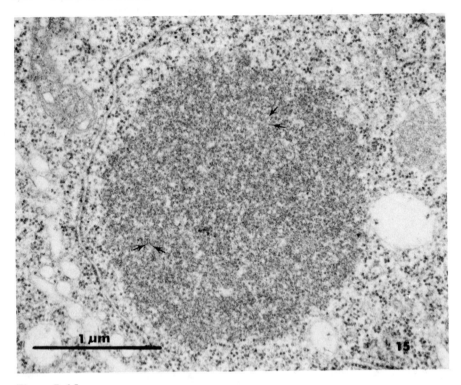

Figure 7-15.
Telophase, persistent nucleolus showing fibrous material (arrows) more compactly
arranged than in nucleoli of metaphase cells.

chromosomes. During prophase, nucleolar material is purported to migrate to and
coat the chromosomes, and in the subsequent telophase this nucleolar material is
believed to contribute to reforming nucleoli (Jacobson and Webb 1952; Kaufmann
et al. 1948). Love (1957), however, interpreted the increase in RNA staining of
chromosomes to be due to RNA of cytoplasmic origin. Autoradiographic studies
that detect RNA associated with mitotic chromosomes (Feinendegen and Bond
1963; Hsu et al. 1965; Kusanagi 1964; LaCour 1963; Taylor 1960; Woodward et
al. 1961), ultrastructural-cytochemical observations of ribonucleoprotein associated
with mitotic chromosomes (Moyne et al. 1974; Papsidero and Braselton 1973),
and biochemical detection of ribosomal or ribosomal-like RNA in preparations of
isolated metaphase chromosomes of animal cell lines (for reviews see Hearst and
Botchan 1970; Mendelsohn 1974) all lend support to the "chromosomal RNA
cycle."

In addition, Stockert et al. (1970) applied several metabolic inhibitors during cell
divisions of *Allium cepa* and showed that the prenucleolar material that appears in
telophase does not consist of newly synthesized material; complete reformation of
the nucleolus, however, does require renewed RNA synthesis. Similarly, experiments using *Amoeba proteus* (Stevens and Prescott 1971) and mammalian cell

cultures (Phillips 1972; Phillips and Phillips 1973), in which actinomycin D was used to prevent ribosomal RNA synthesis in selected stages of the cell cycle, indicated that synthesis of nucleolar RNA in interphase prior to mitosis is necessary for normal reformation of nucleoli in telophase. It is tempting to speculate that nucleolar RNA synthesized just prior to mitosis corresponds to RNA associated with mitotic chromosomes, but we have no direct evidence that this is so. And if and when premitotically synthesized nucleolar RNA and mitotic chromosomal RNA are shown to be one and the same, then controlling mechanisms and metabolic events necessary for nucleolar reformation still must be determined.

To what extent nucleolar behavior during mitosis corresponds to the "chromosomal RNA cycle" occurring in some organisms or how the type of nucleolar behavior influences reformation of nucleoli during telophase are not known. In organisms that have "autonomous" nucleolar behavior such as the alga *Euglena* (Leedale 1967) and the fungus *Sorosphaera veronicae* (Braselton et al. 1975), the nucleolus within each telophase nucleus arises from a large portion of the nucleolus of the parent nucleus. And in the type of persistent behavior expressed in the algae *Chara* (Pickett-Heaps 1967) and *Spirogyra* (Fowke and Pickett-Heaps 1969; Godward and Jordan 1965; Jordan and Godward 1969), a significant percentage of the telophase nucleolus arises from material synthesized prior to division and transported to telophase nuclei via the chromosomes. But the occurrence of persistent nucleoli such as seen in mitosis of mung beans does not necessarily indicate that persistent nucleolar material is going to contribute to reformation of telophase nucleoli, which may occur in some organisms such as was suggested for *Allium sativum* (Hanzely 1974) and the Chinese hamster cell line Dede (Brinkley 1965). As is seen here in mung beans, persistent nucleoli migrate to the poles apparently independently of the chromosomes, do not become incorporated into telophase nuclei, and often can be seen outside a telophase nucleus that contains a new, reforming nucleolus. By the end of telophase persistent nucleoli are no longer seen and it is assumed that the material has been dispersed into the cytoplasm. The ultimate fate of the nucleolar material is not known.

We know that persistent nucleolar behavior is more common in higher plants than generally believed, and we are able to compare and contrast morphological events of persistent nucleolar behavior to the more typical dispersive behavior. But morphological studies are limited to what information can be ascertained from them, and experiments must be devised to more critically determine functional differences (if any) and similarities between the major modes of nucleolar behavior during mitosis. Perhaps further studies to determine relationships of nucleolar cycles to chromosome cycles as was performed on *Allium cepa* (Giménez-Martín et al. 1971) and *Zea mays* (Clowes and de la Torre 1974; de la Torre and Clowes 1972) can yield insight into timing of persistent nucleolar behavior in relation to other aspects of mitosis. Also, refined autoradiographic techniques used in conjunction with RNA polymerase inhibitors may provide information regarding the role(s) or cause(s) of persistent nucleolar components in mitotic divisions. Certainly our understanding of nucleolar organization and function specifically, and of mitosis in general, will be incomplete until we can account for all variations in nucleolar behavior during mitosis in terms of molecular biological concepts. And it is hoped that ultrastructural observations such as those made here on persistent nucleoli of mung beans will provide the necessary background for further studies of variations of nucleolar behavior during mitosis.

ACKNOWLEDGMENT

This study was supported in part by a Sigma Xi Grant-in-Aid of Research

LITERATURE CITED

Allen, R. D. and C. C. Bowen. 1966. Fine structure of *Psilotum nudum* cells during division. *Caryologia* **19**:299–342.

Avanzi, M. G. 1950. Osservazioni sul ciclo nucleolare in *Cassia acutifolia* Delile. *Caryologia* **3**:200–203.

Birnstiel, M. 1967. The nucleolus in cell metabolism. *Ann. Rev. Pl. Physiol.* **18**:25–58.

Braselton, J. P., C. E. Miller, and D. G. Pechak. 1975. The ultrastructure of cruciform nuclear division in *Sorosphaera veronicae* (Plasmodiophoromycete). *Amer. J. Bot.* **62**:349–358.

Brinkley, B. R. 1965. The fine structure of the nucleolus in mitotic divisions of Chinese hamster cells *in vitro*. *J. Cell Biol.* **27**:411–422.

Brown, W. V. and W. H. P. Emery. 1957. Persistent nucleoli and grass systematics. *Amer. J. Bot.* **44**:585–589.

Chouinard, L. A. 1966. Nucleolar architecture in root meristematic cells of *Allium cepa*. *Nat. Cancer Inst. Monogr.* **23**:125–143.

Clowes, F. A. L. and C. de la Torre. 1974. Inhibition of RNA synthesis and the relationship between nucleolar and mitotic cycles in *Zea mays* root meristems. *Ann. Bot.* **38**:961–966.

D'Amato-Avanzi, M. G. 1952. Nuove osservazioni sulla persistenza del nucleolo durante la mitosi nel genere *Cassia*. *Caryologia* **5**:133–135.

De la Torre, C. and F. A. L. Clowes. 1972. Timing of nucleolar activity in meristems. *J. Cell Sci.* **11**:713–721.

Dickinson, H. G. and J. Heslop-Harrison. 1970. The ribosome cycle, nucleoli, and cytoplasmic nucleoloids in the meiocytes of *Lilium*. *Protoplasma* **69**:187–200.

Ehrenberg, L. 1946. Influence of temperature on the nucleolus and its coacervate nature. *Hereditas* **32**:407–418.

Fabbri, F. 1960. Contributo per l'interpretazione della persistenza nucleolare durante la mitosi in *Psilotum nudum* (L.) Beauv. *Caryologia* **13**:297–337.

Feinendegen, L. E. and V. P. Bond. 1963. Observations on nuclear RNA during mitosis in human cancer cells in culture (HeLa-S3), studied with tritiated cytidine. *Exp. Cell Res.* **30**:393–404.

Fowke, L. C. and J. D. Pickett-Heaps. 1969. Cell division in *Spirogyra*. I. Mitosis. *J. Phycol.* **5**:240–259.

Frew, P. E. and R. H. Bowen. 1929. Nucleolar behavior in the mitosis of plant cells. *Quart. J. Microscopical Sci.* **73**:197–214.

Giménez-Martín, G., M. E. Fernández-Gómez, A. González-Fernández, and C. de la Torre. 1971. The nucleolar cycle in meristematic cells. *Cytobiologie* **4**:330–338.

Godward, M. B. E. and E. G. Jordan. 1965. Electron microscopy of the nucleolus of *Spirogyra britannica* and *Spirogyra ellipsospora. J. Roy. Microscop. Soc.* 84: 347–360.

Gori, C. 1951. Persistenza del nucleolo durante la mitosi in *Reseda odorata* L. *Caryologia* 3:294–298.

Gori, C. 1956. Persistenza nucleolare durante la mitosi nel genere *Reseda. Caryologia* 9:45–55.

Håkansson, A. and A. Levan. 1942. Nucleolar conditions in *Pisum. Hereditas* 28: 436–440.

Hanzely, L. 1974. Persistence of nucleolar material during mitosis in root meristematic cells of *Allium sativum. E. M. S. A. Proc.* 32:272–273.

Hearst, J. E. and M. Botchan. 1970. The eukaryotic chromosome. *Ann. Rev. Biochem.* 39:151–182.

Heenan, W. K. and W. W. Nichols. 1966. Persistence of nucleoli in short term and long term cultures and in direct bone marrow preparations in mammalian materials. *J. Cell Biol.* 31:543–561.

Hsu, T. C., F. E. Arrighi, R. R. Klevecz, and B. R. Brinkley. 1965. The nucleoli in mitotic divisions of mammalian cells *in vitro. J. Cell Biol.* 26:539–553.

Hyde, B. B. 1966. Changes in nucleolar ultrastructure associated with differentiation in the root apex. *Nat. Cancer Inst. Monogr.* 23:39–52.

Hyde, B. B. 1967. Changes in nucleolar ultrastructure associated with differentiation in the root tip. *J. Ultrastruct. Res.* 18:25–54.

Hyde, B. B., K. Sankaranarayanan, and M. L. Birnstiel. 1965. Observations on fine structure in pea nucleoli *in situ* and isolated. *J. Ultrastruct. Res.* 12:652–667.

Iyengar, N. K. 1939. Cytological investigations on the genus *Cicer. Ann. Bot.* 3: 271–305.

Jacob, K. T. 1942. Cytological studies in the genus *Sesbania. Bibliograph. Genet.* 13:225–297.

Jacobson, W. and M. Webb. 1952. The two types of nucleoproteins during mitosis. *Exp. Cell Res.* 3:163–169.

Jordan, E. G. and M. B. E. Godward. 1969. Some observations on the nucleolus in *Spirogyra. J. Cell Sci.* 4:3–15.

Kaufmann, B. P., M. McDonald, and H. Gay. 1948. Enzymatic degradation of ribonucleoproteins of chromosomes, nucleoli and cytoplasm. *Nature* 162:814–815.

Kusanagi, A. 1964. Cytological studies on *Luzula* chromosome. VI. Migration of the nucleolar RNA to metaphase chromosomes and spindle. *Bot. Mag.* (Tokyo) 77:388–392.

LaCour, L. F. 1963. Ribose nucleic acid and the metaphase chromosome. *Exp. Cell Res.* 29:112–118.

Lafontaine, J. -G. 1968. Structural components of the nucleus in mitotic plant cells. In *The Nucleus,* eds. Dalton, A. J. and F. Haguenau. Academic Press, New York, pp. 151–196.

Lafontaine, J. -G. 1974. Ultrastructural organization of plant cell nuclei. In *The Cell Nucleus,* Vol. I, ed. H. Busch. Academic Press, New York, pp. 149–185.

Lafontaine, J. -G. and L. A. Chouinard. 1963. A correlated light and electron microscope study of the nucleolar material during mitosis in *Vicia faba. J. Cell Biol.* 17:167–201.

Lafontaine, J. -G. and A. Lord. 1966. Ultrastructure and mode of formation of the telophase nucleolus in various plant species. *Nat. Cancer Inst. Monogr.* **23**:67–75.

Lafontaine, J. -G. and A. Lord. 1969. Organization of nuclear structures in mitotic cells. In *Handbook of Molecular Cytology*, ed. A. Lima-de-Faria. North-Holland Publishing Co., London, pp. 381–411.

Lafontaine, J. -G. and A. Lord. 1974. A correlated light- and electron-microscope investigation of the structural evolution of the nucleolus during the cell cycle in plant meristematic cells *(Allium porrum). J. Cell Sci.* **16**:63–93.

Leedale, G. F. 1967. *Euglenoid Flagellates*. Prentice-Hall, Englewood Cliffs, New Jersey.

Levan, A. 1944. Notes on the cytology of *Dipcadi* and *Bellevalia. Hereditas* **30**:217–224.

Love, R. 1957. Distribution of ribonucleic acid in tumour cells during mitosis. *Nature* **180**:1338–1339.

Mazia, D. 1961. Mitosis and the physiology of cell division. In *The Cell*, Vol. III, eds. J. Brachet and A. E. Mirsky. Academic Press, New York, pp. 77–412.

Mendelsohn, J. 1974. Studies of isolated mammalian metaphase chromosomes. In *The Cell Nucleus*, Vol. II, ed. H. Busch. Academic Press, New York, pp. 123–147.

Mensinkai, S. W. 1939. Cytological studies in the genus *Gladiolus. Cytologia* **10**:59–72.

Moyne, G., J. Garrido, and W. Bernhard. 1974. Localisation ultrastructurale de ribonucléoprotéines au niveau chromosomes mitotiques. *C. R. Acad. Sc. Paris* D **278**:1385–1388.

Noel, J. S., W. C. Dewey, J. H. Abel, Jr., and R. P. Thompson. 1971. Ultrastructure of the nucleolus during the Chinese hamster cell cycle. *J. Cell Biol.* **49**:830–847.

Papsidero, L. D. and J. P. Braselton. 1973. Ultrastructural localization of ribonunucleoprotein on mitotic chromosomes of *Cyperus alternifolius. Cytobiologie* **8**:118–129.

Pathak, G. N. 1940. Studies in the cytology of *Crocus. Ann. Bot.* **4**:227–256.

Perry, R. P. 1969. Nucleoli: The cellular sites of ribosome production. In *Handbook of Molecular Cytology*, ed. A. Lima-de-Faria. American Elsevier Publishing Company, Inc., New York, pp. 620–636.

Phillips, S. G. 1972. Repopulation of the postmitotic nucleolus by preformed RNA. *J. Cell Biol.* **53**:611–623.

Phillips, D. M. and S. G. Phillips. 1973. Repopulation of postmitotic nucleoli by preformed RNA. II. Ultrastructure. *J. Cell Biol.* **58**:54–63.

Pickett-Heaps, J. D. 1967. Ultrastructure and differentiation in *Chara*. II. Mitosis. *Aust. J. Biol. Sci.* **20**:883–894.

Pickett-Heaps, J. D. 1970. The behavior of the nucleolus during mitosis in plants. *Cytobios* **6**:69–78.

Ramanujam, S. 1938. Cytogenetical studies in the Oryzeae. I. Chromosome studies in the Oryzeae. *Ann. Bot.* **2**:107–125.

Rattenbury, J. A. 1952. Specific staining of nucleolar substance with aceto-carmine. *Stain Tech.* **27**:113–124.

Stevens, A. R. and D. M. Prescott. 1971. Reformation of nucleolus-like bodies in the absence of postmitotic RNA synthesis. *J. Cell Biol.* **48**:443–454.

Stockert, J. C., M. E. Fernández-Gómez, G. Giménez-Martín, and J. F. López-Sáez.

1970. Organization of argyrophilic nucleolar material throughout the division cycle of meristematic cells. *Protoplasma* **69**:265–278.

Taylor, J. H. 1960. Nucleic acid synthesis in relation to the cell division cycle. *Ann. N. Y. Acad. Sci.* **90**:409–421.

Tjio, J. H. 1948. Notes on nucleolar conditions in *Ceiba pentandra. Hereditas* **34**: 204–208.

Underbrink, A. G., A. H. Sparrow, A. F. Rogers, and V. Pond. 1967. Observations on the cytology and fine structure of mitosis in the fern, *Ophioglossum petiolatum* Hook. *Cytologia* **32**:489–499.

Wager, H. 1904. The nucleolus and nuclear division in the root apex of *Phaseolus. Ann. Bot.* **18**:29–55.

Williams, E., J. Heslop-Harrison, and H. G. Dickinson. 1973. The activity of the nucleolus organizing region and the origin of cytoplasmic nucleoloids in meiocytes of *Lilium. Protoplasma* **77**:79–93.

Woodward, J., E. Rasch, and H. Swift. 1961. Nucleic acid and protein metabolism during the mitotic cycle in *Vicia faba. J. Biophys. Biochem. Cytol.* **9**:445–462.

Zirkle, C. 1928. Nucleolus in root tip mitosis in *Zea mays. Bot. Gaz.* **86**:402–418.

Membranes in the Spindle Apparatus: Their Possible Role in the Control of Microtubule Assembly

Peter K. Hepler   Stanford University

INTRODUCTION

The study of the formation and organization of spindle microtubules in dividing cells of higher plants has focused largely on the sites of initiation, especially the kinetochore and phragmoplast and to a lesser degree the spindle pole (Bajer and Molė-Bajer 1972; Pickett-Heaps 1969; 1974). Much less attention has been devoted to the conditions necessary, and the structures involved in establishing those conditions, for the formation of the microtubules upon these initiating sites. It is now known that particular levels of divalent cations (Ca^{++}/Mg^{++}) are required for the assembly of microtubules *in vitro* (Borisy et al. 1975; Hepler and Palevitz 1974; Weisenberg 1972). It is also becoming apparent that membranous elements constitute a common component of the spindle apparatus in dividing plant (Hepler and Palevitz 1974) and animal cells (Harris 1975). The idea is developed in this chapter that the membranes are the cellular inclusions which regulate the levels of divalent cations *in vivo* and thus participate in an important way to control the formation of spindle microtubules.

MEMBRANES IN THE SPINDLE APPARATUS

The classic work of Porter and Machado (1960) first made us aware of the presence of the endoplasmic reticulum (ER) in the spindle apparatus. Subsequent work

212

of others has shown this phenomenon to occur in many different species and has provided convincing visual evidence that the ER is aggregated in the region of the spindle pole. Conspicuous ER-spindle associations have been reported in root tip cells of *Hyacinthus, Tradescantia* (Sakai 1969a), *Allium* (Hanzely and Schjeide 1973) and *Triticum* (Burgess and Northcote 1968), mesophyll cells of *Nicotiana* (Esau and Gill 1969), pollen mother cells of *Tradescantia* (Wilson 1970) and *Trillium* (Sakai 1969b), and symmetrically and asymmetrically dividing root epidermal cells of *Hydrocharis* (Cutter and Hung 1972). The ER aggregations are especially prominent in pollen mother cells of *Tradescantia* (Wilson 1970) and *Trillium* (Sakai 1969b) and in the centrifuged root tip cells of *Triticum* (Burgess and Northcote 1968).

Studies in my laboratory have confirmed the findings for dividing root tip cells of *Allium* and have enlarged the list to include pollen mother cells of *Lilium*, root tip cells of *Phleum*, epidermal cells of the stomatal complex, especially asymmetrically dividing subsidiary cells of *Hordeum*, and spermatogenous cells of *Marsilea*. In each case conspicuous clusters of the endoplasmic reticulum are observed in the region of the spindle pole. An example of a symmetrically dividing epidermal cell of *Hordeum* shown in Figure 8-1 reveals dilated cisternae of the ER around almost the entire spindle area. Here, as in other examples, the ER invading the spindle region is largely smooth while towards the peripheral cytoplasm elements of rough endoplasmic reticulum (RER) are evident. The clusters of ER appear much more concentrated in dividing subsidiary cells, where the spindle forms in a cytoplasmic bridge extending between the two side walls of the lateral epidermal cell (Figures 8-2 and 8-3). The spindle apparatus is displaced asymmetrically, being about twice as close to the side wall, common with the guard mother cell, the proximal wall, as it is to the opposite or distal wall (Figures 8-2 and 8-3). ER accumulations are prominent at both poles and in particular at the distal pole they appear to be composed of layers of parallel curved lamellae with their concave surfaces directed towards the spindle (Figures 8-2-8-4). At higher magnification it is frequently observed that membranous elements extend specifically along the kinetochore microtubules (Figures 8-5 and 8-6), often covering the entire distance between the chromosome and the pole (Figure 8-6). Detailed examination reveals that microtubules and membranes are closely apposed to one another and a structural interaction such as crossbridging seems possible (Figures 8-5 and 8-6). Dictyosomes are also observed close to the spindle apparatus and it appears from the micrographs that their vesicles may comprise a part of the membranous elements present (Figures 8-3 and 8-5).

The observations on dividing spermatogenous cells of the water fern *Marsilea* support the above results and extend them to non-flowering vascular plants, in particular to those specific cells in which there is a transformation in spindle structure from anastral to astral, with the formation *de novo* of a blepharoplast, a procentriole-containing spindle pole body (Hepler, 1976). In *Marsilea*, hydrated microspores proceed through a series of nine cell divisions followed by a differentiation phase to yield 32 motile sperm. During the differentiation phase 100-150 basal bodies per cell are required for the growth of the flagella which constitute the motile apparatus for the sperm. The blepharoplast is the source of these basal bodies,

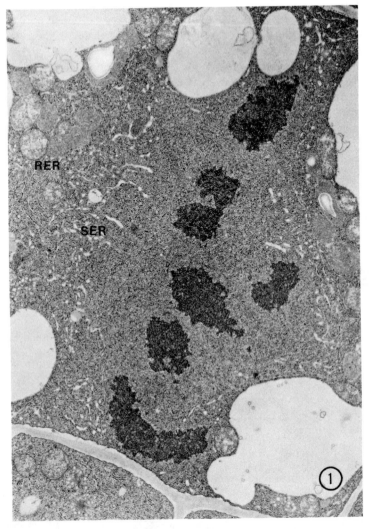

Figure 8-1.
Metaphase in a symmetrically dividing lateral epidermal cell of
Hordeum. The entire spindle region is surrounded and partially
invaded by dilated cisternae of smooth endoplasmic reticulum
(SER). Towards the peripheral cytoplasm, especially on the left side,
elements of rough endoplasmic reticulum (RER) are evident. X
11,000

but it forms prior to differentiation in the division phase and appears to function
initially as a spindle pole body for the last (ninth) division. During early prophase
of the ninth division two young blepharoplasts, which form as a double structure
in telophase of the previous eighth division, separate from one another and move to
opposite poles as spindle tubules grow between them (Hepler, 1976). It is especially

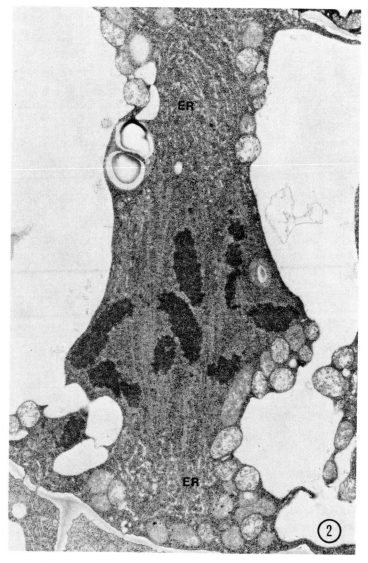

Figure 8-2.
Metaphase in an asymmetrically dividing subsidiary cell of *Hordeum.*
Massive accumulations of ER occur at both pole regions. × 11,000

noticeable at late prophase of the ninth division that they are the focal points for the converging tubules of the spindle apparatus (Figure 8-7). They remain at or close to the poles during the rest of mitosis although the spindle apparatus when fully formed at metaphase is not sharply focused as it had been in prophase (Figure 8-8). Membrane elements form a prominent part of the mitotic apparatus and can be observed at the poles and throughout the spindle (Figures 8-8 and 8-10).

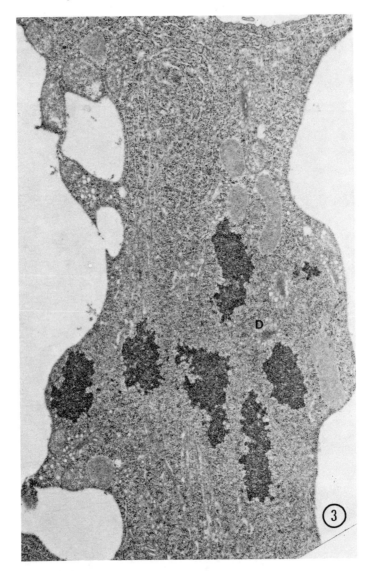

Figure 8–3.
Metaphase in an asymmetrically dividing subsidiary cell of *Hordeum*. The ER of the distal pole (upper portion) forms a large "U" shaped swirl in which the concave side faces the spindle region. The distal pole is shown at higher magnification in Figure 8-4, while a portion of the proximal half spindle is shown in Figure 8-5. Dictyosomes (D) appear within the edge of the spindle region. X 15,000

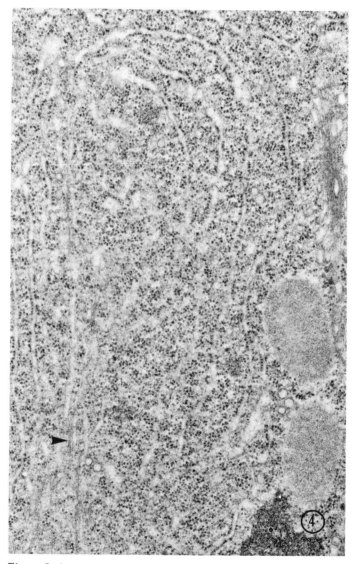

Figure 8-4.
A higher magnification view of the ER at the distal pole of the cell
shown in Figure 8-3. Much of the ER is rough but where it inter-
mingles with microtubules (lower left) it is usually smooth. A close
association of membrane and microtubule is indicated by the arrow
head. × 38,000

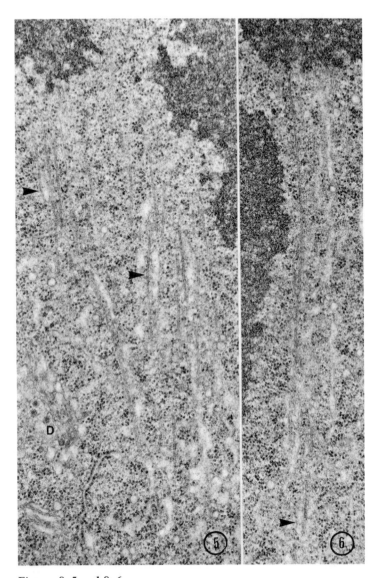

Figures 8-5 and 8-6.
Selected area showing the close association of membranes with the spindle microtubules. Although the profiles of the membranes are not markedly distinct, their presence can be easily determined by their lightly stained, dilated cisternal space. Most of the microtubules shown, especially those on the left side of Figure 8-5 and all of those in Figure 8-6, are thought to be kinetochore tubules. While close membrane-microtubule associations occur frequently, some which are particularly clear are indicated by the arrow heads. ✕ 38,000

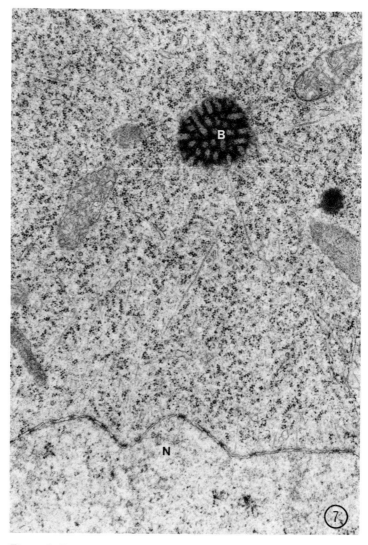

Figure 8-7.
Prophase of the 9th division in a microspore of *Marsilea*. The
blepharoplast (B), a densely stained spherical organelle, is the focal
point for radiating microtubules, many of which impinge upon the
nucleus (N) shown in lower portion of the micrograph. X 30,000

In the preceding (eighth) division no blepharoplast is present and thus it is of considerable interest to compare and contrast its spindle structure with that of the
ninth division. The eighth division spindle is characterized by the presence of large
accumulations of predominantly smooth ER at the spindle pole (Figures 8-9 and
8-11). These membranes also extend throughout the spindle structure, but they

Figure 8-8.
Metaphase of the 9th division in *Marsilea*. Although the blepharoplast is in the region of the pole, the microtubules are not focused
towards it and the spindle, thus, has lost its astral configuration.
Compare Figures 8-7 and 8-8. Membranes can be observed among
the microtubules throughout the spindle apparatus. Chromosomes
(C) always appear lightly stained in *Marsilea*. The outlined area is
shown at higher magnification in Figure 8-10. × 21,000

Figure 8-9.
Metaphase of the 8th division in *Marsilea*. No blepharoplast is present. Membranes occur within the spindle region, and arms of extended metaphase chromosomes (C) are evident. The boundary between spindle and non-spindle cytoplasm can be observed in the upper portion of the micrograph. The spindle pole contains membranes, whereas the non-spindle cytoplasm contains organelles such as dictyosomes (D) and is richer in helical polyribosomes. The outlined area is shown at higher magnification in Figure 8-11. X 21,000

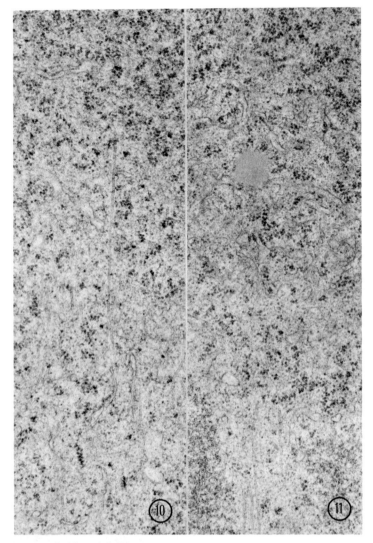

Figures 8-10 and 8-11.
Enlargements of portions of the spindle regions shown respectively
in Figures 8-8 and 8-9. Ninth (Figure 8-10) and 8th (Figure 8-11)
division spindles are similar and both contain numerous elements
of ER, most of which are smooth. X 41,000

end at the poles, and very often a demarcation can be observed at the interface of
the SER-rich spindle pole and the adjacent helical polysome-rich, non-spindle
cytoplasm (Figures 8-9 and 8-11). In many respects the eighth and ninth division
spindles resemble each other, even though the ninth division begins at prophase as
an astral structure with microtubules emanating directly from the region immediately around the blepharoplast (Figure 8-7). In the ninth division, by the time the

nuclear envelope disperses, the pole has defocused and many microtubules are not directed towards the belpharoplast. The role of the blepharoplast in the ninth division thus becomes obscure, and although it serves as the focal point for tubules initially during spindle formation, it loses this property by metaphase. Whether it has an active role in spindle formation cannot be determined yet but its failure to maintain a focused spindle throughout division suggests that factors other than itself are active participants in the maintenance of spindle organization. The presence of membranous elements at the poles denotes an important structural similarity between eighth and ninth division spindles, and it may be these inclusions which ultimately control spindle morphogenesis.

Astral divisions in lower plants, especially the algae and fungi, and in various animals cells have received considerable attention in ultrastructural studies (Hepler and Palevitz 1974; Pickett-Heaps 1974, for reviews). While the emphasis has in the past been primarily directed at the spindle pole body and the microtubules, it is quite apparent from some of these studies that membranes of the ER and the nuclear envelope (NE) frequently surround or invade the spindle region (Marchant and Pickett-Heaps 1973; Pickett-Heaps 1972). In some examples where the nuclear envelope remains intact, except for polar fenestrations, it is noteworthy that extra layers of ER, called the perinuclear envelope, become organized so that they closely surround the nuclear envelope during mitosis (McDonald 1972; Pickett-Heaps 1970).

Marked accumulations of membranes have, in addition, been reported in the mitotic apparatus of dividing animal cells (Harris 1975, for review). In sea urchins, especially when fixed at acid pH with osmium tetroxide alone, the aster consists predominantly of radial arrays of vesicular membrane elements which extend in linear rows along the microtubules. Spindle poles may contain numerous vesicular membranes around the centriole (Harris 1975). Elements of SER have also been observed to change their organization in cells in relation to the different phases of mitosis. Interphase rat hepatic cells have a random array of SER, whereas during prometaphase these membranes become associated with the centriole (Dougherty and Lee 1967). Mouse spermatocytes also possess smooth membranes which change during mitosis and become conspicuously associated with the spindle apparatus (Pleshkewych and Levine 1975).

MEMBRANES DURING SPINDLE FORMATION IN PLANTS

In trying to assess the importance and function of the membranous components of the spindle apparatus it is of value to consider how the spindle and in particular the spindle poles form during prophase. Studies on living, dividing endosperm cells of the African blood lily, *Haemanthus*, with the polarizing light microscope, have revealed that the birefringent spindle elements appear first in a region called the clear zone immediately adjacent to the nuclear envelope (NE) during prophase (Inoué and Bajer 1961). Before the NE disperses the poles become apparent as the focused ends of the birefringent material. Studies by Bajer and Molè-Bajer (1969) show that the spindle may be multipolar during prophase but that evenually two poles emerge as the spindle continues its formation during prometaphase and metaphase. Electron microscopic examination of these cells shows that the tubules first form in a random pattern (Figure 8-12) without a defined pole and later become aligned parallel to one another, forming a sheath around the nucleus (Figure 8-13) at prometaphase. Thus, in *Haemanthus* endosperm cells, the appearance of the poles may be a secondary phenomenon resulting from, rather than producing, the alignment of the perinuclear spindle tubules.

In other types of the cells the clear zone is not strikingly evident during spindle formation. Esau and Gill (1969), examining tobacco mesophyll cells find, in early prophase at a time when the equatorial or preprophase band of tubules is still

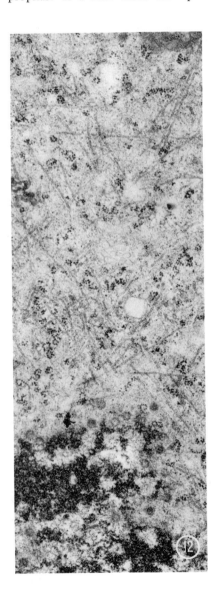

Figure 8-12.
Prophase in an endosperm cell of *Haemanthus*. The section from which this micrograph was made passed tangentially through the edge of the nucleus (lower portion of figure) into the adjacent clear zone, revealing the randomly oriented microtubules of the forming spindle apparatus. X 25,000

present, that the spindle poles appear as discrete aggregates of microtubules and curved elements of SER, referred to as "polar caps." In prometaphase the fragmented nuclear envelope further contributes to the accumulating membranes at the pole. Hepler and Zeiger (unpublished) also observe similar "polar caps" in dividing stomatal subsidiary cells of barley (Figure 8-14). Especially on the side of the asymmetrically dividing nucleus distal to the guard mother cell where the spindle is appreciably focused, we observe at prophase aggregations of vesicular elements at the point where the microtubules converge (Figure 8-14). Microtubules are also found extending along the nucleus immediately adjacent to the nuclear

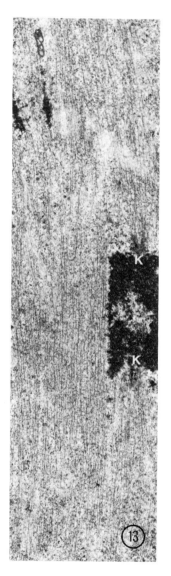

Figure 8-13.
Prometaphase in *Haemanthus*. The section from which this micrograph was photographed occurred in the plane tangential to the surface of the spindle apparatus. Microtubules of the continuous spindle have become parallel to one another and are beginning to invade the region occupied by the chromosomes and intermingle with the kinetochore (K) tubules. × 12,000

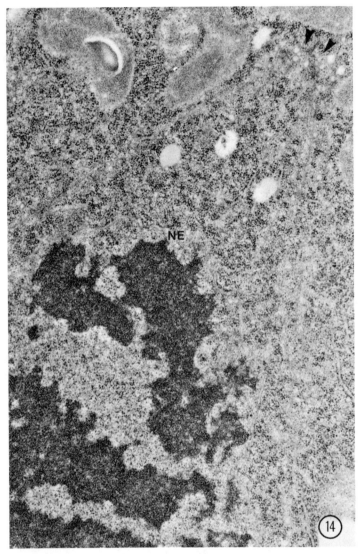

Figure 8-14.
Late prophase in a subsidiary cell of *Hordeum*. The forming spindle pole, indicated by two arrowheads, consists of focused microtubules, together with a small cluster of membrane vesicles. The nuclear envelope (NE) is still intact over portions of the nucleus but it appears to be in the process of breaking down. X 29,000

envelope. In these examples both the nuclear envelope and a membrane rich polar region may participate in spindle formation.

Taken together, the observations from dividing cells which are mitotic or meiotic, anastral or astral, plant or animal show that membranes constitute a common, perhaps universal, component of the spindle apparatus. Furthermore, their specific structural associations with the spindle fibers themselves or their transformations associated with the progression of mitotic events suggest that they may play an important functional role in the formation and/or activity of the mitotic apparatus. While most reports single out the SER as the most ubiquitous component in the spindle it is evident in plants, at least, that the NE, and possibly the Golgi vesicles both comprise a portion of the membrane elements.

FUNCTION OF MEMBRANES IN THE SPINDLE

If membranes of the ER-NE are closely associated with the mitotic apparatus what role might they play in the formation and/or function of the spindle? The membrane-spindle associations may only be fortuitous and it has been suggested that the tubules are the elements which give order to the lamellae of ER. This idea seems unlikely since membrane accumulations appear in some spindle poles, e.g., *Nicotiana* mesophyll cells and *Hordeum* subsidiary cells, at the beginning of spindle formation when there would appear to be too few microtubules to exert a membrane ordering. Burgess and Northcote (1968) favor the idea that the ER serves as a system which transports the tubule subunits to different aggregation sites and in this manner participates in the polymerization process.

Alternatively it has been suggested that the ER functions in a structural capacity by providing an anchor against which the spindle can pull or push (Hanzely and Schjeide 1973). The observations in *Hordeum* (Figures 8-5 and 8-6) which show a close interaction between the kinetochore tubules and the ER might be interpreted as a structural attachment, possibly a crossbridging of microtubules and membranes, or of microtubules with a microfibrillar, "actin" component which in turn is inserted on the ER membranes. The newly emerging results of fluorescent actin antibody (Cande et al. 1975) and fluorescent HMM (Sanger 1975) techniques that indicate the presence of large amounts of actin in the spindle apparatus, especially at the poles, forces us to reexamine the ultrastructure of the spindle and to consider the disposition of presumptive actin fibrils which heretofore have probably been largely obscured or destroyed by the preparative procedures. We simply note at this point that the distribution of ER corresponds closely to the recently observed pattern of actin and therefore that there might be a functional relationship.

Foremost in our thinking about ER-NE function, however, is the idea that these membranes act like the sarcoplasmic reticulum (SR) of muscle to control the Ca^{++}/Mg^{++} ionic milieu of the spindle and thus regulate the polymerization and depolymerization of the spindle microtubules. This idea, which is developed below, has been suggested earlier by Hepler and Palevitz (1974), in brief form, and has been presented more completely in a recent, thoughtful study by Harris (1975) on membranes in the spindle of sea urchins. It is now well established that the divalent cations, Ca^{++} and Mg^{++}, play a crucial role in the control *in vitro* of microtubule assembly (Borisy et al. 1975; Weisenberg 1972), and it seems likely that they also participate in spindle tubule assembly *in vivo*. The ER-NE, because of its proximity to the tubules and the presence of its compartmentalized cisternal space, becomes a candidate for the system which regulates the cytoplasmic levels of these ions. In muscle cells it is well established that the SR, the specialized ER of those cells, contains the necessary proteins, a Ca^{++}-Mg^{++} ATPase and an acid protein "calseques-trin," which respectively pump Ca^{++} ions into the cisternal space of the membrane and retain them through ioinic interaction (Hasselbach 1972). Upon stimulation by the nerve, the SR releases Ca^{++} ions to the muscle cell cytoplasm raising their concentration and permitting interaction between actin and myosin and hense, muscle contraction.

By comparison, it seems possible that the membranes associated with the spindle apparatus contain a Ca^{++}-Mg^{++} ATPase which is capable of pumping Ca^{++} or Mg^{++} into their cisternal space. Given the proper stimulus, these membrane proteins might regulate the ionic levels in accord with the sequential events of the cell cycle and thus control microtubule polymerization and depolymerization. Although the ionic conditions for microtubule polymerization are still under intensive investigation, some recent conclusions have influenced our current thinking. We know that there is an absolute requirement for a divalent cation (Borisy et al. 1975). Mg^{++} (10^{-3} M) is commonly used to stimulate polymerization, whereas Ca^{++} (10^{-3} M) is known to prevent it. In the absence of Mg^{++}, however, small amounts of Ca^{++} (10^{-6} M) will stimulate polymerization (Borisy et al. 1975). Hepler and Palevitz (1974) and others (Harris 1975; Timourian et al. 1974) have suggested that the control for spindle formation may be the result of regulating Ca^{++} levels around 10^{-3} M which, on the higher side, would depolymerize microtubules. However, this possibility seems unlikely since millimolar concentrations of Ca^{++} probably do not exist under physiological conditions within the cell (Borisy et al. 1975). A more reasonable possiblity is that Ca^{++} ions are controlled at the micromolar (10^{-6}) level as they are in the SR of muscle, and that spindle tubule formation is stimulated by increasing the ion concentration (e.g., up to 1×10^{-5} M). In either case it is suggested that the membranes by some mechanism, yet to be explained, respond to the progressive events of the cell cycle and at the proper time (prophase) shift the ion level in such a way that spindle tubules can assemble.

Creating the ionic milieu may only constitute a part of the problem of spindle regulation. Microtubules in dividing cells tend to form at specific sites, for example, at the kinetochore. It is much easier to envision the endoplasmic reticulum as that component creating the proper ionic conditions for microtubles which will shift the equilibrium to polymerization. It is more difficult, however, to envision what it is at the spindle pole to which or from which the microtubules are directed. Inasmuch as the microtubules tend to mutually align, it may only be important in those cells like *Haemanthus* endosperm which do not divide along a predetermined axis to establish the ionic milieu and to then allow the spindle to form in whatever direction it may. However, in cells and spindles where the divisions are polarized, it seems essential that there be some mechanism to establish the pole. Even a localized

ER may be insufficient since due to rapid rates of diffusion it might be impossible to create a localized gradient in cytoplasmic concentration of free Ca^{++}/Mg^{++} ions. Possibly the ER, in addition to regulating Ca^{++}/Mg^{++}, may also contain a microtubule initiating factor or seed, or alternatively maybe the ER, as suggested by Burgess and Northcote (1968), shuttles the tubulin in its cisternal space to specific sites where the polymerization subsequently occurs. Under these conditions microtubule formation could be controlled by the combination of factors, including the proper ionic milieu and a sufficient concentration of tubulin. This aspect of the problem, which is so crucial to cell polarity and cytomorphogenesis in plants, unfortunately eludes us and its clarification awaits a better understanding of the composition and cellular location of the factors controlling microtubule polymerization.

While the postulated ER control of Ca^{++}/Mg^{++} concentration remains highly speculative some important evidence beyond the visualization of the membranes themselves has emerged which supports the idea. Mazia et al. (1972) have identified a Ca^{++} ATPase in dividing sea urchin eggs and have shown the enzyme to be preferentially localized in the mitotic apparatus. Petzelt (1972a,b) has further revealed that the enzyme activity fluctuates during the mitotic cycle with peaks at early prophase when chromosome condensation begins and again at metaphase. Mazia et al. (1972) are cautious in their analysis of these findings since a Ca^{++} ATPase might well be involved in a number of spindle related events including chromosome condensation, the action of the mechanochemical tubules cross bridge, and the control of tubule polymerization and spindle assembly. In a more recent study on the Ca^{++} ATPase in parthenogenetically activated sea urchin eggs, Petzelt and von Ledebur-Villiger (1973) consider the close correlation in appearance of enzyme activity with the time of spindle formation, especially in monasters where no subsequent chromosome motion occurs, and suggest, in general, agreement with the thesis presented in this chapter and by Harris (1975) that the enzyme participates in the control of spindle microtubule assembly by regulating the Ca^{++} ion levels.

Additional studies using the electron microprobe on dividing sea urchin eggs have shown that the Ca^{++} ions themselves are also preferentially localized in the spindle apparatus with an indication that they are even more concentrated in the region of the pole (Timourian et al. 1974). Support for a role of the ER in Ca^{++}/Mg^{++} ion regulation can also be found in the ciliate *Spirostomum* which undergoes extensive contraction of its cell body. At, or immediately preceding contraction, free Ca^{++} around 10^{-6} M, is released presumably from a vesicular membrane system and is resequestered there upon relaxation (Ettienne 1970).

CONCLUSIONS

By suggesting that the ER controls the ionic conditions necessary for spindle formation, we are attempting to integrate our current knowledge of the role of divalent cations for *in vitro* microtubule polymerization with the increasing number of observations which single out membranous elements as common components of the mitotic apparatus. It is recognized that factors basides the ionic levels, for example, the high molecular weight protein associated with tubulin (Burns and Pollard 1974; Gaskin et al. 1974) and the recently discovered tau factor (Weingarten et al. 1975) may be present in the spindle apparatus and participate in the regulation of spindle formation. Nevertheless, it seems highly probable that divalent cations control some aspects of tubule polymerization *in vivo*.

A number of different investigations may now be made to experimentally determine the role of the ER, and of Ca^{++}/Mg^{++}. For example, it will be important to cytochemically localize Ca^{++} within the cell, especially at different phases of the cell cycle. Cell fractionation might also prove valuable, especially in more carefully identifying the fraction to which the spindle Ca^{++} ATPase described by Petzelt (1972a) and Mazia et al. (1972) is associated. Finally, the use of ionophores, e.g., A23187, which is already known to stimulate lymphoytes into mitosis (Luckasen et al. 1974) and chelators, e.g., EGTA-EDTA, to experimentally alter the Ca^{++}/Mg^{++} levels within dividing cells and thereby possibly disturb the polymerization of tubules might bring us closer to an understanding of spindle formation in the cell.

ACKNOWLEDGMENTS

I thank Dr. Patricia Harris, University of Oregon, for discussing some of the above topics with me and for making available to me a preprint of her paper on this subject. I also thank my colleagues at Stanford University for helpful criticisms. This work was supported by grant number BMS 74-15245 from the National Science Foundation.

LITERATURE CITED

Bajer, A. and J. Molè-Bajer. 1969. Formation of spindle fibers, kinetochore orientation, and behavior of the nuclear envelope during mitosis in endosperm. *Chromosoma* 27:448–484.

Bajer, A. and J. Molè-Bajer. 1972. Spindle dynamics and chromosome movements. *Inter. Rev. Cytol.* suppl. 3.

Borisy, G. G., J. M. Marcum, J. B. Olmsted, D. B. Murphy, and K. Johnson. 1975. Purification of tubulin and associated high molecular weight proteins from porcine brain and characterization of microtubule assembly *in vitro. Ann. N. Y. Acad. Sci.* 253:107–132.

Burgess, J. and D. H. Northcote. 1968. The relationship between the endoplasmic reticulum and microtubular aggregations and disaggregation. *Planta* 80:1–14.

Burns, R. C. and T. D. Pollard. 1974. A dynein like protein from brain. *FEBS Letters* 40:274–280.

Cande, W. Z., E. Lazarides, and J. R. McIntosh. 1975. Visualization of actin in functional lysed mitotic cell preparations. *J. Cell Biol.* (abstracts).

Cutter, E. G. and C-Y. Hung. 1972. Symmetric and asymmetric mitosis and cytokinesis in the root tip of *Hydrocharis morsus-ranae* L. *J. Cell Sci.* 11:723–737.

Dougherty, W. J. and M. McN. Lee. 1967. Light and electron microscope studies of smooth endoplasmic reticulum in dividing rat hepatic cells. *J. Ultrastruct. Res.* 19:200–220.

Esau, K. and R. H. Gill. 1969. Structural relations between nucleus and cytoplasm during mitosis in *Nicotiana tabacum* mesophyll. *Can. J. Bot.* 47:581–591.

Ettienne, E. M. 1970. Control of contractility in *Spirostomum* by dissociated calcium ions. *J. Gen. Physiol.* **56**:168–179.

Gaskin, F., S. B. Kramer, C. R. Cantor, R. Adelstein, and M. L. Shelanski. 1974. A dynein like protein associated with neurotubules. *FEBS Letters* **40**:281–286.

Hanzely, L. and O. A. Schjeide. 1973. Structural and functional aspects of the anastral mitotic spindle in *Allium* sativum root tip cells. *Cytobios* **7**:147–162.

Harris, P. 1975. The role of membranes in the organization of the mitotic apparatus. *Exper. Cell Res.* **94**:409–425.

Hasselbach, W. 1972. The sarcoplasmic calcium pump. In *Molecular Bioenergetics and Macromolecular Biochemistry*, ed. H. H. Weber. Springer-Verlag, Berlin, Heidelberg, New York, pp. 149–171.

Hepler, P. K. 1976. The blepharoplast of *Marsilea*: its *de novo* formation and spindle association. *J. Cell Sci.* **21**:361–390.

Hepler, P. K. and B. A. Palevitz. 1974. Microtubules and microfilaments. *Ann. Rev. Plant Physiol.* **25**:309–362.

Inoué, S. and A. Bajer. 1961. Birefringence in endosperm mitosis. *Chromosoma* **12**:48–63.

Luckasen, J. R., J. G. White, and J. H. Kersey. 1974. Mitogenic properties of a calcium ionophore, A23187. *Proc. Nat. Acad. Sci.* **71**:5088–5090.

Marchant, H. J. and J. D. Pickett-Heaps. 1973. Mitosis and cytokinesis in *Coleochaete scutata*. *J. Phycol.* **9**:461–471.

Mazia, D., C. Petzelt, R. O. Williams, and I. Meza. 1972. A Ca^{2+}-activated ATPase in the mitotic apparatus of the sea urchin egg (isolated by a new method). *Exp. Cell Res.* **70**:325–332.

McDonald, K. 1972. The ultrastructure of mitosis in the marine red alga *Membranoptera platyphylla*. *J. Phycol.* **8**:156–166.

Petzelt, C. 1972a. Ca^{2+}-activated ATPase during the cell cycle of the sea urchin *Strongylocentrotus purpuratus*. *Exper. Cell Res.* **70**:333–339.

Petzelt, C. 1972b. Further evidence that a Ca^{2+}-activated ATPase is connected with the cell cycle. *Exper. Cell Res.* **74**:156–162.

Petzelt, C. and M. von Ledebur-Villiger. 1973. Ca^{2+}-stimulated ATPase during the early development of parthenogenetically activated eggs of the sea urchin *Paracentrotus lividus*. *Exper. Cell Res.* **81**:87–94.

Pickett-Heaps, J. D. 1969. The evolution of the mitotic apparatus: an attempt at comparative ultrastructural cytology in dividing plant cells. *Cytobios* **3**:257–280.

Pickett-Heaps, J. D. 1970. Mitosis and autospore formation in the green alga *Kirchneriella lunaris*. *Protoplasma* **70**:325–347.

Pickett-Heaps, J. D. 1972. Cell division in *Cosmarium botrytis*. *J. Phycol.* **8**:343–360.

Pickett-Heaps, J. D. 1974. Plant microtubules. In *Dynamic Aspects of Plant Ultrastructure*, ed. A. W. Robards. McGraw-Hill, London, pp. 219–255.

Pleshkewych, A. and L. Levine. 1975. Phase-contrast and electron-microscopic observations on a membranous complex in primary spermatocytes of the mouse. *J. Cell Sci.* **18**:1–17.

Porter, K. R. and R. D. Machado. 1960. Studies on the endoplasmic reticulum. IV. Its form and distribution during mitosis in cells of onion root tip. *J. Biophys. Biochem. Cytol.* **7**:167–180.

Sakai, A. 1969a. Electron microscopy of dividing cells. II. Microtubules and formation of the spindle in root tip cells of higher plants. *Cytologia* **34**:57–70.

Sakai, A. 1969b. Electron microscopy of dividing cells. III. Mass of microtubules and formation of spindle in pollen mother cells of *Trillium kamtschaticum*. *Cytologia* **34**:593–604.

Sanger, J. W. 1975. Presence of actin during chromosomal movement. *Proc. Nat. Acad. Sci.* **72**:2451–2455.

Timourian, H., M. M. Jotz, and G. E. Clothier. 1974. Intracellular distribution of calcium and phosphorous during the first cell division of the sea urchin egg. *Exper. Cell Res.* **83**:380–386.

Weingarten, M. D., A. H. Lockwood, S-Y. Hwo, and M. W. Kirschner. 1975. A protein factor essential for microtubule assembly. *Proc. Nat. Acad. Sci.* **72**:1858–1862.

Weisenberg, R. C. 1972. Microtubule formation *in vitro* in solution containing low calcium concentrations. *Science* **177**:1104–1105.

Wilson, H. J. 1970. Endoplasmic reticulum and microtubule formation in dividing cells of higher plants—a postulate. *Planta* **94**:184–190.

Interaction of Microtubules and the Mechanism of Chromosome Movements (Zipper Hypothesis) :II. Dynamic Architecture of the Spindle

Andrew S. Bajer

University of Oregon

INTRODUCTION

The crucial tests for any hypothesis lie in its ability to explain already available data, and in the results of experiments specifically designed to test the validity of its predictions. If the hypothesis claims to be universal it should explain a broad spectrum of a selected class of phenomena, both in their normal and abnormal appearance or course. Therefore, any hypothesis must have its limitations. Otherwise it becomes a theoretical description of the event(s) and not an interpretation (example: molecular wind hypothesis; Östergren *et al.* 1960).

The main purpose of this chapter is to draw attention to the significance of the lateral interaction of MTs (microtubules) in chromosome movements. Certain basic, but so far little known, features of the fine structure of the spindle will also be discussed. The role of directional tilt (Couderc 1974) for the analysis of specific features of the spindle fine structure will also be stressed. Finally the limitations and restrictions of the zipper hypothesis will be considered in determining to what extent the zipper hypothesis of chromosome movements (Bajer 1973) meets the criteria of a general hypothesis.

THE ZIPPER HYPOTHESIS

Detailed observations of the rearrangement of the MTs of kinetochore fibers and of the interzonal region during anaphase in endosperm of *Haemanthus katherinae*

233

(see review, Bajer and Malè-Bajer 1972) and coracoid bone tissue culture of the newt *Taricha granulosa granulosa* (Molè-Bajer 1976) led to the development of the zipper hypothesis (Bajer 1973). The data are consistent with the assumption that a specific type of interaction, namely zipping between MTs, supplies the motive force for chromosome movements. Thus, the force is used for the specific elastic deformation of the MTs (bending), and is responsible for the tension within the spindle (p. 249). This assumption should be distinguished from the postulate concerning another type of elastic deformation, i.e., stretching of the MTs (p. 000).

The spindle is formed (assembled) during prophase-prometaphase and is gradually destroyed (disassembled) during anaphase-telophase. Therefore, assembly-disassembly is another important process superimposed on any kind of MTs interaction. Understanding of the functional spindle is impossible without more detailed information concerning assembly-disassembly of MTs.

Only two basic assumptions concerning MTs, in addition to zipping and assembly-disassembly, are needed to account for the many complex types of chromosome movement explained in terms of MT rearrangement. These are:

1. MTs of the spindle neither contract nor stretch. Thus no significant elastic deformation (stretching) occurs during the pull.

2. MTs are anchored by zipping to other MTs and/or by frictional resistance of the spindle matrix. It is assumed that the degree of anchoring of different MTs varies considerably.

The assumption concerning elastic deformation (stretching) applies to any interpretation involving MTs as the force transmitting elements, and will not be further discussed. On the other hand, zipping and anchoring of MTs has several implications, such as overall internal spindle tension, and should be considered in more depth.

LATERAL INTERACTION, ZIPPING, BREAKAGE, AND DISASSEMBLY OF MTs

Several types of lateral interaction or zipping have been mentioned (Bajer 1973), the terms being interchangeable. To simplify further discussion, the terms "lateral interaction" and "zipping" will be defined.

Lateral interaction is defined as sequential and progressive interaction between neighboring MTs, in which no longitudinal shift (sliding) of the interacting MTs occurs. There are several possible types of lateral interaction. For instance, tip interaction (tip zipping–Bajer 1973), in which one or both MTs disassemble during or after zipping, may resemble the interpretation of chromosome movement based on assembly-disassembly (Inoué and Sato 1967; Inoué and Ritter 1975; Inoué et al 1975; Dietz 1972).

"Parallel zipping" is the coming together of two MTs which bond laterally and progressively along their length in such a way that the two ends come closer together. This type of zipping is therefore a specific, more restricted, case of lateral interaction. It is comparatively simple geometrically (Figure 9-1), but certainly not on the molecular level.

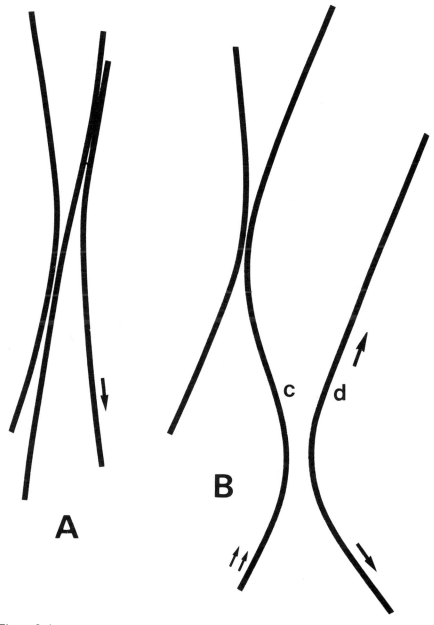

Figure 9-1.
Parallel (A) and touch-and-go zipping. It is assumed that during parallel zipping
the bonds are strong and do not break easily during the pull (arrow). Bonds in
touch-and-go zipping are weak and break up during the pull (arrows). However,
before the break of the bond between MTs c and d, MT c is deformed and this
results in a jerk and a short lasting pull (double arrows). It is possible that these
two types of zipping transform into each other.

"Touch and go" zipping can be defined as parallel zipping followed by breakage of the bond between the zipped MTs rather than of the MTs themselves. This type of zipping will result in jerks if the bonds are short lasting (Figure 9-1). If the bonds last for a longer time the result will be tension within the spindle (see p. 249). According to this representation, one MT may zip in several regions simultaneously, without breakage and disassembly. The final result will be the same as in parallel zipping, i.e., increase of divergence of the kinetochore fiber accompanied by a pulling force.

It is obvious that both parallel and touch-and-go zipping would be even more efficient, if followed by sliding over short distances such as observed in flagella (Warner and Satir 1974). However, the zipper hypothesis assumes that zipping alone, without subsequent sliding, is sufficient to account for pull on the kinetochores.

Both types of zipping, parallel-break and parallel touch-and-go, may occur at the same time. It is impossible at present to discriminate between these two types on the basis of the morphology of MT arrangements. Therefore "zipping" in further discussion will refer to these two types. Both types have some experimental support and thus permit predictions which can be further tested by experiments.

After zipping, MTs are approximately parallel, but do not touch each other. Such coupled MTs are separated by a distance approximately equal to twice the radius of the electron transparent halo found around MTs.

More than two MTs can zip together, as is found in the normal developing phragmoplast, and experiments involving chloral hydrate (Molè-Bajer 1967, 1969), or cold treatment (Bajer and Lambert 1976, Bajer et al. 1975). In the normal half-spindles, units of only two coupled MTs are found most often. During metaphase and anaphase the majority of such segments are not parallel to the direction of movement and are abundant at the periphery of the kinetochore fiber.

The main assumption postulates that zipping provides the motive force for chromosome movements. The force is utilized to bend the MTs, i.e., for elastic deformation of MTs. The most likely candidates for the force producing elements are MTs arms, although less specific interaction is possible (Allen 1975). The arms of MTs are involved in a variety of functions ranging from sliding in cilia and flagella (Satir 1965, 1968), complex bidirectional transport in melanophores (Murphy and Tilney 1974) of tentacles of Tokophrya (Rudzinska 1967, 1970; Tucker 1972), and finally in rigid linkages between MTs (Tucker 1974).

In the zipper hypothesis it is postulated that parallel zipped MTs are not involved in the production of the motive force.

Schematic drawings show how zipping may move chromosomes and simultaneously account for elongation of the spindle, increase in divergence of kinetochore MTs, and the gradual decrease in MT number. Thus, the rate of chromosome movement may not depend at all, or only to some extent, on the number of MTs involved in zipping. Once certain speed of the kinetochore is achieved, increase or decrease in the number of MTs may not influence the rate of the movements. Experimental data of Salmon (1975) involving hydrostatic pressure supports such assumptions. The diagrammatic schema (Figure 9-2) may be much closer to the actual situation than initially suspected. Each zipping interaction can move a kinetochore only a minute distance. Each kinetochore fiber must be involved in hundreds, if not thousands, of interactions to account for the overall movement from the equator to the pole or vice versa.

The principle of zipping shows that at least two MTs are needed to move a kinetochore in a certain direction. The question therefore arises how zipping can explain movement in organisms with only one MT per kinetochore (Heath 1974). At least three explanations are possible:

1. In the zipper hypothesis, peripheral MTs of the kinetochore fiber are mainly responsible for the pull. The core MTs of the kinetochore fiber are parallel and they do not contribute to the pull, but to the elongation of the spindle (Figure 9-2). This suggests a possible interpretation. Single MTs are usually found in cells with numerous small chromosomes. The MTs of such chromosomes may be zipped together and represent, in fact, one or more compound kinetochore fibers. In such cases the whole half-spindle may correspond to the single kinetochore fiber of higher organisms. The motive force would be produced at the periphery of such compound fibers.

MTs which are parallel and close to the kinetochore may bend (diverge) in the polar region and zip with other MTs. Such interaction would result in faster and more efficient movements than the divergence of the whole kinetochore fiber at its base, i.e., at the kinetochore. This might also explain the rapid movements in narrow spindles or movements involving thin, long fibers such as are found during prometaphase.

2. In the "touch-and-go" mechanism a single MT is sufficient to supply the force. Such a single MT may oscillate laterally and zip temporarily with different MTs. This would result in minute jerks in various directions.

3. It is quite conceivable, however, that the mechanism of movement is partially or even totally different in lower organisms. This might be reflected, for instance, in the rate of movement. If it varies by an order of magnitude or more, it would seem unlikely that the same mechanisms would be responsible.

In the functional spindle, zipping is responsible for continuous changes of MT arrangement during the whole course of mitosis. Zipping, and therefore the course of mitosis, depends on the ability of MTs (parallel zipping) and/or bonds (touch-and-go zipping) to break and/or disassemble. Otherwise, all MTs satisfying given geometrical conditions would zip, preventing further rearrangements of MTs, and all movement would come to a complete stop. It is postulated therefore that assembly-disassembly of MTs does not give rise to the motive force but is a factor regulating the rate of zipping and thus the rate of movements. This assumption would also explain why the data on birefringence fit well with the rates of chromosome movements (Dietz 1969, 1972; Inoué et al. 1975).

The existing data (summarized by Inoué and Ritter 1975) suggest that assembly-disassembly (dynamic equilibrium) is in fact fully compatible with any interpretation of chromosome movements on the EM level (Rebhun et al. 1974). It should be realized, however, that in certain cases change of chromosome position (movements) can result from disassembly of the spindle (MTs) alone. Thus, for instance,

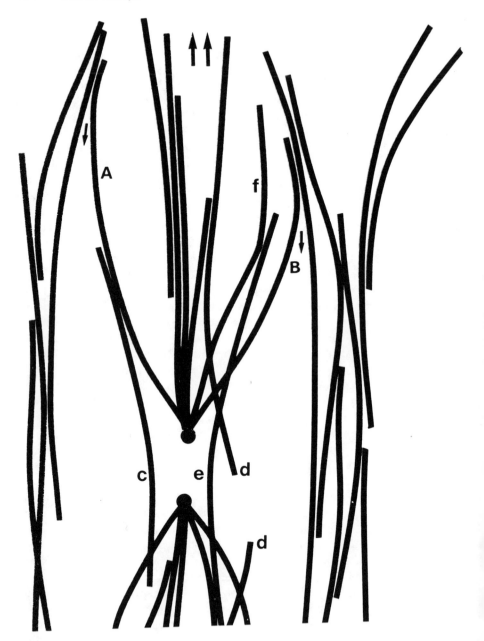

Figure 9-2.
Early anaphase according to the zipper hypothesis. MTs surrounding kinetochore
fiber(s) form a three-dimensional self-supporting "cage." Some of the MTs fan out
due to such interaction (see Figures 9-4 and 9-5). When the MTs (A,B) of the
kinetochore fiber zip progressively toward the plate (thin arrows) with MTs of the
"cage," kinetochores (circles) and the whole kinetochore fiber moves up (double

if the spindle is disrupted by colchicine, chromosomes change their arrangement in a disorderly fashion (Inoué and Ritter 1975). Such movements, however, do not even remotely resemble the movement associated with kinetochores and occur at the time of rapid disintegration of the spindle. Numerous MTs of the spindle and asters may still be present at the time of such movements. It is, therefore, questionable whether the mechanisms are comparable at all.

Some indications that zipping may be directly connected with MT disassembly can be derived from the observation of films taken with the polarizing microscope. Lateral waving of the spindle fiber(s) as seen in polarized light during late prometaphase and metaphase may be explained as evidence of sideways zipping. Such sideways movements have been described as the "northern lights phenomenon" (Inoué and Bajer 1961). Such movements are especially clearly seen in grasshopper spermatocytes (Sato and Izutsu 1974). In some cells, each wave of change of birefringence results in a pronounced decrease of the spindle length, i.e., of spindle disassembly.

If assembly-disassembly of MTs and zipping are not in perfect equilibrium, as is probably the case in the majority of cells, certain predictions can be made. It might be expected, for instance, that if metaphase were experimentally prolonged, or equilibrium shifted toward disassembly, the spindle would decrease in length due to continuous zippings (followed by disassembly). When the spindle would achieve a certain threshold length, anaphase could still proceed but only over a very short distance. With further shortening of the spindle, after chromosomes split at the kinetochore, anaphase would not proceed at all. Effects of glycols on anaphase in *Haemanthus* may be interpreted in this fashion. Anaphase spindles exposed to glycols shorten and finally anaphase is stopped before kinetochores reach the poles. At the same time lateral movements in the phragmoplast are stimulated and formation of phragmoplasts proceeds (lateral movements Molè-Bajer, Unpublished). Therefore, a correlation should exist between the length of the spindle and the maximum distance of chromosome separation. However, a correlation would depend on so many factors that it is doubtful whether precise numerical values could exist.

If the MTs break after zipping, the resulting fragments must be either removed from the spindle (non-kinetochore transport, see p. 250), or disassemble. If this process did not occur, the fragments, if abundant, would make further zipping difficult, if not impossible. The breakage may give rise to shorter or longer fragments of

---

arrows). Such interaction occurs mainly at the periphery of the kinetochore fiber which is seen especially well on slightly oblique sections to the axis of the spindle. Prevailing direction of zipping is determined by the gradient of MT distribution. During the movement of the kinetochore fiber, some MTs are pulled out passively (c) and others are broken (d,d). MTs, before they pull out or break, unzip central, more parallel MTs of the kinetochore fiber. Thus, MT e unzips and splits the kinetochore fiber (see also Figure 9–8). These MTs connect two sister kinetochores by zipping. Their activity results in "bullet-like" shape of the kinetochore fiber (Figure 9–6). Kinetochore fiber is of hybrid nature; i.e., it is composed not only from MTs ending in the kinetochore, but also of some MTs of non-kinetochore origin (see also p. 248).

MTs. Data from *Haemanthus* thin sections (Jensen and Bajer 1973) indicate that MTs are irregularly arranged at the periphery of the kinetochores during anaphase. On the other hand, high voltage electron micrographs (Figure 9-3) indicate that a surprising number of MTs in a single section of a *Haemanthus* cell are over 10 μm long in mid-anaphase and certainly longer in metaphase. This may indicate either that zipped segments persist for a considerable time, or that "touch-and-go" zipping prevails. Thus immediate breaking may not be necessary for chromosome movements especially in long, thin kinetochore fibers where zipping occurs in a comparatively small region of the spindle.

SPINDLE AND KINETOCHORE FIBERS

The technique of directional tilt (Couderc 1974) shows several features of the organization of the spindle and kinetochore fiber as expected on the basis of zipping. In this technique, the micrograph is viewed at 30-40° and rotated in the plane of observation. Such viewing changes the perspective, eliminates the background noise and enables the courvatures of the MTs to be followed easily (Figure 9-3). It demonstrates the following features:

1. MTs of the kinetochore fiber are not straight, but diverge and tend to be arranged in arcs.

2. The majority of MTs which compose the kinetochore fiber end in the kinetochore. Long MTs of the kinetochore fiber are more parallel and more densely arranged (are zipped) close to the kinetochore and in the center of the fiber, than at the edges. Zipping may occur also closer to the polar end of the kinetochore fiber which may have important consequences.

3. Coupled MTs are always found abundantly where expected, i.e., on the periphery of the kinetochore fiber. This is especially clear on sections which are not precisely parallel to the long axis of the kinetochore fiber. MTs are also often zipped together closer to the polar regions in each half-spindle. Thus the diffused polar region of the anastral spindle is, in fact, composed of numerous "mini-poles" or "mini-centers" (Figure 9-3).

4. Non-kinetochore MTs in the half-spindle fan out (Figures 9-4 and 9-5) and intermingle with neighboring kinetochore fibers.

Such organization of the spindle suggests answers to several questions concerning the functional aspects of the spindle and permits several conclusions to be drawn:

1. The kinetochore MTs which form the kinetochore fibers are not the only MTs which are in contact with the chromosomes. Numerous MTs penetrate the chromosomal body (see review, Fuge 1974). In *Haemanthus* such MTs very often form bundles very close to the kinetochores and intermingle with the kinetochore fibers closer to the pole (Jensen and Bajer 1973; Lambert and Bajer 1975), or run through the chromatin of the sister chromosome. The length of such chromosomal MTs is not known. Most of them certainly pass the equator, but probably only a very few, if any, run from pole to pole. Their function may be to unzip the kinetochore fiber, i.e., disrupt parallel arrangement of the MTs in the center of the kinetochore fiber which are not contributing to the movement (Figure 9-9). Consequently during

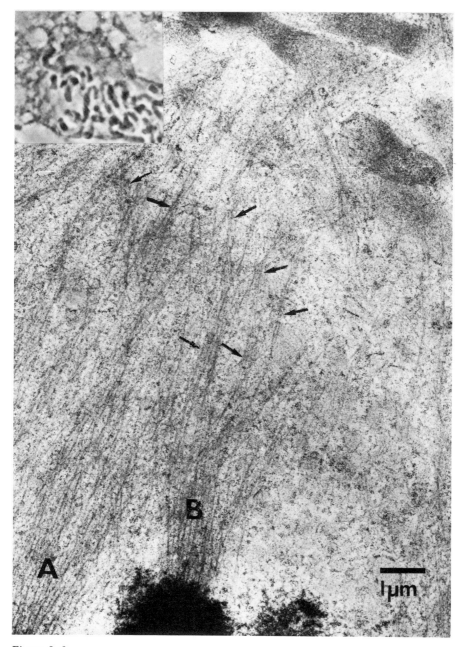

Figure 9–3.
Mid-anaphase. High voltage electron microscope. Arrows mark some of the regions of the spindle where the distance between MTs is sufficient for close contact and interaction. Interdigitation between the two different kinetochore fibers (A and B) is also visible. MTs are very long and form gentle arches. *Haemanthus katherinae.*

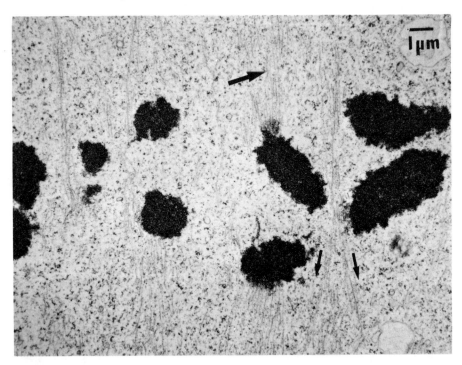

Figure 9-4.
Fanning of non-kinetochore MTs (small arrows). MTs intermingle with kinetochore fibers closer to the pole (see also Figure 9-3). It is interpreted that such fanning out is the result of lateral zipping bending kinetochore and non-kinetochore MTs. See also kinetochore fiber shown in Figure 9-5 marked by the large arrow. Endosperm of *Haemanthus katherinae.*

metaphase and early anaphase the kinetochore fibers are arched (bullet shaped) with the curvature toward the equator (Figure 9-6). Such a shape is consistent with the assumption that the transverse force acting in the spindle tends to expand or disrupt the fiber perpendicular to its long axis (Figure 9-2). This shape would be expected as a result of the touch and go mechanism if the zipped segments persist for a considerable time before breaking or disassembling. The kinetochore fiber fans out during anaphase and the number of MTs decreases. However, few parallel (zipped) MTs still remain in the core of the fiber when the movement ceases at the end of anaphase (Figure 9-7). The same factor may also account for some differences in organization of the spindle between prometaphase and anaphase, i.e., the tendency of parallel arrangement in prometaphase and more divergence in anaphase.

Due to zipping of non-kinetochore MTs with kinetochore MTs, and the existence of the force toward the equator in each half-spindle, the diameter of the metaphase plate increases in metaphase and the spindle shortens. Non-kinetochore and non-chromosomal (continuous) MTs must either break or pull out passively (slide out

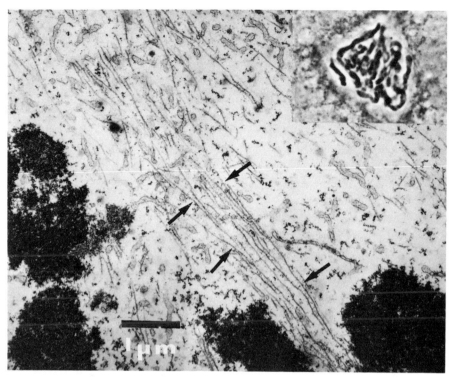

Figure 9-5.
Zipping within fanned out non-kinetochore fiber in prometaphase. Progressive zipping (arrows) leads to fanning out of the non-kinetochore fiber (Figure 9-4) and more parallel arrangement of MTs in the equatorial region. Endosperm of *Haemanthus katherinae*. Bar = 1 μm.

passively) from each other during anaphase. This will necessarily lead to a more parallel arrangement of MTs in the interzonal region between the kinetochores. When points of attachment are lacking, as when the free ends of the MTs are present, a more parallel arrangement is also expected. This may also explain, why, during anaphase, there are so many MTs just below the kinetochore and gradually fewer closer to the equator. It can be concluded, therefore, that no regular and predictable number of MTs would be found on the cross-section of the spindle in anaphase as compared to metaphase. The number may be a little more predictable in small rapidly dividing cells (McIntosh and Landis 1971; Brinkley and Cartwright 1971), although even in such cases precise correlation does not exist. The number and arrangement of MTs within the kinetochore fiber during anaphase depends on the distance of the kinetochore from the pole. The diameter of the kinetochore fiber may change depending on how and where MTs disintegrate during anaphase. The number of MTs should decrease gradually during anaphase in the anastral spindle because the majority of zippings should occur at the periphery of the

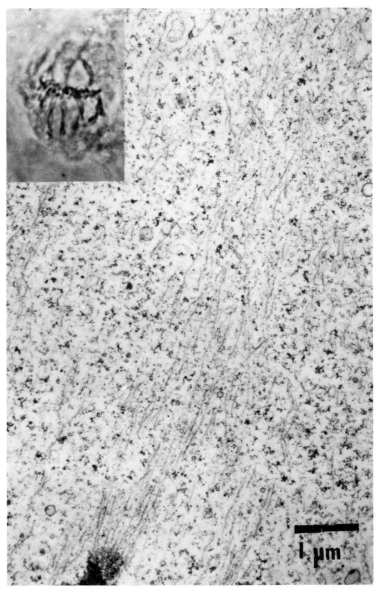

Figure 9-6.
Kinetochore fiber at the start of anaphase. Kinetochore and non-
kinetochore MTs intermingle with each other. This is especially clearly
seen if the micrograph is tilted and is viewed under proper angle and
direction. Endosperm of *Haemanthus katerinae.*

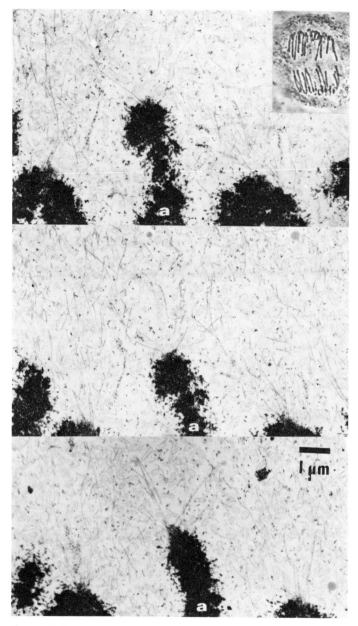

Figure 9-7.
Telophase—three non-consecutive serial sections. MTs of the kinetochore fiber fan out, shorten, and their number decreases in
comparison to earlier stages (see Figures 9-6, 9-8). Straight MTs
persist in the center of the fiber. This is seen in all kinetochores,
but especially clearly in kinetochore a. Such MTs do not contribute to the pull on the kinetochore, but to the elongation of the
spindle (p. 238). Endosperm of *Haemanthus katherinae*.

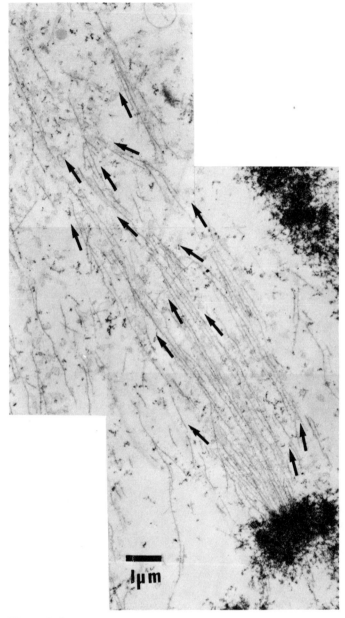

Figure 9–8.
Kinetochore fiber in early anaphase. The kinetochore fiber is
composed of long MTs intermingling with non-kinetochore
MTs. The kinetochore fiber is split into several smaller bundles, especially closer to the poles. MTs and the whole bundle
intersect under different angles (arrows) permitting efficient
pull (see also Figures 9–2 and 9–3.) Endosperm of *Haemanthus katherinae.*

kinetochore fiber. Due to more localized zippings at the poles, the number of the kinetochore MTs in the astral spindle should remain relatively unchanged for a considerable period of anaphase.

2. Depending on how MTs are disassembled under normal and experimental conditions, the paths of the kinetochores and arrangement of MTs (cf., Figures 9-8 and 9-9) will be different. If the MTs disassemble, but before final disassembly, temporarily tend to arrange themselves parallel (Bajer and Lambert 1976), the paths of kinetochores will be straight before they stop. If the MT number decreases uniformly, the paths will change irregularly as a result of the non-uniform structure of the anastral spindle (effects of colchicine). The data on the normal course of division in *Haemanthus* (Jensen and Bajer 1973) and crane fly (Fuge 1974) support these predictions. In the crane fly the autosomes have kinetochore fibers of the astral spindle, while the sex chromosomes, which are distributed when the autosomes have nearly reached the poles, have kinetochore fibers resembling those of the anastral spindle. Consequently the differences between astral and anastral kinetochore fibers should be, and are, revealed in studies of the kinetochore velocity. Variations may result from differences in location of the region of most intensive zipping along the kinetochore fiber. If zipping is localized in a more restricted region closer to the poles, and proceeds at a uniform rate during anaphase, we would expect the plot of kinetochore speed against time to be linear, as in some animal cells (Dietz 1969, 1972). If, however, zipping occurs all along the periphery of the kinetochore fiber one would expect a deceleration, with the slope of the graph decreasing, as in fact it does in flattened cells of *Haemanthus* (Bajer and Molè-Bajer 1963).

3. Interaction and the site of force production are either further or closer to the kinetochore and differently located in various kinetochore fibers. The regions of the kinetochore fiber where MTs are clustered together (Figures 9-3 and 9-8) are interpreted as regions of intensive interaction. There may be several such small centers within each kinetochore fiber, or only one, with a considerable spread. Thus either a large portion of the kinetochore fiber is involved in force production (typically divergent kinetochore fiber), or only the end, or polar region of the fiber (typically straight kinetochore fiber). Non-uniform zipping within and between kinetochore fibers may explain the flaming effect of the kinetochore fiber (oscillation of the kinetochore fiber) described as the northern lights phenomenon in polarized light (Inoué and Bajer 1961).

4. Due to zipping and disassembly, the kinetochore fibers and the whole spindle are composed of interdigitating MTs. Some MTs of the kinetochore fiber may end at a certain distance from the kinetochore. They originate from the breakage of the other MTs and are incorporated into the kinetochore fiber. Such MTs are anchored within the kinetochore fiber by zipping to kinetochore anchored MTs and therefore may perform their function (pull by zipping to others).

The spindle in metaphase is composed of long MTs, although only a few of them run between the poles. Interdigitating MTs in various regions of the spindle are joined by zipping. Thus all MTs form a three-dimensional, elastic, self-supporting "cage" which prevents the collapse of the spindle during chromosome movements and

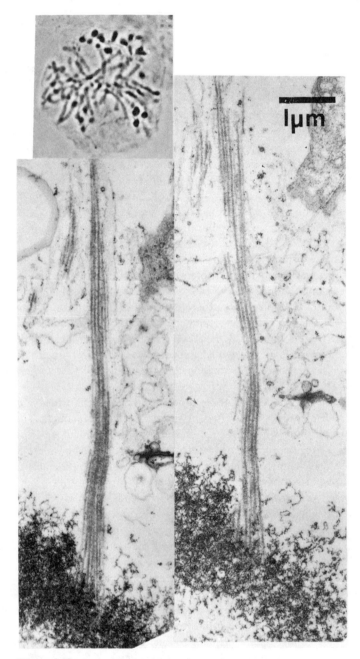

Figure 9-9.
Experimentally zipped kinetochore fibers in early anaphase at low temperature (+2°C). Kinetochore movements are reversibly arrested when the fiber is zipped. The movements start after return to room temperature and fanning out of the fiber. *Haemanthus katherinae.*

makes anchoring of the spindle to the cell wall (membrane) totally unnecessary. The problem of MTs not connected with the chromosomes (continuous and mantle MTs) undoubtedly needs reinvestigation (See also Fig. 9–2).

ANCHORING OF MTs AND SPINDLE TENSION

One assumption of the zipper hypothesis is that all MTs are anchored, but to different extents, by zipping to each other and/or frictional resistance of the spindle matrix. Linkage of MTs by arms or bridges may be sufficient to anchor MTs. The information concerning bridges in the spindle is scarce (McIntosh 1974). However, MTs in several systems are linked by periodic, or non-periodic bridges. Intertubule links may resist considerable tension, as in tentacles of the suctorian *Tokophrya* (Tucker 1974), where they stretch and still hold MTs together. It is quite conceivable that the bridges in the spindle form hundreds, or rather thousands, or links between MTs. Due to gradient of MT distribution (MTs are more dense at the equator than at the poles) formation of links, which tend to zip, will result in bending of MTs into arches and consequently the spindle will assume a "spindle or barrel shape." There is no need to assume that the linkages are very strong. Weak linkages, which can break easily, would be more functional and they could, for instance, account for the "touch and go" type of zipping.

In some systems, for example the tentacles of *Tokophrya* (Tucker 1974), the MTs are anchored firmly in amorphous electron dense material, in which they are very likely assembled. This is not the only way MTs seem to be anchored. MTs may be also connected to membranes by cross-bridges (Franke 1971) or less defined elements (Allen 1975; Bergstrom and Arnold 1974) In both *mnium* (Lambert 1970) and *Haemanthus,* bridges are seen between MTs and the nuclear envelope.

It is therefore postulated that zipping (by means of bridges or other undefined elements) is the major factor in the production of tension and thrust within the spindle. Tension is understood as a force on sections of MTs between the zipped section and the point of their attachment, or between two zipped segments of the same MT. By this definition not all sections of MTs are under tension. The zipping will therefore tend to bend MTs. Thrust will occur due to the tendency of MTs to straighten (elastic deformation). Thus, the motive force for chromosome movements is utilized for bending of MTs. We can only speculate, at present, as to the nature of such force on the molecular level ("attraction" between electron light halos, interaction between dynein type cross-bridges, actin-myosin system, etc.).

Models of the spindle constructed from elastic rods show that comparatively few zippings (bonds) are sufficient to prevent the collapse of a spindle. Such models show several properties of the spindle. They are elastic, but rigid, and can easily "support" kinetochore fibres. Due to tension the structure automatically assumes a spindle shape as a result of the gradient of MT distribution. No special organizing centers are required at the poles, although structures such as asters help considerably to keep MTs closer together and arranged in arches, i.e., under tension.

Increase of diameter of the metaphase plate with simultaneous change of the spindle shape (La Fountain 1972), shortening of the spindle during metaphase, flattening of the endosperm spindles (Molė-Bajer and Bajer 1968), and twisting of the spindle during anaphase (photographed in polarized light, Inoué 1960; Sato and Isutsu 1974), may all be interpreted as direct results of change of tension. Change of tension may occur due to elongation or shortening of the spindle.

The internal tension shortens the whole spindle due to zipping during metaphase because two half-spindles are "joined" at the equatorial plane by MTs running through the chromatin next to kinetochore MTs. When the kinetochore regions are not yet divided, the only way to release the tension is to decrease the diameter of the metaphase plate. This can occur only when the kinetochore regions split at the onset of anaphse. Anaphase can therefore be considered as a general result of the release of tension between MTs within the spindle.

One could argue, however, that the zipping of MTs, leading to release of tension during anaphase, would lead to a more parallel arrangement of MTs during chromosome migration. Such an arrangement would occur only if the zipping was not followed by simultaneous breakage and/or disassembly. Disassembly of MTs during the progress of anaphase does not permit such orderly changes and results in the opposite effect, i.e., disarrangement of the MTs.

The internal tension within the spindle may be directly connected with the control and shift in equilibrium of MT disassembly. It may be speculated for instance that MTs which are zipped, or are under tension, are more stable. The MTs which are not under tension would, depending on conditions, either disassemble or grow. Such an interpretation has some experimental support: zipped MTs seem to be less sensitive to disruption by any experimental treatment, e.g., low temperature (Bajer and Lambert 1976.).

Assuming that all MTs are anchored by zipping to other MTs or to membranes leads to the following conclusion:

Each half-spindle is structurally and functionally an independent unit with the motive force in front of the kinetochores. Anaphase can therefore proceed without any support in the interzonal region, and the presence of MTs between separating chromosome groups is not needed. Theoretically, if the cell were cut into two in the interzonal region during anaphase, chromosome migration toward the poles would continue undisturbed. The spindle of some fungi (Heath 1974) may give support to such speculations. These spindles show very few MTs in the center of the spindle during anaphase but a comparatively disorganized arrangement of MTs in the polar regions.

NON-KINETOCHORE TRANSPORT AND ITS ROLE

Rapid rearrangement of MTs, such as occurs especially during prometaphase and anaphase, must be connected with the localized breakage of MTs or zipped segments, and/or their disassembly. Due to non-kinetochore transport (spindle elimination properties, see Bajer and Molė-Bajer 1972) fragments which do not disassemble must be eliminated toward either the poles of the interzonal region, depending on the stage. Fragments of MTs which accumulate in certain regions of the spindle or interzonal region may serve as seeding elements for the growth of new MTs. Evidence of the relationship between fragments and the formation of new MTs is found during the development of the phragmoplast (Lambert and Bajer 1972). Similarly, large astrospheres contain such fragments. Thus, the formation of the astrosphere may be at least partly due to non-kinetochore transport.

The zipper hypothesis (Bajer 1973) proposes that non-kinetochore transport results directly from the change of MT arrangement from non-parallel to approximately parallel. The direction and intensity of transport depends on the gradient of MT distribution. It should be stressed, however, that unidirectional or even bidirectional transport occurs also in strictly parallel arrays of MTs (nerve cells—Schmitt 1968; Smith 1971; Smith et al. 1975; melanophores—Murphy and Tilney 1974; tentacles of suctorian—Rudzinska 1967, 1970; Tucker 1972).

Further studies are needed to find out whether all transport systems associated with parallel arrays of MTs have the same mechanism. Typical non-kinetochore transport is of the order of chromosome movements (1–3 μm min^{-1}), while suctorian tentacles the rate may be fifty times faster (20 μm s^{-1}—Tucker 1974). Such transport may therefore be due to a different mechanism, although MTs or their arms may produce the motive force in all cases.

Non-kinetochore transport could result also from the growth of MTs toward the poles. Small granules and even chromosome arms, if connected to MTs, could be also pushed to the poles. It is an open question, however, whether such a mechanism exists. For example, in fish melanophores (Murphy and Tilney 1974) the same MTs, or the structures present in the electron light halo around MTs, and not assembly-disassembly, are responsible for bidirectional transport. Some insight concerning non-kinetochore transport has been obtained in experiments involving rapid chilling of endosperm cells in mitosis (Bajer and Lambert 1976). It appeared that when chromosomes are reversibly stopped by such treatment, non-kinetochore also ceases. This is especially clear during prometaphase-metaphase. At the same time the number of MTs decreases and only some kinetochore MTs, which zip together (Figure 9-9), persist. It is evident, therefore, that non-kinetochore transport is connected with the presence of MTs. Therefore, at least two possible explanations for non-kinetochore transport may be proposed:

1. Interaction between MTs: parallel and touch-and-go zipping related directly to the gradient of MT distribution (Bajer 1973).
2. The motive force for non-kinetochore transport may be inherent to the "halo" around MTs, i.e., due to interaction between "halo" and MT, or "halo" and surrounding matrix of the spindle. Several possibilities can be visualized such as actin-myosin interaction, etc.

ZIPPING OF MTs DURING COURSE OF MITOSIS

Analysis of fine structure during normal cell division shows that the basic changes of MT arrangement are similar in all types of mitosis of higher, and many lower, organisms. Modifications of mitosis may result from various factors mentioned previously, including the possibility of existence of different mechanisms (Allen and Bajer, Unpublished).

The spindle cycle in *Haemanthus*, the newt, and certainly most higher organisms, may be characterized by definite stages in the rate of assembly-disassembly and zipping.

1. *Prophase-prometaphase.* Due to zipping MTs become arranged more and more parallel in late prophase. Prometaphase starts with the breaking of the nuclear envelope. The tendency toward parallel arrangement is retained during prometaphase and leads to a gradual slowing down of movement in the majority of cells. Chromosomes with zipping systems working in two opposite directions (two sister kinetochores) must automatically align themselves in the equatorial metaphase position of equilibrium. MT assembly exceeds disassembly throughout this stage.

2. *Metaphase.* This is a comparatively stable stage where the degree of zipping in each half-spindle is approximately the same. MTs zip in both directions (toward the pole and the equator), but the prevailing direction is toward higher density of MT distribution. This leads to a gradual increase of internal tension within the spindle directed toward the equatorial plane, which may be reflected in the decrease of the spindle length and increase in its diameter (*Haemanthus*, crane fly).

3. *Anaphase.* The stable, more parallel arrangement of MTs in metaphase is disorganized during anaphase. Rearrangement of MTs from regular to irregular results in an increase of chromosome velocity as compared to metaphase. The velocity decreases gradually when the spindle is disassembled. Zipping occurs predominantly in one direction and chromosomes move to the poles. Anaphase can be understood as a result of the release of the internal tension in the spindle.

4. *Telophase.* Remnants of MTs left behind the kinetochores zip together and elongate by assembly. This leads to the formation of a new structure such as the phragmoplast or stem body with approximately parallel MTs. Assembly prevails at this stage. There are no bodies (chromosomes) tending to distort the regular parallel arrangement of MTs and they can therefore zip more easily and arrange themselves more regularly than in metaphase. Such an arrangement is more stable; the structure persists for a long time and disassembles slowly. It is more resistant to experimental treatment such as low temperature, or colchicine.

GRADIENT OF MT DISTRIBUTION, DIRECTION OF ZIPPING, AND PROMETAPHASE MOVEMENTS

The MT distribution usually varies in different regions of the spindle. On cross-sections of the *Haemanthus* spindle, there are fewer MTs in the polar regions than at the equatorial plane. Thus a density gradient of MTs exists in each half-spindle. It is generated during prometaphase and plays an important role in establishing the metaphase plate and determining the direction of the chromosome movement.

Three main factors lead to the establishment of such a gradient in prometaphase:

1. The presence of chromosomes with sister kinetochore fibers pointing in opposite directions. The growth of kinetochore fibers must result in breakage of certain bonds, such as those between kinetochore MTs, and an increase in their divergence. The latter is due to the presence of MTs which cross the equator and zip to the kinetochore MTs.

2. The shortening of some MTs by breakage and/or disassembly, a process undoubtedly more active in the polar regions, which in turn may result in shortening of the spindle during prometaphase. It is expected, threfore, that if the kinetochore fiber did not grow continuously during prometaphase-metaphase, the spindle would disassemble completely.

3. The occurrence of zipping between all types of MTs with the prevalent direction being determined by the presence or absence of mechanical obstacles.

If MTs grow from the kinetochore poleward, then the presence of kinetochore fibers on opposite sides of the equatorial plane is a factor which is sufficient to produce a distribution gradient of MTs. More parallel MTs will be found close to the equator while toward the poles the arrangement becomes progressively less and less regular. Such a gradient may determine the degree of zipping toward or away from the equator in various regions of the spindle and consequently the general transport properties of the spindle, including non-kinetochore transport. The zipper hypothesis makes no postulate in regard to polarity of the MTs. It must be assumed therefore that zipping can occur toward both the poles and the equator. In the functional spindle the intensity of the zipping toward the pole must be lower than that toward the equator to account for the prevailing direction of chromosome movement. This may be explained as a direct result of the gradient of MT distribution.

To achieve the equatorial arrangement of metaphase, zipping of each sister kinetochore fiber must proceed predominantly in the direction of the metaphase plate. The pulling force due to zipping, acting on the two sister kinetochores, is usually in equilibrium when the kinetochores reach the equator. If there is no density gradient, or MTs do not break (disassemble), equilibrium in the spindle is stable and the metaphase plate may persist for a very long time. This may actually occur in the eggs of some marine organisms where the density of MTs is uniform throughout the spindle (Sato 1969).

CONCLUSIONS

When the zipper hypothesis was proposed (Bajer 1973) no experimental evidence for existence of the active zipping in the spindle was available. At that time most of the evidence was derived from studies on the phragmoplast (Lambert and Bajer 1972) and not the spindle. Experiments on the effects of low temperature (Bajer and Lambert 1976) and glycols (Molè-Bajer, Unpublished) are considered as sufficient evidence that such active zipping exists both in the half-spindles (Figure 9-9) and the phragmoplast. The possibility that the active zipping between MTs results from experimental treatment and is not present in the normal untreated spindle seems to be only remotely possible.

The basic question is, therefore, whether the existence of the active zipping is sufficient proof that zipping is the motive force for the chromosome movements. Although the results of the experiments are fully consistent with the assumptions of the zipper hypothesis, an affirmative answer cannot yet be given, before the role of zipping in the following processes is established and understood. At least three groups of processes are involved:

1. Active zipping may be necessary to hold the MTs of the spindle together. Zipped segments of the MTs may serve as dynamic and rapidly changing "joints," permitting relative freedom of movements.

2. Several possible mechanisms acting between the matrix of the spindle and MTs may exist. Such mechanisms in the presence of active zipping may produce similar structural rearrangements of MTs as observed, if the zipping was the motive force for chromosome movements. For instance, the "molecular pump hypothesis" (Östergren et al. 1960) does in fact assume that there is sliding (shearing force) between the kinetochore fiber and the matrix of the spindle (kinetochore fibers are pushed by the molecular wind toward the pole).

3. Non-kinetochore transport (spindle elimination property—Bajer and Molè-Bajer 1972) and transport of particles or states (Allen et al. 1969) may in fact play a more important role in the mechanism of chromosome movement than anticipated.

Further data concerning these mechanisms are necessary for the understanding of how MT rearrangement is related to the mechanism of chromosome movement. There is little doubt, however, that the active MT rearrangement and their assembly-disassembly are two distinct, but certainly closely related aspects of the functional spindle. Thus one of the most, if not the most, important problem of studies on mitosis today is to find out how the rearrangement of MTs is related to their assembly-disassembly.

There is enough evidence that assembled MTs are in equilibrium with the pool of available tubulin. Rebhun et al. (1974) stress that the temperature determines the proportion of the tubulin assembled in MTs, and the cycling process between MTs and tubulin may not be related at all to any mechanism of chromosome movement. It may be, therefore, compatible with any interpretation(s). The zipper hypothesis assumes that the rate of disassembly determines the rate of zipping. Therefore, the data on assembly must be in full agreement with the measurements of the rate of chromosome movements.

These two main postulates (zipping and assembly-disassembly) lead to the speculation that it should be possible to induce experimental conditions where the equilibrium between zipping and assembly-disassembly is considerably disturbed. This would lead to pronounced changes in MT arrangement and would be reflected in chromosome movement. Theoretically such conditions should exist for only a short time under the influence of several chemical and physical factors, if applied for proper time and at concentrations which do not lead to the total destruction of the spindle.

In conclusion, let us summarize some advantages of the zipper hypothesis as the temporary basis for the analysis and interpretation of the spindle structure and function.

1. The zipper hypothesis draws attention to the aspects of the spindle which were considered so far non-essential, such as: oblique MTs and their role, change of spindle shape during prometaphase, and relation of chromosome movements to MTs rearrangements.

2. Only according to the zipper hypothesis, "spindle shape" of the spindle is func-

tional. Such shape is not functional, for instance, when it is assumed that any kind of parallel arrangements are responsible for the movement.
3. Zipper hypothesis explains in a simple way anchorage of MTs in the spindle.
4. Zipping combined with assembly-disassembly explains some phenomena the purpose and function of which are not clear at all according to other interpretations. These are, for instance: northern lights phenomena, twisting of the spindle in late anaphase (many spermatocytes), movements of chromosomes in chloral hydrate treated cells, etc.

SUMMARY

The purpose of this second theoretical paper on the zipper hypothesis is (a) to introduce postulates which restrict the hypothesis and therefore permit experimental tests; and (b) to draw attention to several predictions and features concerning fine structure and chromosome movements, which can be checked in the future on a wider range of material. Two main postulates of the hypothesis are:

1. The movement of kinetochores is the result of multiple zippings, each (or rather a group of them) being responsible for a minute displacement of the kinetochore. It is postulated that zipping alone is a sufficient factor to account for the internal tension within the spindle. Zipping in the normal conditions occurs due to the action and/or formation of possibly weak and transitory linkages between MTs themselves and membranes. Arms or bridges or less defined structures may be the structural links. Pulling of MTs together or toward the membranes will result in their bending. The motive force for chromosome movement (i.e., the force responsible for zipping between MTs or MTs and membranes) is utilized for bending of MTs perpendicular to the long axis of the spindle and results in arching of the MTs. Pulling force for chromosome movement is understood as the combination of tension and thrust within the spindle. The existence of the thrust results in release of the tension due to the tendency of bent MTs to straighten. An assumption concerning the bending of MTs and resulting internal tension and thrust within the spindle, suggests an interpretation concerning anchorage of the MTs in the spindle. The problem of anchorage and its variation depending on the type of the spindle (astral, anastral) is discussed in detail. The spindle is understood as a self-supporting structure (no attachment of poles to the cell membrane is necessary). Each half-spindle is an autonomous unit and the mechanism(s) moving kinetochores is located in front of the moving kinetochore.

2. Zipping is followed by MT breakage and/or disassembly. Otherwise the movements would stop even before metaphase. Anaphase can be explained only by zipping followed by disassembly. Two types of zipping—parallel and touch-and-go are discussed. The hypothesis assumes that basically two MTs are needed to pull a single kinetochore. In this connection, possible mechanisms of chromosomes with one MT per kinetochore are considered.

Simple geometrical relations between MTs permit various movements of chromosomes and account for the differences between astral and anastral spindle. This permits predictions of expected differences between wide and thin spindles and of organisms with large and small chromosomes.

It is further suggested that assembly-disassembly is not the motive force for chromosome movements, but it regulates the rate of MT interaction, i.e, zipping. Therefore, the quantitative data on birefringence fits well with the change of rate of chromosome movements.

The gradient of MT (density) distribution in various regions of the spindle is discussed. It is postulated that such a gradient is present in all typical spindles (astral and anastral) in stages when movement takes place. The gradient regulates the direction of zipping and does not require any assumptions concerning the polarity of MTs.

The role of non-kinetochore transport is discussed and attention drawn to the complexity of the problem.

ACKNOWLEDGMENTS

The work was partly supported by NIH Research Grant, GM 21741 (A. Bajer). The John Simon Guggenheim Fellowship (1975) and the CRNS appointment (1974/1975) were instrumental in carrying out this work. Figure 9-3 was made with the cooperation of Dr. H. Ris of the University of Wisconsin (Madison).

REFERENCES

Allen, R. D. 1975. Evidence for firm linkages between microtubules and membrane-bound vesicles. *J. Cell Biol.* **64**:497–503.

Allen, R. D., A. Bajer, and J. LaFountain. 1969. Poleward migration of particles or states in spindle fiber filaments during mitosis in *Haemanthus*. *J. Cell Biol.* **43**:4a.

Bajer, A. S. 1973. Interaction of microtubules and the mechanism of chromosome movement (zipper hypothesis). 1. General principle. *Cytobios.* **8**:139–150.

Bajer, A. and J. Molè-Bajer. 1963. Cine analysis of some aspects of mitosis in endosperm. In *Cinematography in Cell Biology*, ed. G. G. Rose. Academic Press, New York, pp. 357–409.

Bajer, A. S. and J. Molè-Bajer. 1972. Spindle dynamics and chromosome movements. *Int. Rev. Cytology*, Suppl. 3:1–271. Academic Press, New York.

Bajer, A. S. and A. M. Lambert. 1976. Reversible arrest of chromosome movements and microtubule (MT) rearrangements at low temperature (LT). *J. Cell Biol.* **70**:128a.

Bajer, A. S., J. Molè-Bajer, and A. M. Lambert. 1975. Lateral interaction of microtubules and chromosome movements. In *Microtubules and Microtubule Inhibitors*, Eds. M. Borgers and M. de Brabander. North-Holland Pub. Co. Amsterdam, pp. 393–423.

Bergstrom, B. H. and J. M. Arnold. 1974. Nonkinetochore association of chromatin and microtubules. *J. Cell Biol.* **62**:917–920.

Brinkley, B. R. and J. Cartwright, Jr. 1971. Ultrastructural analysis of mitotic spindle elongation in mammalian cells *in vitro*: Direct microtubule counts. *J. Cell Biol.* **50**:416–431.

Couderc, H. 1974. Anamorphoses des images D'objets biologiques tridimensionnels observes au M.E.B. *Jeol News* 12:11-15.

Dietz, R. 1969. Bau und Funktion des Spindelapparats. *Naturwissenschaften* 56: 237-248.

Dietz, R. 1972. Die Assembly-Hypothese der Chromosomenbewegung und die Veranderungen der Spindellange während der Anaphasel in Spermatocyten von *Pales ferruginea* (Tipulidae, Diptera). *Chromosoma* 38:11-76.

Forer, A. 1966. Characterization of the mitotic traction system and evidence that birefringent spindle fibres neither produce nor transmit force for chromosome movement. *Chromosoma* 19:44-98.

Franke, W. W. 1971. Cytoplasmic microtubules linked to endoplasmic reticulum with cross bridges. *Exp. Cell Res.* 66:486-489.

Fuge, H. 1974. Ultrastructure and function of the spindle apparatus. Microtubules and chromosomes during nuclear division. *Protoplasma* 82:289-320.

Heath, I. P. 1974. Mitosis in the fungus *Thraustotheca clavata*. *J. Cell Biol.* 60: 204-220.

Inoué, S. 1960. Birefringence in dividing cells. Time lapse film, available from G. W. Colburn Laboratories, Inc., Chicago, Ill.

Inoué, S. and A. Bajer. 1961. Birefringence in endosperm mitosis. *Chromosoma* 12:48-63.

Inoué, S., J. Fuseler, E. D. Salmon, and G. W. Ellis. 1975. Functional organization of mitotic microtubules. Physical chemistry of the in vivo equilibrium systems. *Biophy. J.* 15:725-744.

Inoué, S. and H. Ritter. 1975. Dynamics of mitotic spindle organization and function. In *Molecules and Movement*, eds. S. Inoué and R. E. Stephens, Raven Press, New York, pp. 3-30.

Inoué, S. and H. Sato. 1967. Cell motility by labile association of molecules. *J. Gen. Physiol.* 50:259-292.

Jensen, C. and A. Bajer. 1973. Spindle dynamics and arrangement of microtubules. *Chromosoma* (Berl.) 44:73-89.

La Fountain, J. R., Jr. 1972. Spindle shape changes as an indicator of force production in crane fly spermatocytes. *J. Cell Sci.* 10:79-83.

Lambert, A. 1970. Etude de structures cinetiques en rapport avec la rupture de la membrane nucleaire en dubut de meiose chez *Mnium hornum* L. Organisation des centromeres. *C. R. Acad. Sc.* (Paris) 270:ser. D. 481-484.

Lambert, A. M. and A. S. Bajer. 1972. Dynamics of spindle fibers and microtubules during anaphase and phragmoplast formation. *Chromosoma* 39:101-144.

Lambert, A. M. and A. J. Bajer. 1975. Dynamics and fine structure of prometaphase spindle. *J. de Micr. Biol.* 23:181-194.

McIntosh, J. R. 1974. Bridges between MTs. *J. Cell Biol.* 61:166-187.

McIntosh, J. R. and S. C. Landis 1971. The distribution of spindle microtubules during mitosis in cultured human cells. *J. Cell Biol.* 49:468-497.

Molè-Bajer, J. 1967. Chromosome movements in chloral hydrate treated endosperm cells *in vitro*. *Chromosoma* 22:465-480.

Molè-Bajer, J. 1969. Fine structural studies of apolar mitosis. *Chromosoma* 26: 427-448.

Molè-Bajer, J. and A. S. Bajer. 1968. Studies of selected endosperm cells with the light and electron microscope. The technique. *La Cellule* 67:257-265.

Murphy, D. B. and L. G. Tilney. 1974. The role of microtubules in the movement of pigment granules in teleost melanophores. *J. Cell Biol.* **61**:757–779.

Östergren, G., J. Molè-Bajer, and A. Bajer. 1960. An interpretation of transport phenomena at mitosis. *Ann. N. Y. Acad. Sci.* **90**:381–406.

Rebhun, L. I., M. Mellon, D. Jemiolo, J. Nath, and N. Ivy. 1974. Regulation of size and birefringence of the *in vivo* mitotic apparatus. *J. Supramol. Struct.* **2**:466–485.

Rudzinska, M. A. 1967. Ultrastructures involved in the feeding mechanism of suctoria. *Trans. N. Y. Acad. Sci.* **29**:512–525.

Rudzinska, M. A. 1970. The mechanism of food intake in *Tokophrya infusionum* and ultrastructural changes in food vacuoles during digestion. *J. Protozool.* **17**:626–641.

Salmon, E. D. 1975. B. Spindle microtubules: thermodynamics of *in vivo* assembly and role in chromosome movement. *Ann. N. Y. Acad. Sci.* **253**:383–406.

Satir, P. 1965. Studies on cilia. II. Examination of the distal region of the ciliary shaft and the role of the filaments in motility. *J. Cell Biol.* **26**:805–834.

Satir, P. 1968. Studies on cilia. III. Further studies on the cilium tip and a "sliding filament" model of ciliary motility. *J. Cell Biol.* **39**:77–94.

Sato, H. 1969. Analysis of form birefringence in mitotic spindle. *Amer. Zool.* **9**:592.

Sato, H. and K. Izutsu. 1974. Birefringence in mitosis of the spermatocyte of grasshopper *Chrysochraon japomicus*. Motion Picture. Available from G. W. Colburn Lab. Inc. Chicago, Ill. 60606.

Schmitt, F. O. 1968. II. The molecular biology of neuronal fibrous proteins. *Neuroscience Res. Prog. Bull.* **6**:119–144.

Smith, D. S. 1971. On the significance of cross-bridges between microtubules and synaptic vesicles. *Philos. Trans. Roy. Soc. Lond.* Ser. B. **261**:395–405.

Smith, D. S., U. Jarlfors, and B. F. Cameron. 1975. Morphological evidence for the participation of microtubules in axonal transport. *Ann. N. Y. Acad. Sci.* **253**:472–506.

Tucker, J. B. 1972. Microtubule-arms and propulsion of food particles inside a large feeding organelle in the ciliate *Phascolodon vorticella*. *J. Cell Sci.* **10**:883–903.

Tucker, J. B. 1974. Microtubule arms and cytoplasmic streaming and microtubule bending and stretching of intertubule links in the feeding tentacle of the suctorian ciliate Tokophrya. *J. Cell Biol.* **62**:424–437.

Warner, F. D., and P. Satir. 1974. The structural basis of ciliary bend formation. *J. Cell Biol.* **63**:35–63.

Part III. Mechanisms of Cell Division

Cell Division in Higher Plants

G. Giménez-Martín, C. de la Torre, and
J. F. López-Sáez

Instituto de Biología Celular, C.S.I.C.

INTRODUCTION

"When an experimentalist begins to be worried about the way specific terms are used . . . his colleagues are apt to shake their heads and conclude that his days as an experimentalist are numbered" (Harris 1974).

In spite of this, we still dare to consider cell division, in *sensu stricto*, as the process by which one cell gives rise to two cells. This process includes nuclear division, also called mitosis or karyokinesis, and cytoplasmic division or cytokinesis. However, *sensu lato*, cell division is identified with the whole process of cell proliferation, including interphase growth and division.

In this review we shall be dealing with cell proliferation in the root meristems of higher plants, placing particular stress on its dissection by means of chemicals.

The ever increasing body of inhibitors has made possible our understanding of some of the control mechanisms operating during mitosis, especially in prophase, and the nucleolar reorganization stage during telophase. By their use, different phases in plant cytokinesis have also been discerned.

ADVANTAGES OF ROOT MERISTEMS IN CYCLE CONTROL STUDIES

First of all, the principal metabolic pathways in plants are similar to those in other eukaryotic cells, so that, in principle, the findings in root meristems can be generalized to other plant and animal cells.

Root meristems are easily handled, for the seed or bulb itself provides the nutrients for its growth, so that the culture medium may be as simple as aerated tap water, a simple mineral medium, or a moist chamber. Besides, they do not require sterile conditions, which is an advantage compared with organ or animal cell cultures, and roots are made up of non-green cells, i.e., with no chloroplasts, and their metabolism is not light dependent.

Under fixed conditions steady-state root growth can be obtained. These conditions are achieved by simply controlling the main environmental factors (e.g., temperature, moisture, and oxygen tension) throughout the experiment. Steady state kinetics is characterized by a constant growth rate, a constant population of cells in proliferation with a constant mean cell cycle duration and constant cycle parameters (mitotic index, frequency of cells in the different cell compartments, growth fraction, etc.).

Being the plant's absorption organ, physiologically speaking, the root has high permeability which makes the penetration and elimination of solutes easy and facilitates drug testing.

The simple medium in which the root may be grown has the additional advantage of restricting the interactions between any substance under investigation and the medium.

Indeed, all the above characteristics have made root tips a classical material for studies of the physiology of nuclear phenomena, in addition to their value in chromosome cytology. For instance, it was with growing root tips of *Vicia faba* that Howard and Pelc (1953) measured cycle parameters for the first time by means of radioactive DNA precursors, thereby discovering the three interphase compartments G_1, S, and G_2.

The small number of large chromosomes characteristic of many plants makes root tips a good material for studying the effects of various agents on chromosomes.

Quite apart from these advantages, *Allium* bulbs are readily available in the market most of the year. A crown of 20–50 roots sprout nearly synchronously from each bulb, providing a fair sized population of genetically identical meristematic cells.

Moreover the cytology of *Allium cepa* cells is very simple, squashing is easy, and the DNA content, 33.5 pg per telophase nucleus, is high enough to allow easy cytophotometric studies.

As to nucleolar morphology, *A. cepa* shows two large nucleoli sometimes fused into one large nucleolar mass, with the nucleolar organizing region (NOR) located in a submetacentric pair of chromosomes of its complement (2n = 16).

INHIBITORS AS TOOLS

The application of inhibitors in solving biological problems has led to a decisive insight into the structure and function of the living state. The living state is characterized by a series of enzymatic reactions taking place at the cellular level, some running in parallel, others in sequence. When genetic dissection of the interrelationship between them is difficult, inhibitors seem to be the right tools to use.

The specific inhibitors of a given process, especially those whose molecular action mechanisms are known, are very valuable in establishing the role that any process plays in cellular physiology.

Even inhibitors whose action mechanism is imperfectly known can still be useful in detecting specific regions or periods characterized by their high sensitivity. The bugbears of drug side effects can be avoided by using various inhibitors whose main action is common and whose side effects differ. Besides, the study of the action of an inhibitor is, very often, an interesting biological problem itself.

As the pharmacological saying goes "the value of a drug depends on the worker's ability to use it." Of course, inhibitors may be incorrectly employed, either in the planning of experiments or in the conclusion drawn, but we do not intend to discuss, in the present review, the well known "use of inhibitors by uninhibited workers."

POLYNUCLEATE CELLS AS A CYCLE DISSECTING TOOL

The use of mutants in the study of animal proliferation appears to be a promising tool in the dissection of regulating controls. However, it has still to be exploited in higher plants.

Polynucleate cells, homokaryons or heterokaryons, offer great possibilities for the study of those mechanisms which govern gene regulation in eukaryotic cells. A successful approach for studying cell cycle controls has indeed been afforded by the fusion of heterophasic animal cells (Johnson and Rao 1971).

In plants, the use of polynucleate cells also seems to be an adequate tool to dissect proliferation mechanisms, especially those involved in nucleus-nucleus and nucleus-cytoplasm interactions. The discovery that xanthic bases exercise an inhibitory effect on cytokinesis has opened the way to inducing binucleate and polynucleate cells in root meristems at will (Mangenot and Carpentier 1944; Kihlman 1955; Giménez-Martín et al. 1965).

Under steady-state kinetics the meristem population of a root is uniformly distributed throughout the cell cycle. Caffeine inhibits cytokinesis by blocking cell-plate formation in those spontaneously synchronous cells traversing telophase during treatment, thus rendering them binucleate. This binucleate population constitutes a labeled subpopulation and it can be easily distinguished from the rest of meristem cells by morphological criteria. This synchronous population then, has been obtained by selection synchrony and not by induction synchrony, and it displays a cycle similar to that of mononucleate cells (Giménez-Martín et al. 1968).

In order to induce polynucleate cells, the binucleate population can be subjected to another caffeine incubation during its next telophase (bi-telophase). This second caffeine treatment gives rise to cells with an 8n chromosome complement, which can be distributed among 2, 3, or 4 nuclei depending on the number of nuclear fusions (Figure 10-1).

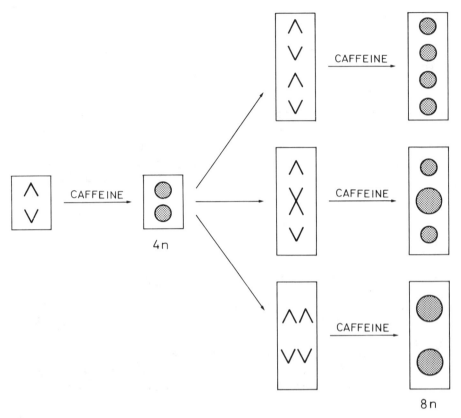

Figure 10-1.
Diagram showing the induction of polynucleate cells in a meristem. Caffeine inhibits cytokinesis in cells progressing through telophase, rendering them binucleate. The newly formed binucleate cells (2n-2n) initiate the interphase of their next cycle a few hours later, depending on growth conditions; the binucleate cell population enters mitosis but overlapping waves of prophase, metaphase, anaphase, and telophase can be recorded in the binucleate population. If a second pulse of caffeine is given at the time when the wave of bitelophases is due, 8n tetranucleate cells are obtained. The nuclei of these 8n cells occasionally fuse and besides the original tetranucleate cells (2n-2n-2n-2n), tri- (2n-4n-2n), and binucleate cells (4n-4n) can be obtained.

For instance, at 25°C, a 3-hour caffeine treatment induces 7.8% binucleate cells in the onion meristem. A second caffeine treatment for one hour (between the 14th and 15th hours after the first caffeine treatment) gives rise to 8n cells representing roughly 1-2% of the cells in a meristem. The first prophases in this 8n polynucleate population will appear 9 hours after this second caffeine treatment (From Giménez-Martín et al. 1968, by permission).

PREMITOTIC REQUIREMENTS

For a cell to divide it must have fulfilled a series of previous requirements. These requirements are connected with the exact duplication of the cell's genetic information concurrent with a rough duplication of its other components.

The newly initiated meristem cell must mature before it is ready to divide. This sort of development at the individual cell level takes place during interphase.

Replication

There is general agreement that chromosomes need to duplicate prior to their equal distribution into the two daughter cells resulting from mitosis.

The already classic paper by Howard and Pelc (1953) mentioned earlier showed that DNA replication is a periodic event in the cell cycle, taking place in a discrete portion of interphase (S period), preceded and followed by stages (G_1 and G_2) when replication does not occur. This periodic pattern is virtually general for all eukaryotic cells.

Initiation of DNA Replication

The initiation of DNA replication can take place immediately after completion of telophase in cap initial cells of maize roots so that G_1 requirements must be met in the earlier cycle in this particular case (Clowes 1967).

The initiation of replication was seen to be concurrent with the capacity of X-ray induced micronuclei to organize nucleoli (Das 1962), so that some relationship seems to exist between these two events, at least in onion and bean cells.

Studies of polynucleate cells have shown that the different nuclei sharing a common cytoplasm are either unlabeled or isolabeled when given a ^{3}H-thymidine pulse at the G_1/S transition so that the anisolabeling frequency was very low (Table 10-1). This fact implies the synchronous initiation of replication in these nuclei and supports the hypothesis that there is a cytoplasmic factor controlling the G_1/S progression, which would then seem to be an inducible event.

Thompson and McCarthy (1968) proved *in vitro* stimulation of nuclear DNA and RNA synthesis by means of cytoplasm extracts. Fusion of animal cells (Yamanaka and Okada 1966; Johnson and Harris 1969; Johnson and Rao 1970; see Harris 1970, 1974) and nuclear transplants (De Terra 1960; Gurdon 1967) also support the hypothesis of a positive factor or factors controlling the initiation of DNA synthesis.

The Course of DNA Replication

The incorporation of DNA precursors by binucleate cells points to differences in the replication rate during S period, with a depression about half way through (Figure 10-2). Differential sensitivity to deoxyadenosine (a DNA synthesis inhibitor) can be paralleled with the different rates of DNA replication (c.f., Figures 10-2 and 10-3).

Table 10-1

Frequency of asynchrony at G_1/S and S/G_2 transitions (aniso-labeling) and at the G_2/prophase boundary, in the different nuclei of 8n polynucleate cells. $15°C$

Cell type	G_1/S anisolabeling[1]	S/G_2 anisolabeling[2]	G_2/prophase asynchrony[3]
4n-4n	0.000	0.197	0.204
2n-4n-2n	0.001	0.700	0.361
2n-2n-2n-2n	0.001	0.404	0.389

1. Unpublished data. They correspond to the 9th hour after the second caffeine treatment which rendered the cells polynucleate (see Figure 10-1). At this time roughly 50% of polynucleate cells have entered S period (see control Figure 10-7).
2. Adapted from table 2 González-Fernández et al. (1971b). The polynucleate cells were treated with ^{3}H-thymidine from 26th to 30th hours after caffeine, and then fixed.
3. Unpublished data. They correspond to the 26th hour after the second caffeine treatment, when roughly 40% of the poly-nucleate cells are in prophase.

A lack of uniformity in the DNA synthesis rate was likewise observed in *Vicia faba* by Howard and Dewey (1961), who found that, after treatment with ^{3}H-thy-midine, two peaks appeared in the labeled mitosis curve, with an interval of 3–4 hours between. Kasten and Strasser (1966) showed that DNA replication in syn-chronized cultures of human epithelial adenocarcinoma also occurred in two waves separated by a short interval about midway through S when replication almost ceased.

Apart from this, microspectrophotometric measurements of the DNA content during interphase in binucleate cells have revealed some lack of synchronism in the replication rate of two nuclei sharing a common cytoplasm in onion (Fer-nández-Gómez 1968). Benbadis et al. (1974) estimated that the lack of synchronism in the replication of both nuclei in *A. sativum* meristems extended to about 20% of the binucleate cell population.

A specific ordered pattern of DNA replication for individual chromosomes and for different chromosome segments throughout the S period has repeatedly been proved in plants, e.g., in *Crepis capillaris* (Taylor 1958), in *Tradescantia paludosa* (Wimber 1961), in *Vicia faba* (Evans 1964), in *Scilla campanulata* (Evans and Rees 1966), in *Puschkinia libanotica* (Barlow and Vosa 1969), as well as in animal cells. Hence, the timing of the whole S period is but the sum of the replication time of the different genome portions.

Termination of DNA Replication

In order to ascertain whether the several nuclei in each polynucleate cell ended the S period synchronously, they were given ^{3}H-thymidine at the S/G_2 boundary.

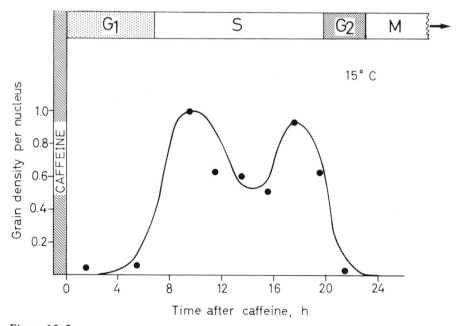

Figure 10-2.
Rate of DNA synthesis during the interphase of binucleate cells. The length of interphase is represented by the bar in the upper part of the figure. At 15°C, 23 hours after treatment with caffeine the first biprophases appear so that 23 hours corresponds to the minimum interphase or interphase of the fastest cells in the binucleate population.

1-hour treatments with ^{3}H-thymidine were given at different times after caffeine and the meristems were all fixed and studied at 24th hour after caffeine. The number of grains per nucleus is represented as a fraction of the maximum number at hour 10 after caffeine, which was taken as density 1.0. Between the 13th and the 14th hours of the interphase there is an area where the rate of DNA synthesis drops considerably (From Fernández-Gómez 1968, by permission).

Table 10-1 shows the frequency of polynucleate cells with anisolabeling in the different nuclei sharing the same cytoplasm. These frequencies point to the fact that termination of the S period in the different nuclei of a polynucleate cell is not necessarily a synchronous event and that the nuclei do not apparently interact with each other.

The degree of asynchrony is greater in tetranucleate cells (2n-2n-2n-2n) than in binucleate tetraploid cells (4n-4n). The greatest degree of asynchrony is found in cells with nuclei of different ploidy in a common cytoplasm (2n-4n-2n).

The arrangement of the nuclei in the cytoplasm seems to be of considerable importance, since termination of replication occurred earlier in nuclei having a larger cytoplasmic area. Generally, those nuclei at the ends of a polynucleate cell are the earliest to complete the S period. It has been postulated that under certain conditions, cells can pass from late S and/or early G_2 into next G_1. Endoreduplication is the term for this mitotic bypass. In it, no chromosome cycle is visible and,

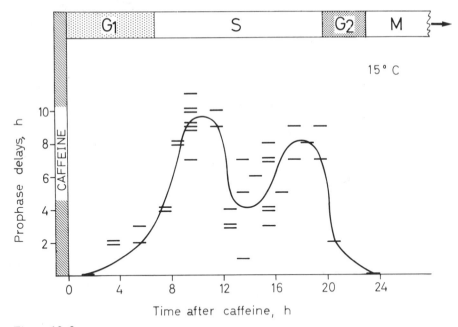

Figure 10-3.
Differential sensitivity to a DNA synthesis inhibitor (deoxyadenosine = AdR) during interphase of binucleate cells. Minimum G_1, S, and G_2 under these steady state growth conditions are shown in the upper bar of the figure. These minimum durations for interphase stages correspond to those found in the fastest binucleate cells (representing around 10% of the binucleate population).

The AdR-sensitivity follows a pattern resembling that of the rate of DNA synthesis (c.f. Figure 10-2), since it also drops around mid S.

Horizontal bars represent the 1-hour treatments with 5×10^{-5} M AdR in different experiments (Adapted from Fernández-Gómez et al., 1970, by permission).

after another S period, tetraploid mitosis with diplochromosomes is seen. The appearance of diplochromosomes indicates an initial centromere malfunction.

Endoreduplication often takes place in differentiating plant tissue in a spontaneous fashion (D'Amato 1952). The experimental induction of endoreduplication has been described in pea root meristems during recovery from 8-azaguanine treatment (Nuti-Ronchi et al. 1965). Lin and Walden (1974) described induction of endoreduplication by hydroxylamine sulfate in maize roots, also during recovery from such a treatment. In animal cells many different chemicals induce endoreduplication, e.g., colchicine (Rizzoni and Palitti 1973).

The existence of such a mechanism highlights the fact that chromosome duplication and mitosis are cellular processes that are or can be made independent.

The last portion of the S period and/or early G_2 seems to be highly sensitive to 5-aminouracil (5-AU). This chemical has been used successfully for synchronizing meristem cell populations (Duncan and Woods 1953; Clowes 1965; Wagenaar, 1966). Socher and Davidson (1971) proposed that 5-AU arrests the cells at the

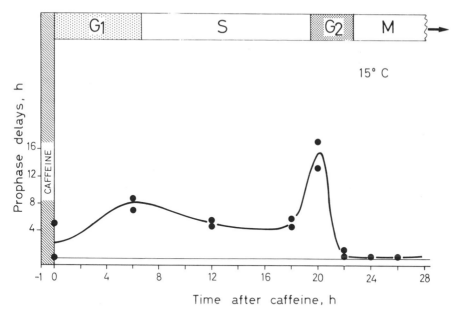

Figure 10-4.
Sensitivity to 5-aminouracil during interphase of binucleate cells. Minimum G_1, S, and G_2 for the fastest cells of this population is shown. The 6 hour treatments with 0.5 mM 5-AU start in all cases at the time shown by the points.

An increased sensitivity to 5-AU is shown towards S/G_2 boundary (From Díez et al. 1976, by permission).

S/G_2 transition, while Scheuermann and Klaffke-Lobsien (1973) postulated a 5-AU preferential action on a DNA polymerase responsible for heterochromatin replication. Experiments with *Allium* support this last interpretation, since a high sensitivity region can be detected by late S (Figure 10-4), and a strong and preferential depression in the transit rate through late S has been measured (Díez et al. 1976).

Transcription

Das (1963) and Das and Alfert (1963) showed that transcription was continuous throughout the interphase of meristem cells, stopping only at mitosis.

Up to now, the use of inhibitors as tools for studying transcriptional controls in the cell cycle has not been amply exploited due to lack of reliable evaluated inhibitors for plant material. We lack the counterpart of actinomycin on plants, since, at least in onion meristems, it is not effective at reasonable concentrations.

Nucleolar Transcription

Transcription in the meristem cell nucleolus represents most nuclear transcription at any given moment of the cell cycle, as shown by autoradiography.

Jakob (1972) found that the bulk of RNA synthesis preceded the bulk of DNA synthesis during G_1 and early S of the first cell cycle in germinating root meristems of *Vicia faba*. At least part of this RNA synthesis seemed necessary for the normal rate of DNA replication. One of the rapidly labeled RNA species then synthesized corresponded to the molecular weight reported for rRNA precursor.

As far as we know there are no studies on the role of nucleolar transcription in cycle progression for meristems growing under study-state conditions.

Extranucleolar Transcription

By employing α-amanitin (an inhibitor of extranucleolar RNA polymerase, Jacob et al. 1971) in meristem cells at a dosage which slows down without blocking the progress of interphase, protraction of G_1 and G_2 occurs while S period hardly lengthens at all (Table 10-2). These differential sensitivities are mooted to be due to differences in enzyme activity, and could be correlated with data on mammalian cells, because two peaks in RNA synthesis, situated in G_1 and G_2, were found in HeLa cells (Kim and Pérez 1965).

Translation

Although protein synthesis is a continuous process during interphase, it is known that the proteins produced at different interphase times differ qualitatively as do their rates of synthesis.

By inhibiting protein synthesis in meristem cells it is shown that there is no initiation of DNA replication without previous protein synthesis (Figure 10-5).

The fact that the initiation of replication depends on a previous synthesis of proteins(s) in G_1 seems to be a principle that applies to all proliferating cells. This could also explain the arrest of plant cells in G_1 induced by carbohydrate starvation because protein synthesis is energy-dependent (Webster and Van't Hof 1970). And, moreover, it is in agreement with the results of Everhard and Prescott (1972) and Highfield and Dewey (1972) for Chinese hamster cells.

This protein(s) could be one of those described by Salas and Green (1971) that

Table 10-2
Duration of G_1, S and G_2 in control and α-amanitin treated meristems (in hours) $15°C^1$

Interphase stage	Control	α-amanitin
G_1	7.2	13.6
S	13.1	14.5
G_2	5.9	10.2

1. Adapted from Table 4, De la Torre et al. (1974).

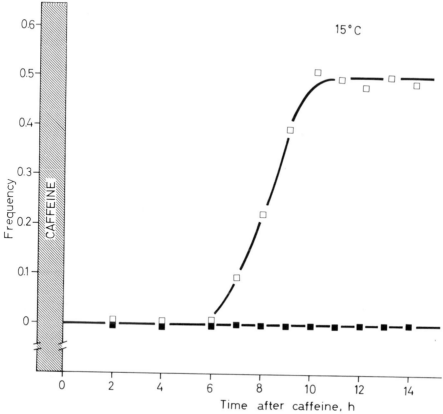

Figure 10–5.
Frequency of binucleate cells initiating their chromosome replication (□ – □, control conditions; ■ – ■, in cells where protein synthesis was simultaneously inhibited). After the caffeine treatment, the bulbs were subjected to 10-minute pulses of ^{3}H-TdR at hourly intervals. The minimum G_1 takes around 7 hours in controlled growth conditions; only 50% of the induced binucleate population enters S period, as would be expected from meristems with a constant number of cells and where 50% of the newly produced cells are in the differentiation zone immediately after mitosis.

Initiation of S period depends on protein synthesis. Inhibition of protein synthesis was brought about by $1^{\mu g}$ ml^{-1} anisomycin, in continuous treatment after caffeine. Roots were immediately fixed at the end of each pulse and incorporation was sought in the binucleate cells. No binucleate cell initiates S period when protein synthesis is inhibited (Adapted from González-Fernández et al. 1974, by permission).

has a certain affinity for DNA and that, according to their suggestion, might control the replication of the genetic material. Moreover, this protein synthesis in G_1 might be connected with the passage of the cells from the "indeterministic" G_1 state into the "real" G_1, in which the cells are already predetermined to go through a new cycle, if we accept Smith and Martin's hypothesis (1973).

Another interphase period, when protein synthesis is required for cell progression towards mitosis, was timed in binucleate cells following the experimental scheme shown in Figure 10-6. It was located between hours 19 and 20 in an interphase 23-hours long (interphase of the fastest binucleate cells at 15°C). This period seems to be located in early G_2, since thymidine labeling shows that about 10% of the cells from the binucleate population (the fastest cells) have ended replication by then. This protein synthesis period is quite possibly the same one whose product was subsequently distributed over the entire chromosome complement in onion (Bloch et al. 1967).

The work of Johnson and Rao (1970) showed, by hybridization, that cells in metaphase were capable of inducing mitotic chromosomal condensation in interphase nuclei, suggesting that a protein present in cytoplasm could be the factor determining chromatin condensation (see Douvas and Bonner, this volume).

Van't Hof (1973) also showed that in the dormant seed and in carbohydrate

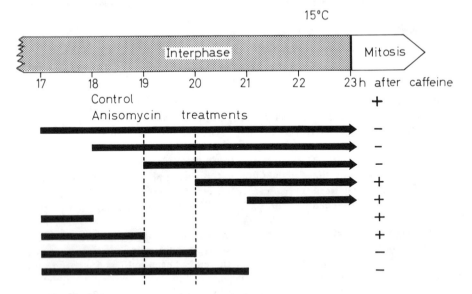

Figure 10-6.
Diagram of the treatments carried out to inhibit protein sunthesis covering different portions of the last part of binucleate cell interphase. The appearance of biprophases was delayed with respect to controls (23rd hour) whenever the protein synthesis inhibition covered the 19th–20th hour period. Initiation of prophase appears to depend on protein synthesis taking place in late S/early G_2 of these cells (Data from González-Fernández et al. 1974, by permission).

starved root meristems the potentially proliferative cells arrest in either G_1 or G_2 owing to a lack of proteins. Accordingly, he postulated the existence of two principal control points in interphase progression (see Rost, this volume). Evans and Van't Hof (1973, 1974) characterized a factor from the cotyledons which promotes selective cell arrest in G_2; this factor is operative in shoots and roots of *Pisum* and in roots of *Vicia*.

TIMING MODIFIERS FOR PREMITOTIC REQUIREMENTS

Internal Factors

There are internal factors modifying the fulfilment time of the different premitotic requirements for a cell to divide. First of all, the genetic make-up of a cell is important.

Östergren and Östergren (1966) found a mutation which eliminates at least the chromosome replication requirement for mitosis to occur, so that unreplicated chromosomes were able to divide. This confirms that genome duplication and mitosis can be made independent, as evident from endoreduplication.

Van't Hof and Sparrow (1963) and Van't Hof (1965, 1975) have shown that, as a rule the higher the DNA content the longer the duration of the cell cycle. Studies of interphase time in polynucleate cells confirm the cycle to be somewhat longer when either the ploidy or the number of nuclei contained by the cell increase (Table 10–3).

Results obtained by Evans et al. (1968) in autopolyploid monocotyledons also show that the degree to which the mitotic cycle lengthens by increased DNA content is less in polyploids than in diploids.

Giménez-Martín et al. (1966) and Alfert and Das (1969) postulated that replication rate depends on the nuclear surface : nuclear volume ratio, according to their observation of a higher rate of replication in binucleate diploid cells (2n-2n) than in tetraploid cells (4n).

Table 10–3

Duration of interphase and mitosis (hours) in relation to DNA content and its distribution in mononucleate and polynucleate cells *Allium cepa* L. root meristems. $25°C$[1]

DNA content	Cell type	Minimum Interphase	Mitosis	Synchronization time
2n	2n	9.2	2.3	—
4n	2n-2n	9.5	2.5	0.4
8n	4n-4n	9.1	2.9	0.4
	2n-4n-2n	10.1	3.0	1.8
	2n-2n-2n-2n	9.8	3.2	1.9

1. From table 2, Giménez-Martín et al. (1968).

The presence of accessory chromosomes causes a disproportionate increase in the duration of the mitotic cycle (Ayonoadu and Rees 1968; Evans et al. 1968). On the other hand, Nagl (1974) showed that the amount of heterochromatin per genome is related to the reduction of cycle times in annual plants with large DNA content.

Apart from the genetic make-up, positional information seems to lead to important changes in cycle time, as Clowes (1961, 1967) proved when comparing the cycle of the cap initials, quiescent center cells, and cells from two other different regions of the root meristem of *Zea mays*, L. A study of seventeen regions of maize root meristems showed that the main difference between them was in the mean duration of G_1 (Barlow and McDonald 1973). Interaction between different meristem regions can be made evident by removal of the root cap, which leads to a change in the rate of mitosis throughout the meristem (Clowes 1972; Barlow 1974).

Contradictory reports about hormone action on cycle progression are no doubt related to contradictory responses at successive hormone thresholds. It was possible to measure directly a 4-hour shortening in a 23-hour-long interphase produced by kinetin plus indoleacetic acid (González-Fernández et al. 1972). This is consistent with MacLeod's (1968) finding that the rate of DNA synthesis increased under kinetin while G_1 was shortened and G_2 lengthened.

Environmental Factors

As is to be expected, plant cells, which may be considered as "poikiloterm" creatures, are strongly influenced by temperature, though they can grow at any temperature from 0 to 35°C.

The fact that plant cells are permissive to such a range of growth temperatures is no doubt a fortunate thing (for researchers) since the timing of a particular process can be conveniently conditioned to the desired experimental conditions.

The progression of their whole cycle is dependent on temperature. Figure 10-7 shows the rates at which cells go through the cycle (cycle rate) at different temperatures. The cycle rate is but the inverse of the duration of the whole cycle. The highest cycle rate is seen at 30-35°C, and the rate diminishes with decreasing temperature.

Figure 10-8 is a schematic representation of the duration of the cycle and its compartments in onion meristem cells in the 10-30°C range. Different cycles are represented by means of a number of concentric circles, whose radii are in each case proportional to absolute cycle time at each temperature. Cycle time increases with decreasing temperature, but the relative duration of cycle phases is preserved at all temperatures, as is evidenced by the constancy of cell frequencies in the different cycle compartments (López-Sáez et al. 1966a; González-Fernández et al. 1971a). These results agree with those of Van't Hof and Ying (1964) in *Pisum*, and Mackenzie et al. (1966) in *Tetrahymena*.

On the other hand, Burholt and Van't Hof (1971, in *Helianthus*), Wimber (1966,

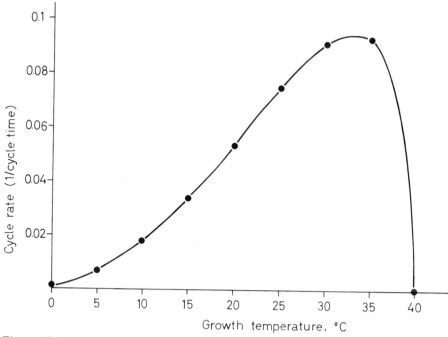

Figure 10-7.
Cycle rate in onion meristems at different growth temperatures. Cycle rate is the inverse of cycle time in hours. The cycle rate is a linear function of growth temperature. The maximum cycle rate is seen at 30–35°C. There is no cell progression at 40° but there is some small cycle rate at 0°C (Data from López-Sáez et al. 1966 a, by permission).

in *Tradescantia*), and Murin (1966, in *Vicia*) found that cell cycle compartments are temperature dependent but that the proportion of the cycle occupied by them changes disproportionately. The disagreement between both pieces of evidence may be related either to the difference in material or to the environment being inadequate for steady state kinetics.

The proportionality of cycle compartments at different temperatures can hardly be explained by similar temperature coefficients for most enzyme reactions, because roots that have been grown at either 15°C or 35°C can achieve their own particular steady state; nevertheless, when they are shifted from one temperature to another they undergo a thermal shock, reflected in differential lengthening of their interphase. Moreover this sensitivity is cycle-position dependent (Figure 10-9).

With respect to other environmental factors, apart from temperature, we will only comment that the rate at which cells go through the cycle (cycle rate) decreases with increasing osmotic pressure but the quantitative effect is not very high. However, cycle rate is very much conditioned by oxygen tension, and its rate increases with increasing oxygen concentration in air-bubbling up to 20% (Figure 10-10).

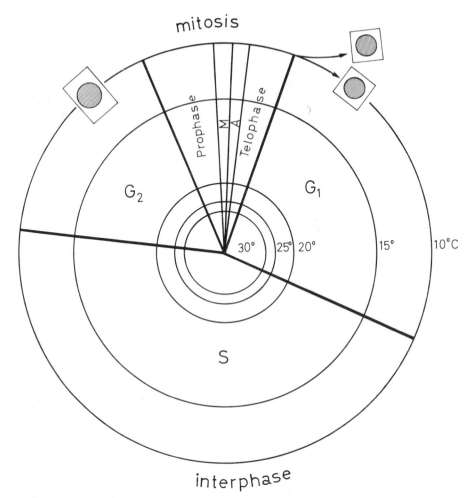

Figure 10–8.
Diagram showing the constant distribution of cell cycle phases in *Allium cepa* meristems growing at different temperatures (10–30°C). The radii of the different circles are proportional to the different cycle times. Cycle time lengthens with decreasing growth temperature. For instance, the cycle lasts 13.5 hours at 25°C, 29.8 hours at 15°C, and 54.6 hours at 10°C.

The circle sections are proportional to the stage frequency in meristem, which were similar at all temperatures tested. Hence, temperature modifies cycle duration without disturbing the relative duration of the different interphase and mitotic stages (Data from López-Sáez et al. 1966a, by permission; and from González-Fernández et al. 1971a, by permission).

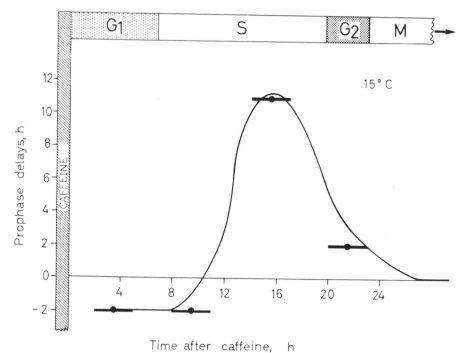

Figure 10-9.
Interphase sensitivity to 35°C thermal shocks in binucleate cells growing at 15°C.
Bars represent thermal shock duration. The shocks in G_1 or early S shorten interphase duration, as would correspond to a faster cycle rate at a higher temperature.
However, mid S is a highly thermosensitive stage, roughly corresponding to that
within S period where replication rate diminishes (c.f. Figure 10-2) and sensitivity
to DNA synthesis inhibition drops (c.f. Figure 10-3). G_2 is moderately thermosensitive (Adapted from De la Torre et al. 1971, by permission).

Cytoplasmic Regulation of G_2 Timing

Binucleate and polynucleate cells served as a basis for studying the asynchronous
development which takes place at the initiation of mitosis by different nuclei
present in the same cytoplasm of animal cells (Oftebro and Wolf 1967; Johnson
and Rao 1970).

Cells with a chromosome complement of 8n distributed in either four nuclei
(2n-2n-2n-2n), three nuclei (2n-4n-2n), or two nuclei (4n-4n) were used to study
the initiation of mitosis in onion meristem cells. The frequency of prophase asynchrony between the different nuclei sharing a common cytoplasm is recorded in
Table 10-1. These frequencies are similar to those of asynchrony at S/G_2 transition
except for cells with three nuclei (2n-4n-2n) where asynchrony frequency diminished
(González-Fernández et al. 1971b).

As seen in Figure 10-11B and C, the morphological pattern of prophase asynchrony also resembles that at S/G_2, so that nuclei surrounded by a larger cytoplasm

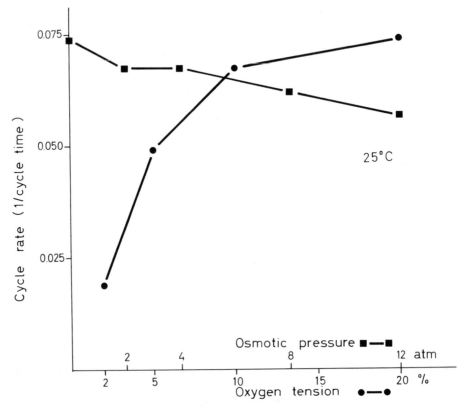

Figure 10-10.
Cycle rate at different osmotic pressure and at different oxygen tension. Cycle rate increases with increasing oxygen tension in the range tested (0–20%), while it diminishes with increasing osmotic pressure in the range 0–12 atm (adapted from González-Bernáldez et al. 1968, by permission; López-Sáez et al. 1969, by permission).

enter prophase sooner. The asynchronous initiation of mitosis seems to reflect the S/G_2 asynchrony as may be directly visualized when ^{3}H-thymidine is given to polynucleate cells on the S/G_2 boundary and later studied at the G_2/prophase boundary.

"Fast" nuclei already in prophase appear unlabeled, i.e., they are the first to reach prophase and were also ahead at G_2, when the pulse was given. On the contrary, "slow" nuclei which are still in interphase are seen to be labeled, so they were already behind (i.e., still in S) when the pulse was given.

An important fact emerging from the following experiment is that the asynchrony between the several nuclei of a polynucleate cell on the S/G_2 boundary is partially balanced out or compensated during G_2.

G_2 can be either shortened or lengthened by the cytoplasm environment. A minimum G_2 of about 4 hours exists in onion meristem mononucleate cells, at 15°C, measured by recording ^{3}H-thymidine labeled prophases. However, in (2n-4n-

2n) polynucleate cells, "fast" nuclei show a G_2 of 6 hours (two hours longer than the fastest G_2 in a mononucleate cell), while "slow" nuclei show a G_2 of only 3 hours (González-Fernández et al. 1971b).

The shortening of the G_2 period in "slow" nuclei seems to indicate that when they reach this stage some of the metabolic processes that they should go through have already been passed by the "fast" nuclei. This fact, as well as the lengthening of G_2 period in "fast" nuclei of a polynucleate cell, could be explained by a common mechanism. The "fast" nuclei in a polynucleate cell take longer over their G_2 period than mononucleate cells, probably because they have to interact with a larger volume of cytoplasm to prepare for mitosis, while the other nuclei are still completing their DNA synthesis. But "slow" nuclei take less time for G_2 than mononucleate cells since the interaction with the cytoplasm has been partially effected by the "fast" nuclei. Similar conclusions may· also be made from the experiments, on animal cells we mentioned earlier (Oftebro and Wolf 1967; Johnson and Rao 1970).

PROPHASE SYNCHRONIZATION

The study of mitosis in the polynucleate cell population shows that at the beginning of a mitotic wave all prophases are asynchronous, while the frequency of asynchrony progressively diminishes with time. (Giménez-Martín et al. 1968).

From metaphase onwards all the nuclei are invariably synchronized (Figure 10-11E,F). This suggests two possibilities: (1) that "fast" nuclei wait for the "slow" ones to catch up before beginning metaphase, or (2) that the "slow" ones speed up their process of development so as to catch up with "fast" ones.

An increase in synchronization time takes place as the number of nuclei per cell increases (Table 10-3) suggesting that "fast" nuclei wait for the rest during prophase. "Slow" nuclei may, perhaps, produce an inhibitor preventing "fast" ones from progressing towards metaphase, this inhibitor being able to act on other nuclei via cytoplasm. This prophase synchronization mechanism may be similar to that postulated for animal cells (De Terra 1960; Oftebro and Wolf 1967; Krishan and Ray-Chaudhuri 1969; Johnson and Harris 1969; Rao and Johnson 1974).

What is the biological significance of the control we are referring to in a mononucleate cell? If each chromosome or chromosome group can produce an inhibitor and put it out of action when it has completed interphase, such a mechanism would explain why all the chromosomes end prophase simultaneously, even if the previous duplication was not synchronized and even given asynchronous condensation of chromosomes during prophase (Kuroiwa 1974).

The same mechanism could account for the synchrony in groups of cells connected by cytoplasmic bridges, such as in many animal and plant meiocytes (Treub 1879; Fawcett et al. 1959; Heslop-Harrison 1964; Erickson 1964; Risueño et al. 1969).

MACROMOLECULAR BIOSYNTHESIS DURING PROPHASE

The majority of macromolecule biosynthesis in the cell cycle occurs during interphase and, in fact, declines in mitosis, being virtually arrested at metaphase. In both

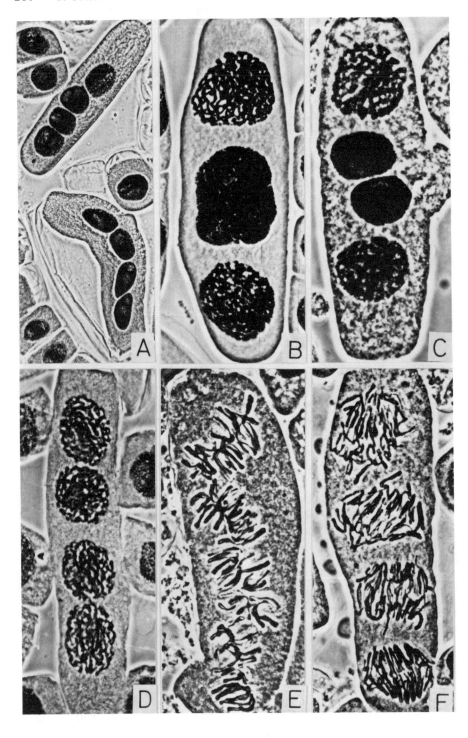

plant and animal cells a number of authors have shown that RNA and protein syntheses also take place during prophase and telophase (Scharff and Robbins 1965; Donnelly and Sisken 1967; Nešković 1968).

The occurrence of macromolecule syntheses during mitosis constitutes circumstantial evidence that does not explain their possible role in the mitotic cell's physiology. What is important is to show whether these syntheses are required for the occurrence of particular cytological events.

Protein Synthesis

By inhibiting protein synthesis in a binucleate cell population of onion meristems from the time when its fastest cells are in the initial third of prophase, it appears that the prophase cells are unable to progress towards metaphase or to remain at prophase. They must jump to interphase (upper part, Figure 10-12).

In an attempt to confirm this, colchicine was used in a parallel experiment. The weak metaphase accumulation seen in the lower part of Figure 10-12, accompanied

---

Figure 10-11.

Different stages of the cycle of 8n polynucleate cells. These cells have been induced in *Allium cepa* L. root meristems by two consecutive caffeine treatments, as displayed in Figure 10-1 (see legend).

A. Two polynucleate cells (2n-2n-2n-2n) as they appear in a meristem squash (From figure 3, Giménez-Martín et al. 1968, by permission).

B. Polynucleate cell (2n-4n-2n) showing asynchrony in the G_2/prophase boundary, since nuclei at the extremes have reached prophase while the central tetraploid nucleus is still in interphase. (From Figure 9, Giménez-Martín et al. 1968, by permission).

When ^{3}H-thymidine is provided at the S/G_2 boundary of the polynucleate cell population and the label pattern is sought in cells such as this, the "slow" tetraploid nucleus is invariably the one which incorporates ^{3}H-thymidine, i.e., the one which was also behind at the S/G_2 boundary. Hence, the nuclear asynchrony in G_2/prophase progression reflects the S/G_2 asynchrony.

The reported asynchrony in G_2/prophase progression is more frequent in this cell type (2n-4n-2n) than in the other type (2n-2n-2n-2n). (See values in Table 10-1).

C. G_2/prophase asynchrony in a (2n-2n-2n-2n) polynucleate cell. (From Figure 1, González-Fernández et al. 1971b, by permission). Nuclei at extremes are usually faster than those in central position.

D. Prophase in a tetranucleate cell. The high frequency of asynchrony in G_2/prophase passage during early hours of the prophase wave of the polynucleate population, progressively disappears so that, later on, all nuclei in the polynucleate cells are in prophase. From prophase onwards, all nuclei in each polynucleate cell are always synchronous.

E. Metaphase in a tetranucleate cell. (From Figure 4, González-Fernández et al. 1971b, by permission).

F. Synchronous anaphase. (From Figure 1, González-Fernández et al., 1971b.)

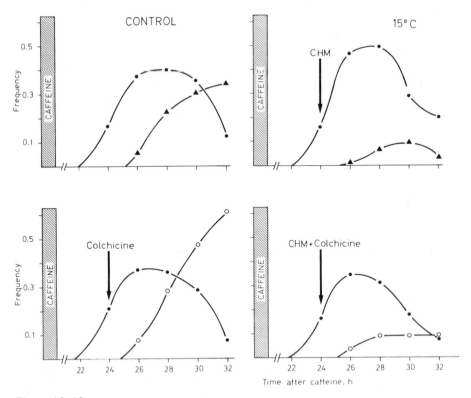

Figure 10–12.

Prophase and metaphase waves in the binucleate cell population (upper left part). The fastest binucleate cells reach prophase at the 23rd hour and metaphase at the 26th hour after caffeine, at 15°C.

Normal prophase progression towards metaphase depends on a prophase protein synthesis, since when cycloheximide-mediated inhibition of protein synthesis is started at the 24th hour (upper right part) a similar prophase wave occurs but no metaphase wave follows it. This suggests that prophase cells pass to interphase without further progress through metaphase.

By using colchicine in the parallel experiments shown in the lower part of the figure, this fact is confirmed.

Anisomycin inhibition of protein synthesis gives a similar picture. (Adapted from García-Herdugo et al. 1974, by permission).

by diminishing prophases when protein synthesis is inhibited indeed indicates their passage into interphase (García-Herdugo et al. 1974).

The biprophases seem to return to G_2 and not to bypass mitosis by jumping to the next G_1, since (1) during recovery from an 8-hour inhibition of protein synthesis, a new wave of biprophases occurs though it is much less steep because of the loss of synchrony, and (2) biprophases with tetraploid nuclei (4C–4C) were sought but not found (García-Herdugo, personal communication). On the other hand, the entrance of binucleate cells into prophase for at least 2 hours in the presence of the inhibitor confirms that synthesis of proteins essential for mitosis initiation does not occur in late G_2.

As far as we know, the studies carried out with animal material indicate that all proteins necessary for successfull cell division are produced prior to the onset of mitosis (Scharff and Robbins 1965; Tobey et al. 1966; Buck et al. 1967). In *Physarum polycephalum*, however, Cummins et al. (1966) observed that cycloheximide produced complete prophase blockage, when the cells were treated during prophase, just before the nucleoli dissolved. They concluded that proteins necessary for mitosis were synthesized up to prometaphase, and that after that time no essential proteins were synthesized.

In generative nuclei of pollen grains of *Ornithogalum virens* cultured in vitro, Niitsu et al. (1972) observed that protein synthesis inhibition at prometaphase brought about a remarkable delay in the times of entrance into new stages, especially into metaphase.

To sum up, protein synthesis taking place in plant prophase is essential for further mitotic progression since a lack of synthesis induces a return from prophase to interphase.

RNA Synthesis

In the normal progression of mitosis, nuclear envelope rupture marks the initiation of prometaphase. From that moment up to telophase, when the nuclear membrane is reconstituted, the chromosomes remain free in the cytoplasm.

Microtubules from the spindle are thus able to attach to the centromeres, ensuring their subsequent division and the separation of each chromosome half towards each cell pole by late mitosis. The Bajers have made beautiful films on these processes in the endosperm of *Haemanthus*, where the first trace of spindle formation is seen a few hours before the breakdown of the nuclear envelope. In these time-lapse films, nuclear envelope rupture is preceded by changes resembling "boiling," especially at polar regions where the envelope first breaks (Bajer and Molè-Bajer 1972).

As we shall see, in plant cells nuclear envelope breakdown at the end of prophase depends on RNA synthesis occurring in prophase itself. When binucleate cells that are initiating prophase are treated with inhibitors of transcription, e.g., ethidium bromide (EB), or 3′deoxyadenosine (3′AdR), it was found that cell progression towards metaphase is prevented as is nuclear envelope breakdown.

When 3′AdR is given to a binucleate cell population when it reaches prophase (24 hours after caffeine), it is demonstrated that entrance into prophase is not prevented by the treatment, while nuclear envelope rupture is inhibited and an endomitotic process is initiated. The chromosomes continue their cycle up to an advanced stage of condensation, greater than in normal metaphase (Figure 10-13 A), but inside the intact nuclear membrane (Giménez-Martín et al. 1971b), which resembles animal cell endomitosis as described by Geitler (1939).

Endomitosis itself makes it evident that (1) the mitotic chromosome cycle is independent of any concurrent RNA synthesis, (2) metaphase chromosome contraction is independent of nuclear membrane rupture and (3) apparent division

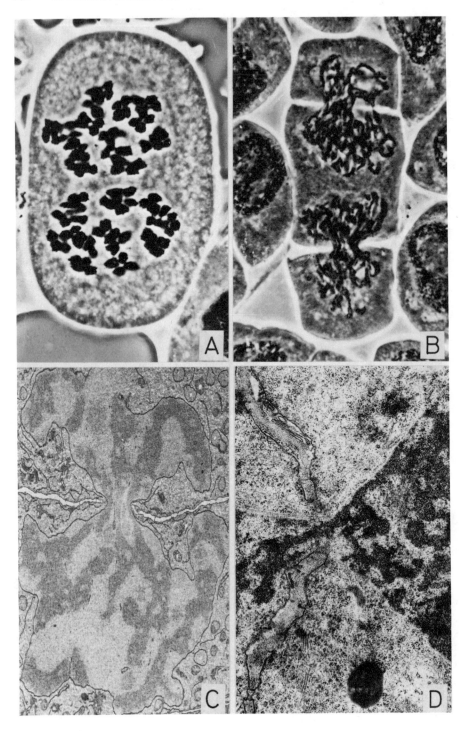

and separation of daughter centromeres can occur inside an intact nuclear envelope. 3'AdR-induced endomitosis only differs in this last feature from the endonuclear c-mitosis which organic mercury compounds produce in *Allium* (Levan 1971). Endonuclear c-mitosis reveals that all features of colchicine-treated chromosomes can be obtained by chromosomes inside an intact nuclear envelope.

Much earlier induction of endomitosis seems to take place spontaneously in some differentiating plant cells (D'Amato 1964; Nagl 1970, 1972, 1973; Tschermark-Woess 1971). In these cells, endomitosis is characterized by the initial decondensation of nuclear chromocenters (the dispersion stage during very early prophase) followed by a recondensation of the now cleaved chromocenters, while further chromosome cycle and nuclear membrane rupture are lacking (Nagl 1970, 1972, 1973).

The use of ethidium bromide (EB) in binucleate cells shows that this intercalant drug can induce an amitotic process in prophase cells. Observations under the light and the electron-microscope of meristem cells in the presence of EB have shown how this chemical inhibits the nuclear membrane from breaking down. The chromosomes retain a prophase-like condensation, though cytokinesis takes place (González-Fernández et al. 1970a,b; Risueño et al. 1971; De la Torre et al. 1973). Cytokinesis usually ends with an unbalanced genome distribution since chromosomes are randomly partitioned (Figure 10-13B,C,D).

From EB-induced amitosis we learn that cytoplasmic division can take place with-

---

Figure 10-13.
The inhibition of RNA synthesis in prophase produces inhibition of nuclear envelope breakdown. *Allium cepa* root meristems.

A. Induction of endomitosis by 3'deoxyadenosine (3'AdR), a nucleotide analogue which inhibits transcription. When the inhibition starts at the initiation of the prophase wave of the binucleate cell population, the chromosomes continue their condensation cycle inside an intact nuclear envelope. The centromeres of the chromosomes appear divided and separated, the only feature in which induced endomitosis differs from the endonuclear c-mitosis produced by organic mercury componds (Levan 1971). (From Figure 3, Giménez-Martín et al. 1971b, by permission.)

B. When ethidium bromide (an intercalant inhibitor of transcription) is given at the time when the prophase wave is initiated in the binucleate cell population, an amitotic process follows. In amitosis, the chromosomes do not attain significant condensation. The corresponding cytokinetic plates are formed so that unbalanced distribution of chromosomes takes place when cytokinesis is complete. Induced amitosis is evidence of the independence of nuclear and cytoplasmic division controls. (From Figure 1 f, González-Fernández et al. 1970b).

C. Ethidium bromide-induced amitosis, as seen at the electron microscope level. The permanganate fixation shows the growing cell plate "strangling" the cell nucleus. With this fixation only the membrane system is well preserved.

D. Ethidium bromide-induced amitosis. The conventional fixation (glutaraldehyde-osmium) preserves other cell components apart from the membrane system. Spindle microtubles are seen. (From Figure 1, Risueño et al. 1971, by permission).

out a previous nuclear division. We may thus make the following generalizations: (1) there is a certain independence of nuclear from cytoplasmic division controls; the induction of binucleate cells (nuclear without cytoplasmic division) demonstrates the same independence, (2) cytokinesis appears to be triggered in parallel to mitosis, (3) cytokinesis is independent of mitotic RNA synthesis, (4) cytokinesis is also independent from nuclear membrane rupture and from chromosome cycle progression, and (5) formation of the mitotic spindle is independent of both mitotic RNA synthesis and nuclear envelope rupture.

Induction of endomitosis and amitosis processes, as well as the return to interphase by inhibiting gene expression in prophase cells, shows that prophase is a decisive period for cell division and/or differentiation in plant cells, since mitotic progression depends on protein and RNA syntheses which take place during prophase itself, and mitotic alternatives involved in normal plant differentiation can then be triggered.

Timing of Protein and RNA Syntheses During Prophase

As we saw within prophase, RNA and protein syntheses are required for the normal progression towards metaphase. The timing of RNA and protein syntheses could enlighten the relationship between both processes.

With a view to locating, within prophase, the relative position of both processes, the experiments displayed in Figure 10-14 were carried out. The results suggest

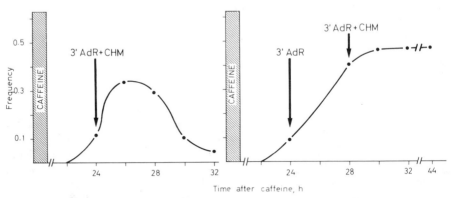

Figure 10-14.
Mapping of prophase region sensitivity to protein and RNA synthesis inhibition. Under simultaneous inhibition of protein and RNA synthesis from the 24th hour after caffeine treatment, the normal wave of biprophases is recorded (left part of the figure).

However, if the simultaneous inhibition of RNA and protein synthesis starts at the 28th hour after caffeine (four hours later than the beginning of RNA synthesis inhibition) biprophases remain accumulated. Hence, the period sensitive to protein synthesis inhibition is earlier than that sensitive to RNA synthesis inhibition in prophase, (Adapted from García-Herdugo et al. 1974.)

that the protein synthesis period occurs earlier than the period sensitive to RNA synthesis inhibitors. Therefore the prophase protein synthesis does not appear to be a translation of the RNA transcribed during prophase itself.

FROM METAPHASE TO TELOPHASE

The aspect of c-mitotic substances which we will just touch on here is the fact that, through them, it is possible to conclude that by mid-mitosis the chromosome cycle is independent of the existence of any mitotic apparatus. The discovery of actin and myosin in *Physarum* nuclei, if extended to other eukaryotic cells, could explain this phenomenon (Jockush et al. 1974).

Hypoxia accumulates cells in metaphase suggesting probably a high energy dependence of the beginning of anaphase. Nevertheless chromosome cycle and spindle formation occur under this low energy production situation. The metaphase chromosomes under hypoxia appear highly condensed but, in contrast with c-mitosis, centromeres appear double and separated as in normal metaphase. When roots recover from hypoxia all the metaphases develop into anaphases within two hours. The process takes place slowly, so that it is possible to see the early emigration of sister centromeres while telomeres still appear close together.

TELOPHASE

During normal telophase, nuclear reconstruction is achieved by a process roughly the reverse of prophase. Two main morphological occurrences can be easily discerned, i.e., nuclear membrane formation, and chromosome decondensation. It is thought they are the structural basis for realigning the transcription and replication patterns of the cycle, for they will lead the newly formed cell into its fully active interphase condition.

As Wolfe and Rodríguez (1974) have shown by using dithiothreitol, two processes can be resolved in the reconstruction of the nuclear envelope in sea urchin embryos; first, each individual chromosome becomes more or less surrounded by nuclear membrane and then all the membrane-coated chromosome vesicles fuse to form a continuous envelope containing the full chromosome complement.

To our knowledge, dissociation of both stages has not been found in plant material, but lagging chromosomes or chromosome segments in plant telophase are apt to form micronuclei. It is possible that the rough endoplasmic reticulum participates in nuclear membrane (NM) reconstruction as it does in the repair of NM injury in *Amoeba* (Flickinger 1974). The reconstruction process seems to be favored by alkaline pH (pH 8.0) as shown in Chinese hamster cells (Obara et al. 1973). Most probably this is generally true for all eukaryotic cells. Macromolecular biosynthesis in mid-mitosis can hardly be involved in the process of nuclear membrane reconstruction, because protein synthesis is very low in mid-mitosis and there is no detectable transcription. As far as we know, nuclear membrane formation and chromosome decondensation show no sensitivity to simultaneous inhibition of macromolecule synthesis.

The chromosome decondensation taking place in telophase is the starting point for the initiation of new nuclear activity in the newly formed cell. In general, it is thought that gene transcription is not compatible with chromosome condensation (Frenster 1969). However, DNA replication can occur on condensed chromatin, because Clowes (1967) showed that telophase chromosomes replicate in maize cap initial cells and Nagl (1970) found incorporation of DNA precursors on chromo-centers of *Allium carinatum* cells, in late S.

The earliest detectable transcription in onion meristem cells takes place at approximately telophase and is confined to the nucleolar region, as shown by the *in situ* assay for endogenous RNA polymerase (unpublished results). This seems to be in opposition to the findings of Simmons et al. (1973) in HeLa cells, where resumption of nuclear RNA synthesis occurs in nucleoplasm, at or near the reforming nuclear envelope, and represents extranucleolar RNA synthesis.

The relationship between chromosome decondensation and nucleologenesis (a process linked to rRNA transcription) is not close. The latter can take place either in early telophase or in late telophase/early G_1 or even well into G_1, as was seen in different cell populations of the same maize root (De la Torre and Clowes 1972). This fact suggests that the end of the chromosome decondensation cycle is not, in itself, the trigger of transcription.

DURATION OF MITOTIC PHASES

In a fashion similar to interphase, mitotic time is clearly modified by internal and external factors. Studies on the relationship between cycle duration and DNA content suggest that there is a linear correlation between them, all the mitotic phases being proportionally modified (Van't Hof and Sparrow 1963; Van't Hof 1965).

Kaltsikes (1971), by studying the duration of the mitotic phases in a triticale and its parents, found that the timing of prophase and other mitotic phases were not completely parallel to the cell's DNA content (Figure 10–15).

Timing and, moreover, completion of prophase is fully dependent on macromolecule synthesis taking place in prophase itself. The mean duration of metaphase and anaphase in meristems increases or decreases twofold for quiescent center and cap initial cells, respectively (Clowes 1961), which confirms that a cell's metabolic status can modify the duration of these phases. Apart from the synchronization time during prophase, the last half of mitosis was uniformly prolonged over metaphase, anaphase, and telophase in polynucleate meristematic cells. On the other hand, metaphase-to-telophase progression in onion was undisturbed by simultaneous RNA and protein synthesis inhibition. Both facts suggest that cell metabolism before but not after metaphase could somehow predetermine the progression rate of mitosis.

The study of mitotic and phase indices in onion roots, at different growth temperatures, in steady-state kinetics, showed the constancy of the relative proportion of each mitotic phase, as represented in Figure 10–8. Therefore we are obliged to admit that the duration of all the morphological stages is affected to the same

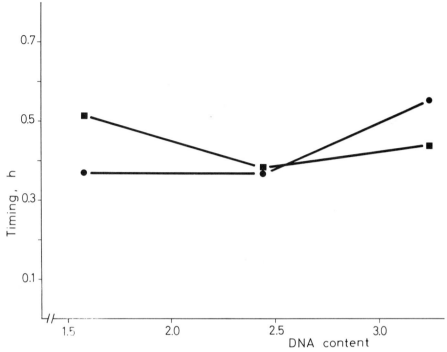

Figure 10-15.
Duration of prophase and other mitotic phase (metaphase, anaphase, and telophase) in a triticale and its parents with different DNA content. (Adapted from Kaltsikes 1971, by permission.)

extent as that of the whole cycle at each temperature and that their relative duration remains constant. According to data in Table 10-4, the length of mitosis can be as short as 1.3 hours (at 30°C) or as long as 15.3 hours (at 5°C) in *Allium cepa* and, for instance, anaphase can take from 8.4 minutes to 1.5 hours without modifying its relative duration in mitosis.

Lastly, osmotic pressure shows negligible effects on mitosis time while low oxygen tension relatively increases metaphase duration, without significantly affecting the other phases.

CYTOKINESIS

In plant cells, cytokinesis is characterized by the formation of the cell plate separating the two daughter cells. Whaley and Mollenhauer (1963) and Frey-Wyssling et al. (1964) have shown that the cell plate is formed by the coalescence of vesicles produced by Golgi bodies. The membranes of these vesicles make up the plasma membrane of the new cell surfaces; and the content of these vesicles gives rise to the amorphous matrix of the new wall.

Cytokinesis appears to be a continuous process in which it is difficult to separate

Table 10-4

Duration of interphase and mitosis at different temperatures in *Allium cepa* root meristems (hours)[1]

	Temperature, $^{\circ}C$					
	5	10	15	20	25	30
Interphase	114.1	48.2	26.2	16.6	12.0	9.7
Prophase	7.2	3.1	1.7	1.0	0.7	0.6
Metaphase	2.0	0.8	0.5	0.3	0.2	0.2
Anaphase	1.5	0.6	0.3	0.2	0.1	0.1
Telophase	4.6	1.9	1.1	0.7	0.5	0.4
Mitosis	15.3	6.4	3.6	2.2	1.5	1.3

1. Durations have been estimated for a mean mitotic index of 11.9 (percentage of meristem cells in mitosis) at all temperatures tested. The frequency of cells at different mitotic stages, taken from original data supporting fig. 5 in López-Sáez et al., 1966a. The duration of the cycle at different temperatures taken from table 3, ibid. The absolute duration of any phase was simply obtained by using the equation :phase duration = phase frequency X cycle duration. This equation applies to meristems growing under steady state conditions, since linear kinetics occur.

the various mechanisms that operate as one harmonious whole. The large number of treatments producing binucleate cells constitute a valuable arsenal of means by which the dissection of cytokinesis may be effected. The production of binucleate cells is the common denominator, while the mechanism at work might be different. Thanks to these treatments, it has proved possible to divide cytokinesis into a number of phases overlapping like tiles on a roof (López-Sáez et al. 1966b; Risueño et al. 1968). The phases of cytokinesis are (1) production of Golgi vesicles, (2) vesicle accumulation, (3) vesicle arrangement, and (4) vesicle coalescence.

Production of Golgi Vesicles

This stage involves the formation, in large quantities, of vesicles from Golgi bodies. At anaphase and early telophase the Golgi bodies accumulate on the border of the phragmoplast and the vesicles congregate at the equatorial region.

Observations under the electron microscope show that a heat shock (38°C) causes a marked drop in the production of vesicles and even disorganizes them by dissociating their flattened sacs (at 40°C). After 1-hour treatment, the anaphases show a very low proportion of vesicles compared with those of untreated cells. After 1-hour recovery, the telophases show an apparently normal arrangement but with fewer vesicles. Separate groups of vesicles form a few isolated fragments of cell plate, which fail to join. At the beginning of interphase these fragments of cell plate disperse about the cytoplasm.

This seems to show that the formation of the cell plate requires a threshold number of vesicles, and if their production is blocked, the normal development of cytokinesis is impeded. These experiments also show that Golgi bodies are thermolabile systems.

Vesicle Accumulation

This stage reaches its highest point at early telophase. The continuous supply of new vesicles facilitates their accumulation at the central equatorial region (Figure 10-16 A) as well as the centrifugal spreading of the accumulation cloud.

C-mitotic substances (colchicine or γ-hexachlorocyclohexane) also inhibit the formation of the cell plate. During the first 3 hours of treatment, Golgi bodies produce vesicles in apparently normal quantities, and the number of these vesicles in the cytoplasm increases. When the chromosomes are enclosed within the nuclear membrane, a polyploid nucleus is formed, while the vesicles remain dispersed about the cytoplasm. Hence, c-mitotic substances do not inhibit the production of Golgi vesicles, but they block cytokinesis insofar as they prevent the vesicles from accumulating in the equatorial region.

Later studies have confirmed the role of microtubules in cell plate formation. They appear to be involved in the accumulation, arrangement, and fusion of vesicles (Hepler and Newcomb 1967; Lambert and Bajer 1972). Since these processes take place in sequence, microtubule depolymerization should lead to blockage of the accumulation phase, as happens effectively when c-mitotic substances are tested.

Vesicle Arrangement

During telophase, the arrangement of the cloud of Golgi vesicles, previously congregated, begins in the central part of the equatorial plane and proceeds centrifugally (Figure 10-16B). At the beginning of interphase the accumulation of vesicles in the equatorial region disappears gradually, and they lie more or less uniformly distributed all over the cytoplasm.

Under caffeine, production and accumulation of the Golgi vesicles develop apparently in a normal way. However, in mid-telophase the vesicles accumulated in the equatorial region do not line up on the equatorial plane.

Cells under 400 atm hydrostatic pressure show a similar pattern; cytokinesis also seems to be blocked in the arrangement of the vesicles. They form a cloud which remains stationary between the daughter nuclei when the interphase has already begun, until the treatment is discontinued, whereupon the vesicles move away from the equatorial region.

Vesicle Coalescence

In late telophase the coalescence of Golgi vesicles starts in the inner parts of the equatorial plane (Figure 10-16C,D). Coalescence, arrangement, and accumulation can be observed, respectively, in the inner, intermediate, and marginal parts of the cell plate. This observation clearly reflects the course of the overlapping phases

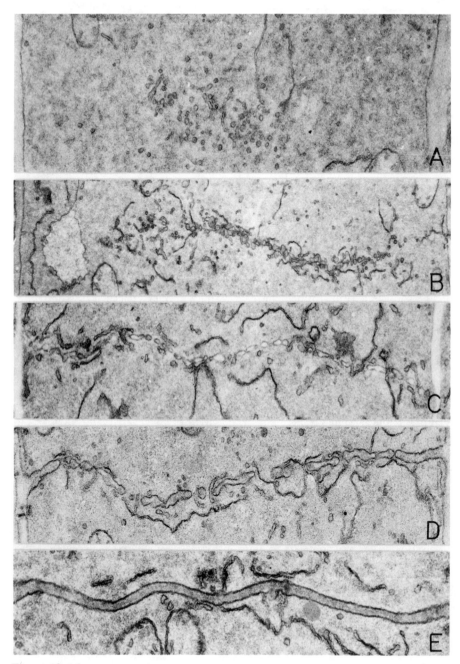

Figure 10–16.
Sequential phases in plant cytokinesis, where the cell plate is formed by the coalescence of vesicles produced by Golgi bodies. The vesicle *production* stage is a thermosensitive process, since a heat shock blocks it.

which make up cytokinesis. Finally, the longitudinal walls of the mother cell are reached by the growing cell plate and cytokinesis is complete.

In a recent study, Paul and Goff (1973) investigated the action mechanism of caffeine in the prevention of Golgi vesicle coalescence. They propose that this inhibitor might act directly on membrane fusion capacity by blocking Ca^{++} binding sites. In fact, López-Sáez et al. (1966b) and Paul and Goff (1973) agree in claiming for caffeine an action on vesicle fusion, but while the former authors postulate that caffeine blocks vesicle arrangement and subsequently fusion, the latter postulate that a direct action on vesicle coalescence occurs.

The completion of cell plate is seen in Figure 16E.

THE NUCLEOLAR CYCLE

By staining nucleoli, we can distinguish four types of cells in a meristem population:

1. interphase cells with two fully organized nucleoli (sometimes fused in one larger nucleolus) (Figure 10-17E);
2. prophase cells whose nucleoli take on an irregular shape (Figure 10-17A);
3. metaphase and anaphase cells with no apparent nucleoli (Figure 10-17B);
4. telophase cells with an evenly distributed number of "prenucleolar bodies" scattered along the chromosomes (Figure 10-17C and D) (Stockert et al. 1969; Giménez-Martín et al. 1971a).

These cellular types represent different stages of the nucleolar cycle found in a meristem population.

The nucleolus is not a permanent organelle in the life of a cycling cell since it disappears during mitosis. It is visible during those periods of the cycle in which the nuclear membrane is present. When the nucleolus begins to disappear in *Allium cepa* L., it loses its spherical shape (Figure 10-17A) and assumes a multilobulated form such as has been described by Lafontaine and Chouinard (1963) in the prophase of *Vicia faba*. In onion, it assumes an appearance resembling segregation, i.e., granular moieties acquire a peripheral location while fibrillar moieties concentrate in the center (Figure 10-17A). Guttes et al. (1968) made an ultrastructural study of nucleolar disorganization in *Physarum* and found that this also occurs before metaphase, and that the remnants of the nucleolus can be seen during the middle

---

A. *Accumulation* of vesicles in equatorial region. Late anaphase. C-mitotic chemicals inhibits this stage.

B. *Arrangement* of vesicles on the equatorial plane. Telophase. Caffeine inhibits this stage.

C. *Coalescence* of vesicles whose membranes fuse giving rise to the new plasma membranes which will separate daughter cells. Mid telophase.

D. The end of the coalescence phase. Late telophase. The new cell wall reaches both edges of the mother cell wall.

E. Cell wall is now completely formed. Daughter cells are fully independent. (From Risueño et al. 1968, by permission.)

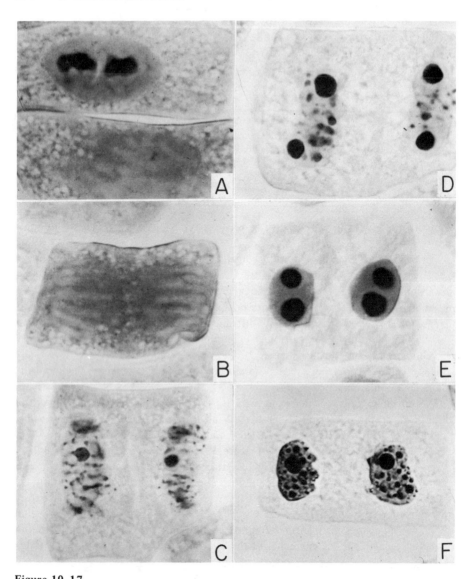

Figure 10–17.
Different nucleolar stages in onion meristem cells. A protein component of the nucleolus is highly contrasted by silver impregnation. (Fernández-Gómez et al. 1969, by permission.)

A. On the left, nucleoli in disorganization in a prophase cell. Nucleoli adopt a lobed appearance and they gradually diminish in size as prophase advances. On the right, cell in prometaphase. Chromosomes appear bleached under this stain. Two paired argentophylic spots close to the metaphase plate are probably the nucleolar organizer regions (NOR). (From Figure 1b, Giménez-Martín et al. 1971a, by permission).

stages of mitosis. In plant material Chardard (1962) and Das and Alfert (1963) advanced the postulate that the nucleolar material is dispersed about the cytoplasm at the end of prophase.

Studies at the ultrastructural level show that nucleolar remnants are transported on mitotic chromosomes, and are, most probably, the origin of the prenucleolar bodies which appear in telophase. When a preferential stain for nucleolar protein is used, nucleolar-like material is seen on metaphase and anaphase chromosomes. In telophase, the prenucleolar bodies are also highly contrasted by this technique.

From the data in Table 10-5, we can infer how the nucleolus develops in com-

Table 10-5

Frequency of meristem cells in the different stages of chromosome and nucleolar cycles and absolute timing (in hours) for the different phases, at $15°C$ growth temperature[1]

Chromosome cycle	prophase	metaphase-anaphase	telophase
Frequency	0.061	0.029	0.039
Time	1.8	0.9	1.2
Nucleolar cycle	disorganization	dispersion	reorganization
Frequency	0.047	0.038	0.074
Time	1.4	1.1	2.2

1. Adapted from Giménez-Martín et al. (1971a).

B. Anaphase. No nucleolar remnant is seen. (From Figure 2a, Giménez-Martín et al. 1971a.)

C. Nucleologenesis at telophase. Prenucleolar material seems to be scattered along chromosome arms. One incipient nucleolus is seen at each nucleus.

D. Telophase in a more advanced stage of nucleologenesis. Prenucleolar material is still dispersed, while the 2 nucleoli in formation are seen on NORs. The symmetric position of daughter chromosomes with NORs is clearly seen. (From Figure 3a, Giménez-Martín et al. 1971a, by permission.)

E. Daughter cells in early G_1. Nucleologenesis has finished and the two fully organized nucleoli of each nucleus are seen. This is the normal interphase situation for functional nucleoli. (From Figure 3b, Giménez-Martín et al. 1971a, by permission. On the average, at $15°C$, nucleologenesis is finished by 1.5 hours after the end of telophase.

F. Inhibition of RNA synthesis is freshly formed binucleate cells (by 3'deoxyadenosine or ethidium bromide) prevents the completion of nucleologenesis. For instance, at $15°C$, nuclei still show this image 12 and 24 hours after the end of telophase.

CHROMOSOME   CYCLE

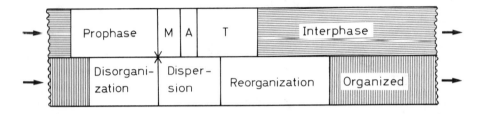

NUCLEOLAR   CYCLE

Figure 10-18.
Chromosome and nucleolar cycles in onion root meristems. The length of any phase in the diagram is proportional to its frequency in the meristem and hence proportional to its duration in time units (given the steady growth kinetics in roots). Matching of both cycles is possible since the prophase/metaphase boundary is the same as the nucleolar disorganization/dispersion boundary in this meristem. As shown in Table 10-5 the absolute duration of the whole nucleolar cycle in mitosis takes 4.7 hours in a 29.8 hour long cycle 15°C. (From Giménez-Martín et al. 1971a, by permission.)

parison with the mitotic cycle, some lack of correlation being present between the stages of the nuclear and nucleolar cycles.

Matching of Chromosome and Nucleolar Cycles

In matching nucleolar cycle to chromosome cycle in onion and maize cells, we capitalized on the fact that all prophase nuclei contain nucleoli and that no metaphase nuclei have them. The prophase/metaphase boundary is therefore taken to occur at the same time as the disorganization/dispersion phase occurs.

The matching of both cycles in onion root meristems is shown in Figure 10-18. The length of each phase in the diagram is proportional to its frequency found in squashed cells of meristems. Disorganization of nucleoli starts later than the prophase condensation cycles and reorganization occupies the last half of telophase and a portion of early G_1 similar in duration to telophase.

Role of Macromolecular Biosynthesis in Nucleolar Cycle

Nucleolar disorganization, though somehow modified in its morphological features, takes place in 3'AdR-induced endomitosis (Giménez-Martín et al. 1972). Hence, completion of nucleolar disorganization is (1) independent of simultaneous RNA synthesis and (2) independent of nuclear membrane rupture.

Six hours after the beginning of protein synthesis inhibition, more than 50% of the prophases show fully organized nucleoli, while the number of them with disorganizing nucleoli is very low compared to control meristems (Fernández-Gómez et al. 1972). This suggests that inhibition of protein synthesis induces a return to prophase from the early disorganization stage, but it does not tell us whether this is independent or not of the reverse chromosome cycle.

To study the dependence of nucleologenesis on simultaneous biosynthesis, the kinetics of the process were studied in freshly formed binucleate cells (Figure 10-19). Fifty percent of these cells, at 15°C, had completed nucleologenesis 1.6 hours after G_1 initiation (in a 7.8-hour-long G_1). Under all treatments, the appearance of prenucleolar bodies was normal in morphology and timing was comparable to controls. That is to say prenucleolar bodies are not immediate products of transcription or translation in the telophase/early G_1 cell. Moreover, the appearance of prenucleolar bodies is independent of the intranuclear presence of a nucleolar organizer region, as made evident in induced micronuclei in plant cells (Das 1962; Stockert et al. 1969).

Figure 10-19 also shows that nucleologenesis depends on simultaneous nucleolar transcription since its inhibition leads to a permanent interruption of nucleologenesis (Figure 10-17F). Induced micronuclei confirm this conclusion, since nucleologenesis depends on the intranuclear presence of a nucleolar organizer region, intracellular presence being insufficient (Das 1962).

The dependence of nucleologenesis on simultaneous transcription of ribosomal RNA genes may be the rule in eukaryotic cells. Phillips (1972) found similar results in human tumor cells, while in two other cell lines, under simultaneous RNA inhibition, she believed that re-formation of nucleoli occurred. However, the normal karyotypic condition of these cell lines is not sufficiently known and non-clustered ribosomal cistrons can lead to the interphase coexistence of numerous small nucleoli which are difficult to distinguish on a morphological basis from prenucleolar bodies.

This seems to be the case in *Amoeba*, where nucleolar material remains dispersed throughout its whole cell cycle, and nucleolus-like bodies indeed appear when simultaneous RNA synthesis is inhibited (Stevens and Prescott 1971).

Figure 10-19 shows, moreover, the unexpected fact that nucleologenesis speeds up when simultaneous protein synthesis is blocked. Prenucleolar bodies seem to group themselves together to form fully developed nucleoli at almost double the normal rate when protein synthesis is prevented. Fakan (1971) postulates that this inhibition uncouples 45S rRNA processing and hence back-inhibits rRNA transcription. Our data could better be explained if protein synthesis inhibition increased the rate of RNA synthesis as has been reported (see Stenram 1973; Foury and Goffeau 1973).

The perfect timing correlation between the disappearance of prenucleolar bodies and the growth of nucleoli in normal as well as in accelerated or delayed nucleologenesis seems to suggest that prenucleolar bodies really are incorporated into the re-forming nucleolus even though some workers believe the opposite (Chouinard 1971). Both Phillips' (1972) finding that new nucleoli do actually contain RNA synthesized before mitosis, and the proof that subnucleolar particles are associated with metaphase chromosomes (Fan and Penman 1971), are in line with this theory.

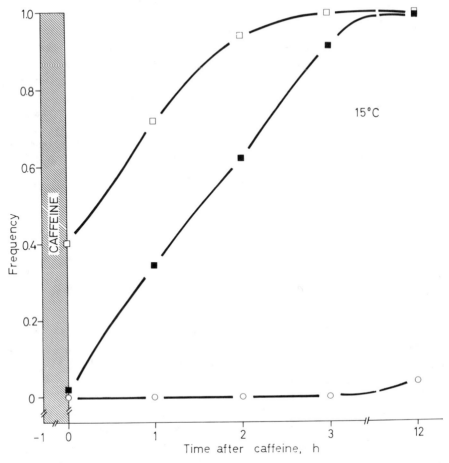

Figure 10-19.
Nucleologenesis in newly formed binucleate cells. Metabolic inhibitors were used following the experimental scheme: caffeine + inhibitor for 1 hour and then inhibitor in continuous treatment. Nucleologenesis does not occur when simultaneous RNA synthesis is inhibited (by 3' deoxyadenosine or ethidium bromide). An extranucleolar RNA synthesis inhibitor (α-amanitin), though lengthening interphase by 1.5 times, allowed a normal nucleologenesis rate. Simultaneous protein inhibition by cycloheximide or anisomycin speeds up nucleologenesis compared to controls. (Adapted from Giménez-Martín et al. 1974, by permission.)

bb - basal bodies	n - nucleus
c - chromosomes	nc - nucleolus
cn - canal	ne - nuclear envelope
cv - contractile vacuole	p - paramylon granule
f - flagellum	r - reservoir
i - interzonal spindle	t - microtubules

CONCLUSIONS

The use of chemicals and of induced polynucleate cells, among other experimental approaches, permits the dissection of various underlying mechanisms involved in cycle progression.

The cycle rate in meristems is inversely proportional to temperature (from 5°C to 30°C), while the relative duration of the different cycle compartments remains constant under steady-state kinetics.

The initiation of genome replication depends on a previous protein synthesis, and initiation is an inducible event for nuclei having replication capacity. The replication capacity appears to be coordinated with the intranuclear presence of a nucleolar organizer region. Chromosome duplication can be uncoupled from chromosome distribution, since there can be either mitosis without previous chromsome duplication or chromosome duplication without subsequent mitosis.

Another protein synthesis, in early G_2 for onion meristems, is needed for prophase chromosome condensation to occur. Synthesis taking place in prophase itself is essential for further mitotic progression, since its lack induces the return of cells to interphase.

The nuclear envelope breakdown at the end of prophase depends on RNA synthesis occurring in prophase itself. If this prophase RNA synthesis is prevented, endomitosis or amitosis can be initiated.

In prophase, the developmental stage sensitive to protein synthesis inhibition was found to be earlier than the stage sensitive to RNA synthesis inhibition.

Apparently, the mitotic chromosome cycle can take place independently of any concurrent RNA synthesis, nuclear membrane rupture, and the presence of mitotic apparatus.

Metaphase to telophase progression and the mechanism of chromosome decondensation and nuclear membrane reconstruction in telophase are processess independent of any concurrent synthesis of informational molecules.

Plant cytokinesis is characterized by the formation of the cell plate through the coalescence of vesicles produced by Golgi bodies and it can be separated into phases by means of different inhibitory treatments.

The induction of binucleate cells on the one hand, and of amitosis on the other, reveals the independence of nuclear from cytoplasmic division controls.

Lastly, nucleologenesis depends on the intranuclear presence of a nucleolar organizer region and of simultaneous nucleolar transcription. Unexpectedly, the rate of nucleologenesis increases when simultaneous protein synthesis is inhibited.

In short, the apparent continuity of the cell cycle is sustained by a series of controls, some running in sequence, others in parallel. Their experimental separation is but an open window towards future understanding of cell proliferation as a whole.

ACKNOWLEDGMENTS

We wish to thank Dr. P. W. Barlow, Dr. F. A. L. Clowes, and Dr. W. Nagl for their suggestions and comments on the manuscripts.

This review has been partially supported by the III Plan de Desarrollo (Spain) and by a grant of the Foundation Rodríguez Pascual (Spain).

We also acknowledge Mr. Christropher Dove's help in correcting the English in this chapter, and to Mrs. M. C. Partearroyo, Miss M. L. Martínez, Mrs. A. Partearroyo, and Miss O. Partearroyo for their valuable cooperation in dealing with secretarial and technical matters.

LITERATURE CITED

Alfert, M. and N. K. Das. 1969. Evidence for control of the rate of nuclear DNA synthesis by the nuclear membrane in eukaryotic cells. *Proc. Natl. Acad. Sci.* **63**:123-128.

Ayonoadu, U. A. U. and H. Rees. 1968. The regulation of mitosis by B-chromosomes in rye. *Exptl. Cell Res.* **52**:284-290.

Bajer, A. S. and J. Molè-Bajer. 1972. Spindle dynamics and chromosome movements. Intern. *Review of Cytol.* suppl. 3.

Barlow, P. W. 1974. Regeneration of the cap of primary roots of *Zea mays*. *New Phytol.* **73**:937-954.

Barlow, P. W. and P. D. M. Macdonald. 1973. An analysis of the mitotic cycle in the root meristem of *Zea mays*. *Proc. R. Soc.* Lond. B. **183**:385-398.

Barlow, P. W. and C. G. Vosa. 1969. The pattern of DNA replication in the chromosomes of *Puschkinia libanotica*. *Chromosoma* **28**:457-467.

Benbadis, M. C., M. Ribsztejn, and G. Deysson. 1974. The mode of nuclear DNA synthesis in experimentally induced binucleate cells of root meristems. *Chromosoma* **46**:1-11.

Bloch, D. P., R. A. Macquigg, S. D. Brack, and J. R. Wu. 1967. The synthesis of deoxyribonucleic acid and histone in the onion root meristem. *J. Cell Biol.* **33**:451-468.

Buck, C. A., G. A. Granger, and J. J. Holland. 1967. Initiation and completion of mitosis in HeLa Cells in the absence of protein synthesis. *Currents in Mod. Biol.* **1**:9-13.

Burholt, D. R. and J. Van't Hof. 1971. Quantitative thermal-induced changes in growth and cell population kinetics of *Helianthus* roots. *Amer. J. Bot.* **58**:386-393.

Chardard, R. 1962. Recherches sur les cellules mères des microspores des Orchidées. Etude au microscope électronique. *Rev. Cytol. et Biol. végét.* **24**:1-148.

Chouinard, L. A. 1971. Behaviour of the structural components of the nucleolus during mitosis in *Allium cepa*. *Advances in Cytopharmacology* **1**:69-87.

Clowes, F. A. L. 1961. Duration of the mitotic cycle in a meristem. *J. Exp. Bot.* **12**:283-293.

Clowes, F. A. L. 1965. Synchronization in a meristem by 5-aminouracil. *J. Exp. Bot.* **16**:581-586.

Clowes, F. A. L. 1967. Synthesis of DNA during mitosis. *J. Exp. Bot.* **18**:740-745.

Clowes, F. A. L. 1972. Regulation of mitosis in roots by their caps. *Nature New Biol.* **235**:143-144.

Cummins, J. E., J. C. Blomquist, and H. P. Rusch. 1966. Anaphase delay after inhibition of protein synthesis between late prophase and prometaphase. *Science* **154**:1343-1344.

D'Amato, F. 1952. New evidence on endopolyploidy in differentiated plant tissues. *Caryologia* 4:121–144.

D'Amato, F. 1964. Endopolyploidy as a factor in plant tissue development. *Caryologia* 17:41–52.

Das, N. K. 1962. Synthetic capacities of chromosome fragments correlated with their ability to maintain nucleolar material. *J. Cell Biol.* 15:121–130.

Das, N. K. 1963. Chromosomal and nucleolar RNA synthesis in root tips during mitosis. *Science* 140:1231–1233.

Das, N. K. and M. Alfert. 1963. Silver staining of a nucleolar fraction, its origin and fate during the mitotic cycle. *Ann. Histochim.* 8:109–114.

De la Torre, C. and F. A. L. Clowes. 1972. Timing of nucleolar activity in meristems. *J. Cell Sci.* 11:713–721.

De la Torre, C., M. E. Fernández-Gómez, G. Giménez-Martín, and A. González-Fernández. 1974. Sensitivity of the cell division cycle to α-amanitin in *Allium* root tips. *J. Cell Sci.* 14:461–473.

De la Torre, C., A. González-Fernández, G. Giménez-Martín, and J. F. López-Sáez. 1971. The effect of thermal shock on the division cycle of meristematic cells. *Cell and Tissue Kinetics* 4:569–575.

De la Torre, C., M. C. Risueño, and G. Giménez-Martín. 1973. Ethidium bromide and its effect on the nuclear membrane cycle. *Protoplasma* 76:363–371.

De Terra, N. 1960. A study of nucleo-cytoplasmic interactions during cell division in *Stentor coeruleus. Exptl. Cell Res.* 21:41–48.

Díez, J. L., A González-Fernández, and J. F. López-Sáez. 1976. On the mechanisms of the synchronization induced by 5-aminouracil in cell cycle. *Exptl. Cell Res.* 98:79–89.

Donnelly, G. M. and J. E. Sisken. 1967. RNA and protein synthesis required for entry of cells into mitosis and during the mitotic cycle. *Exptl. Cell Res.* 46:93–105.

Duncan, R. E. and P. S. Woods. 1953. Some cytological aspects of antagonism in synthesis of nucleic acids. *Chromosoma* 6:45–60.

Erickson, R. O. 1964. Synchronous cell and nuclear division in tissues of the higher plants. In *Synchrony in Cell Division and Growth*, ed. E. Zeuthen. Interscience, New York, pp. 11–27.

Evans, H. J. 1964. Uptake of ^{3}H-thymidine and patterns of DNA replication in nuclei and chromosomes of *Vicia faba. Exptl. Cell Res.* 35:327–341.

Evans, H. J. and H. Rees. 1966. The pattern of DNA replication at mitosis in the chromosomes of *Scilla campanulata. Exptl. Cell Res.* 44:150–160.

Evans, H. J., H. Rees, C. L. Snell, and S. Sun. 1968. The relationship between nuclear DNA amount and the duration of the mitotic cycle. In *Chromsomes Today*. v. 3, pp. 24–31.

Evans, L. and J. Van't Hof. 1973. Cell arrest in G_2 in root meristems: A control factor from the cotyledons. *Exptl. Cell Res.* 82:471–473.

Evans, L. and J. Van't Hof. 1974. Promotion of cell arrest in G_2 in root and shoot meristems in *Pisum* by a factor from the cotyledons. *Exptl. Cell Res.* 87:259–264.

Everhard, L. P. and D. M. Prescott. 1972. Reversible arrest of Chinese hamster cells in G_1 by partial deprivation of leucine. *Exptl. Cell Res.* 75:170–174.

Fakan, S. 1971. Inhibition of nucleolar RNP synthesis by cycloheximide as studied by high resolution radioautography. *J. Ultrastruct. Res.* 34:586–596.

Fan, H. and S. Penman. 1971. Regulation of synthesis and processing of nucleolar components in metaphase arrested cells. *J. Mol. Biol.* **59**:27–42.

Fawcett, D. W., S. Ito, and D. Slautterback. 1959. The occurrence of intercellular bridges in groups of cells exhibiting synchronous differentiation. *J. biophys. biochem. Cytol.* **5**:453–460.

Fernández-Gómez, M. E. 1968. Rate of DNA synthesis in binucleate cells. *Histochemie* **12**:302–306.

Fernández-Gómez, M. E., C. de la Torre and G. Giménez-Martín. 1972. Accelerated nucleolar reorganization with shortened anaphase and telophase during cycloheximide inhibition of protein synthesis in onion root cells. *Cytobiologie* **5**:117–124.

Fernández-Gómez, M. E., J. C. Stockert, A. González-Fernández, and J. F. López-Sáez. 1970. Delays in prophase induced by adenosine 2-deoxyriboside and their relation with DNA synthesis. *Chromosoma* **29**:1–11.

Fernández-Gómez, M. E., J. C. Stockert, J. F. López-Sáez, and G. Giménez-Martín. 1969. Staining plant cells nucleoli with AgNO₃ after formalin-hidroquinone fixation. *Stain Technol.* **44**:48–49.

Flickinger, C. J. 1974. The role of endoplasmic reticulum in the repair of Amoeba nuclear envelopes damaged microsurgically. *J. Cell Sci.* **14**:421–437.

Foury, F. and A. Goffeau. 1973. Stimulation of yeast RNA synthesis by cycloheximide and 3', 5' cyclic AMP. *Nature New Biol.* **245**:44–47.

Frenster, J. H. 1969. Biochemistry and molecular biophysics of heterochromatin and euchromatin. In *Handbook of molecular Cytology*, ed. Lima-de-Faria. North-Holland Pub. Co., pp. 251–276.

Frey-Wyssling, A., J. F. López-Sáez, and K. Muhlethaler. 1964. Formation and development of the cell plate. *J. Ultrastruct. Res.* **10**:422–432.

Garcia-Herdugo, G., M. E. Fernández-Gómez, J. Hidalgo, and J. F. López-Sáez. 1974. Effects of protein synthesis inhibition during plant mitosis. *Exptl. Cell Res.* **89**:336–342.

Geitler, L. 1939. Die Entstehung der polyploiden Somakerne der Heteropteren durch Chromosomenteilung ohne Kernteilung. *Chromosoma* (Berl.) **1**:1–22.

Giménez-Martín, G., C. de la Torre, M. E. Fernández-Gómez, and A. González-Fernández. 1972. Effect of the cordycepin on the nucleolar cycle. *Caryologia* **25**:43–58.

Giménez-Martín, G., C. de la Torre, M. E. Fernández-Gómez, and A. González-Fernández. 1974. Experimental analysis of nucleolar reorganization. *J. Cell Biol.* **60**:502–507.

Giménez-Martín, G., M. E. Fernández-Gómez, A. González-Fernández, and C. de la Torre. 1971a. The nucleolar cycle in meristematic cells. *Cytobiologie* **4**:330–338.

Giménez-Martín, G., A. González-Fernández, and J. F. López-Sáez. 1966. Duration the division cycle in diploid, binucleate and tetraploid cells. *Exptl. Cell Res.* **43**:293–300.

Giménez-Martín, G., A. González-Fernández, C. de la Torre, and M. E. Fernández-Gómez. 1971b. Partial initiation of endomitosis by 3'deoxyadenosine. *Chromosoma* **33**:361–371.

Giménez-Martín, G., A. González-Fernández, and J. F. López-Sáez. 1965. A new method of labeling cells. *J. Cell Biol.* **26**:305–309.

Giménez-Martín, G., J. F. López-Sáez, P. Moreno, and A. González-Fernández. 1968. On the triggering of mitosis and the division cycle of polynucleate cells. *Chromosoma* 25:282–296.

González-Bernaldez, F., J. F. López-Sáez, and G. Garcia-Ferrero. 1968. Effect of osmotic pressure on root growth, cell cycle and cell elongation. *Protoplasma* 65:255–262.

González-Fernández, A., J. L. Díez, G. Giménez-Martín, and C. de la Torre. 1972. Direct measurement of interphase shortening produced by kinetin plus indol acetic acid in meristematic cells of *Allium cepa* L. *Experientia* 28:247.

González-Fernández, A., M. E. Fernández-Gómez, J. C. Stockert, and J. F. López-Sáez. 1970a. Effect produced by inhibitors of RNA synthesis on mitosis. *Exptl. Cell Res.* 60:320–326.

González-Fernández, A., G. Giménez-Martín, and C. de la Torre. 1971a. The duration of the interphase periods at different temperatures in root tip cells. *Cytobiologie* 3:367–371.

González-Fernández, A., G. Giménez-Martín, J. L. Díez, C. de la Torre, and J. F. López-Sáez. 1971b. Interphase development and beginning of mitosis in the different nuclei of polynucleate homokaryotic cells. *Chromosoma* 36:100–111.

González-Fernández, A., G. Giménez-Martín, M. E. Fernández-Gómez, and C. de la Torre. 1974. Protein synthesis requirements at specific points in the interphase of meristematic cells. *Exptl. Cell Res.* 88:163–170.

González-Fernández, A., G. Giménez-Martín, and J. F. López-Sáez. 1970b. Cytokinesis at prophase in plant cells treated by ethidium bromide. *Exptl. Cell Res.* 62:464–467.

Gurdon, J. B. 1967. On the origin and persistence of a cytoplasmic state inducing nuclear DNA synthesis in frog's eggs. *Proc. Nat. Acad. Sci.,* 58:545–552.

Guttes, S., E. Guttes, and R. A. Ellis. 1968. Electron microscope study of mitosis in *Physarum polycephalum. J. Ultrastruct. Res.* 22:508–529.

Harris, H. 1970. Cell fusion. *The Dunham Lectures.* Clarendon Press, Oxford.

Harris, H. 1974. *Nucleus and Cytoplasm*, 3rd edition. Clarendon Press, Oxford.

Hepler, P. K. and E. H. Newcomb. 1967. Fine structure of cell plate formation in the apical meristem of *Phaseolus* roots. *J. Ultrastruct. Res.* 19:498–513.

Heslop-Harrison, J. 1964. Cell walls, cell membranes and protoplasmic connections during meiosis and pollen development. In *Pollen Physiology and Fertilization*, ed. H. F. Liskens. North-Holland Publ. Co., Amsterdam.

Highfield, D. P. and W. C. Dewey. 1972. Inhibition of RNA synthesis in synchronized Chinese Hamster cells treated in G_1 or early S phase with cycloheximide or puromycin. *Exptl. Cell Res.* 75:314–320.

Howard, A. and D. L. Dewey. 1961. Non-uniformity of labelling rate during DNA synthesis. *Exptl. Cell Res.* 24:623–624.

Howard, A. and S. R. Pelc. 1953. Synthesis of deoxyribonucleic acid in normal and irradiated cells and its relation to chromosome breakage. *Heredity* (suppl.) 6: 261–274.

Jacob, S. T., E. H. Sajdel, and H. N. Munro. 1971. Mammalian RNA polymerase and their selective inhibition by amanitin. *Adv. Enzyme Regul.* 9:169–181.

Jakob, K. M. 1972. RNA synthesis during the DNA synthesis period of the first cell cycle in the root meristem of germinating *Vicia faba. Exptl. Cell Res.* 72:370–376.

Jockusch, B. M., M. Becker, I. Hindennach, and H. Jockusch. 1974. Slime mould actin : homology to vertebrate actin and presence in the nucleus. *Exptl. Cell Res.* **89**:241–246.

Johnson, R. T. and H. Harris. 1969. DNA synthesis and mitosis in fused cells. I. HeLa homokaryons. *J. Cell Sci.* **5**:645–697.

Johnson, R. T. and P. N. Rao. 1970. Mammalian cell fusion : induction of premature chromosome condensation in interphase nuclei. *Nature* **226**:717–722.

Johnson, R. T. and P. N. Rao. 1971. Nucleo-cytoplasmic interactions in the achievement of nuclear synchrony in DNA synthesis and mitosis in multinucleate cells. *Biol. Rev.* **46**:97–155.

Kaltsikes, P. J. 1971. The mitotic cycle in amphiploid (Triticale) and its parental species. *Can. J. Genet. Cytol.* **13**:656–662.

Kasten, F. H. and F. F. Strasser. 1966. Nucleic acid synthetic patterns in synchronized mammalian cells. *Nature* **211**:135–140.

Kihlman, B. 1955. Chromosome breakage in *Allium* by 8-ethoxycaffeine and X-rays. *Exptl. Cell Res.* **8**:345–368.

Kim, J. H. and A. G. Perez. 1965. Ribonucleic acid synthesis in synchronously dividing populations of HeLa cells. *Nature* **207**:974–975.

Krishan, A. and R. Ray-Chadhuri. 1969. Asynchrony of nuclear development in cytochalasin induced multinucleate cells. *J. Cell Biol.* **43**:618–621.

Kuroiwa, T. 1974. Fine structure of interphase nuclei. III. Replication site analysis of DNA during the S period of *Crepis capillaris*. *Exptl. Cell Res.* **83**:387–398.

Lafontaine, J. -G. and L. A. Chouinard. 1963. A correlated light and electron microscope study of the nucleolar material during mitosis in *Vicia faba*. *J. Cell Biol.* **17**:167–209.

Lambert, A. M. and A. S. Bajer. 1972. Dynamics of spindle fibers and microtubules during anaphase and phragmoplast formation. *Chromosoma* **39**:101–144.

Levan, A. 1971. Cytogenetic effects of hexyl mercury bromide in the *Allium* test. *J. Indian Bot. Soc. Golden Jubile.* **50A**:340–349.

Lin, M. S. and D. B. Walden. 1974. Endoreduplication induced by hydroxylamine sulfate in *Zea mays* root tip nuclei. *Exptl. Cell Res.* **86**:47–52.

López-Sáez, J. F., G. Giménez-Martín, and A. González-Fernández. 1966a. Duration of the cell division cycle and its dependence on temperature. *Zeits. Zellforsch.* **75**:591–600.

López-Sáez, J. F., F. González-Bernaldez, and G. Garcia-Ferrero. 1969. Effect of temperature and oxygen tension on root growth, cell cycle and cell elongation. *Protoplasma* **67**:213–221.

López-Sáez, J. F., M. C. Risueño, and G. Giménez-Martín. 1966b. Inhibition of cytokinesis in plant cells. *J. Ultrastruct. Res.* **14**:85–94.

Mackenzie, J. B., G. E. Stone, and D. M. Prescott. 1966. The duration of G_1, S and G_2 at different temperatures in *Tetrahymena pyriformis*. *J. Cell Biol.* **31**:633–635.

MacLeod, R. D. 1968. Changes in the mitotic cycle in lateral root meristems of *Vicia faba* following kinetin treatment. *Chromosoma* **24**:177–187.

Mangenot, G. and S. Carpentier. 1944. Sur les effets mitoclassiques de la caféine et de la thérophylline. *C. R. Soc. Biol.* **138**:232–233.

Murin, A. 1966. The effect of temperature on the mitotic cycle and its time parameters in root tips of *Vicia faba*. *Naturwissenschaften* **53**:312–313.

Nagl, W. 1970. The mitotic and endomitotic nuclear cycle in *Allium carinatum*. II. Relations between DNA replication and chromatin structure. *Caryologia* **23**: 71–78.

Nagl, W. 1972. Molecular and structural aspects of the endomitotic chromosome cycle in Angiosperms. In *Chromosomes Today*, vol. 3, eds. C. D. Darlington and K. R. Lewis. Hafner, New York, pp. 17–23.

Nagl, W. 1973. The mitotic and endomitotic nuclear cycle in *Allium carinatum*. IV. ^{3}H-uridine incorporation. *Chromosoma* **44**:203–212.

Nagl, W. 1974. Role of heterochromatin in the control of cell cycle duration. *Nature* **249**:53–54.

Nešković, B. A. 1968. Development phases in intermitosis and the preparation for mitosis of mammalian cells *in vitro*. *Int. Rev. Cytol.* **24**:71–97.

Niitsu, T., A. Hanaoka, and K. Ochiyana. 1972. Reinvestigations of the spindle body and related problems. II. Effects of cycloheximide upon the development of kinetochore fibers studies in the pollen mitosis of *Ornithogalum virens in vivo*. *Cytologia* **37**:143–154.

Nuti-Ronchi, V., S. Avanzi, and F. D'Amato. 1965. Chromosome endoreduplication (endopolyploidy) in pea root meristems induced by 8-azaguanine. *Caryologia* **18**:599–617.

Obara, Y., H. Yoshida, L. S. Chai, H. Weinfeld, and A. A. Sandberg. 1973. Contrast between the environmental pH dependencies of prophasing and nuclear membrane formation in interphase-metaphase cells. *J. Cell Biol.* **58**:608–617.

Oftebro, R. and J. Wolf. 1967. Mitosis of bi- and multinucleate HeLa cells. *Exptl. Cell Res.* **48**:39–52.

Östergren, G. and K. Östergren. 1966. Mitosis with undivided chromsomes. III. Inhibition of chormosome repreduction in *Tradescantia* by specific mutations. In *Chromosomes Today*, vol. I, eds. C. D. Darlington and K. R. Lewis. Hafner, New York, pp. 128–130.

Paul, D. C. and Ch. W. Goff. 1973. Comparative effects of caffeine, its analogues and calcium deficiency on cytokinesis. *Exptl. Cell Res.* **78**:399–413.

Phillips, S. G. 1972. Repopulation of the postmitotic nucleolus by preformed RNA. *J. Cell Biol.* **53**:611–623.

Rao, P. N. and R. T. Johnson. 1974. Regulation of cell cycle in hybrid cells. In *Control of Proliferation in Animal Cells*. Cold Spring Harbor Laboratory, pp. 785–800.

Risueño, M. C., G. Giménez-Martín, and A. González-Fernández. 1971. Ultrastructural analysis of cytokinesis in plant cells blocked at prophase. *J. Microscopie* **10**:331–336.

Risueño, M. C., G. Giménez-Martín, and J. F. López-Sáez. 1968. Experimental analysis of plant cytokinesis. *Exptl. Cell Res.* **49**:136–147.

Risueño, M. C., G. Giménez-Martín, J. F. López-Sáez, and M. I. Rodriguez-Garcia. 1969. Connexions between meiocytes in plants. *Cytologia* **34**:262–272.

Rizzoni, M. and F. Palitti. 1973. Regulatory mechanism of cell division. I. Colchicine-induced endoreduplication. *Exptl. Cell Res.* **77**:450–458.

Salas, J. and H. Green. 1971. Proteins binding to DNA and their relation to growth in cultured mammalian cells. *Nature New Biol.* **229**:165–169.

Scharff, M. D. and E. Robbins. 1965. Synthesis of ribosomal RNA in synchronized HeLa cells. *Nature* **208**:464–466.

Scheuermann, W. and G. Klaffke-Lobsien. 1973. On the influence of 5-aminouracil on the cell cycle of root tip meristems. *Exptl. Cell Res.* **76**:428–436.

Simmons, T., P. Heywood, and L. Hodge. 1973. Nuclear envelope-associated resumption of RNA synthesis in late mitosis of HeLa cells. *J. Cell Biol.* **59**:150–164.

Smith, J. A. and L. Martin. 1973. Do cells cycle? *Proc. Natl. Acad. Sci.* **70**:1263–1267.

Socher, S. H. and D. Davidson. 1971. 5-aminouracil treatment. A method for estimating G_2. *J. Cell Biol.* **48**:248–252.

Stenram, U. 1973. Relationship between nucleolar size and the synthesis and processing of pre-ribosomal RNA in the liver of rat. In *Biochemistry of Cell Differentiation*, vol. 24, ed. A. Monroy and R. Tsanov. Academic Press, Inc. New York, pp. 131–141.

Stevens, B. J. and D. M. Prescott. 1971. Reformation of nucleus-like bodies in the absence of postmitotic RNA synthesis. *J. Cell. Biol.* **48**:443–454.

Stockert, J. C., M. E. Fernández-Gómez, G. Giménez-Martín, and J. F. López-Sáez. 1969. Organization of argyrophilic nucleolar material throughout the division cycle of meristematic cells. *Protoplasma* **69**:267–278.

Taylor, J. H. 1958. The mode of chromosomal duplication in *Crepis capillaris*. *Exptl. Cell Res.* **15**:350–357.

Thompson, L. R. and B. J. McCarthy. 1968. Stimulation of nuclear DNA and RNA synthesis by cytoplasmic extracts *in vitro*. *Biochem. biophys. Res. Commun.* **30**:166–172.

Tobey, R. A., E. C. Anderson, and D. F. Petersen. 1966. RNA stability and protein synthesis in relation to the division of mammalian cells. *Proc. Natl. Acad. Sci.* **56**:1520–1527.

Treub, M. 1879. Sur la pluralité des noyaux dans certaines cellules végétales. *C. R. Hebd. Séanc. Acad. Sci.* (Paris) **89**:494–505.

Tschermak-Woess, E. 1971. Endomitose. In *Handbuch der allgemeinen Pathologie* 2/II/1. Berlin-Heidelberg-New York. Spring, pp. 569–625.

Van't Hof, J. 1965. Relationship between mitotic cycle duration, S period duration and the average rate of DNA synthesis in the root meristem cells of several plants. *Exptl. Cell Res.* **39**:48–58.

Van't Hof, J. 1973. The regulation of cell division in higher plants. *Brookhaven Symp. Biol.* **25**:152–165.

Van't Hof, J. 1975. The duration of chromosomal DNA synthesis, of the mitotic cycle, and of meiosis of higher plants. In *Handbook of Genetics*, vol. 2, ed. R. C. King. Plenum Press, London.

Van't Hof, J. and A. H. Sparrow. 1963. A relationship between DNA content, nuclear volume and minimum mitotic cycle time. *Proc. Natl. Acad. Sci.* **49**:897–902.

Van't Hof, J. and W. K. Ying. 1964. Relationship between the duration of the mitotic cycle, the rate of cell production and the rate of growth of *Pisum* roots at different temperatures. *Cytologia* **29**:399–406.

Wagenaar, E. B. 1966. High mitotic synchronization induced by 5-aminouracil in root cells of *Allium cepa* L. *Exptl. Cell Res.* **43**:184–190.

Webster, P. L. and J. Van't Hof. 1970. DNA synthesis and mitosis in meristems : requirements for RNA and protein synthesis. *Amer. J. Bot.* **57**:130–139.

Whaley, W. G. and H. H. Mollenhauer. 1963. The Golgi apparatus and cell plate formation. A postulate. *J. Cell Biol.* **17**:216–221.

Wimber, D. E. 1961. Asynchronous replication of DNA in root tip chromosomes of *Tradescantia paludosa. Exptl. Cell Res.* **23**:402–407.

Wimber, D. E. 1966. Duration of the nuclear cycle in *Tradescantia* root tips at three temperatures as measured with ³H-thymidine. *Amer. J. Botany* **53**:21–24.

Wolfe, J. and L. Rodríguez. 1974. Inhibition of nuclear reconstruction by dithiothreitol. *J. Cell Biol.* **60**:497–502.

Yamanaka, T. and Y. Okada. 1966. Cultivation of fused cells resulting from treatment of cells with H. V. J.: II. *Exptl. Cell Res.* **49**:461–469.

Cell Division in
Euglena and *Phacus*
I. Mitosis

Jeremy D. Pickett-Heaps   University of Colorado

Kenneth L. Weik   Lake Forest College

INTRODUCTION

The euglenoid flagellates comprise a widespread and diverse group of algae, particularly familiar to biologists; *Euglena* especially has been used in all manner of biological investigations (e.g., see Buetow 1968a,b). Much electron microscopy has been carried out on these interesting organisms, particularly *Euglena,* which has revealed various unusual features (e.g., their elastic or rigid pellicle composed of interlocking proteinaceous strips, complex reservoir and canal, and eyespot situated in the cytoplasm); this ultrastructural work has been excellently and thoroughly reviewed by Beutow (1968c) and Leedale (1967, 1970).

Cell division in this group has also invoked considerable interest. Light microscopy has given us numerous accounts of the sequence of events (e.g., Leedale 1968, and references cited in this chapter). However, at the ultrastructural level, the information available is far less complete. Sommer and Blum (1965), in their short account of the mitotic nucleus in *Astasia,* showed that the spindle was closed and contained fibers which we now identify as microtubules; they could not detect any kinetochores. They confirmed the intimate association of the dividing nucleus with the flagellar bases, and suggested that the spindle fibers briefly came in contact with the kinetosomes (basal bodies). Leedale (1968, 1970) also described the closed nature of the spindle, the existence of spindle microtubules, and the apparent lack of kinetochores. He also stressed that as the endosome (the term commonly applied to the euglenoid nucleolus) elongates, microtubules surround it. These and several other provocative observations (e.g., the most unusual, prolonged duration of anaphase, the resistance of mitosis to many inhibitors of normal mitosis such as colchicine, the staggered anaphase movement of chromatids which in some species seem to be in *singlets* at metaphase; Leedale 1967, 1968) have all led to a general feeling that euglenoid mitosis in unconventional, and that a "normal" spindle is not present. Leedale (1967, 1968) discusses the possibility that chromosomal movement may be autonomous, arising from "repulsion" of daughter chromatids, and therefore that the mechanisms of mitosis could be more primitive than in many other eucaryotic organisms; he emphasizes that detailed electron microscopy would be needed to clarify these puzzling features. Since that date, no such study has been published to our knowledge, and so we decided to attempt it, being further stimulated by an increasing interest in the whole topic of the evolution of the spindle (e.g., Pickett-Heaps 1974a; Kubai 1975). The lack of information concerning these flagellates has led, for example, to much unresolved and perhaps unfounded speculation as to whether the mitotic chromosomes could be moved by attachment to the nuclear envelope, as now appears to be the case in several dinoflagellates (Kubai and Ris 1969; Soyer and Haapala, 1974).

MATERIALS AND METHODS

Euglena gracilis var. *gracilis* was grown in Huntner's medium at *p*H 3.5 (Miller and Staehelin 1973), in two liter flasks with constant illumination (approximately 75 ft. candle) at 20°C. The cells were always fixed during an exponential growth phase.

Phacus longicaudus, one of the largest species in this genus, was collected locally and identified by K. L. Weik; our cultures were cloned from a single cell, and thereafter maintained in biphasic soil/water medium with calcium carbonate added, at 18°C with a 15:9 hour light:dark cycle supplied by cool white fluorescent lamps.

Transmission Microscopy

E. gracilis was fixed in 1% glutaraldehyde made up in the culture medium, for 1 hour, washed in the medium several times and then post-fixed in 1% osmium tetroxide, (OsO_4) also made up in the culture medium, for 1/2 hour. *P. longicaudus* was similarly fixed using soil/water medium in the fixative, or else 1/2% glutaraldehyde in 0.1 M cacodylate buffer. The cells were washed several times, the *Euglena* post-stained in 1% aqueous uranyl acetate overnight and then all were dehydrated slowly in cold acetone before being embedded in Spurr's (1969) resin. Preselected dividing cells were remounted and sectioned; the sections were collected on Formvar-coated grids, stained with 1% methanolic uranyl acetate and lead citrate in the normal way. For high voltage electron microscopy, thicker sections (1/4-1/2 μm) were cut from these blocks, mounted on slightly thicker Formvar films and stained as above but for appreciably longer (the actual times varied, but normally, uranium staining lasted about 10 minutes, and lead, 5 minutes). These grids were then coated on one or both sides with carbon before being examined in a JEM 1000 at 1Mev.

Light Microscopy

The orientation of the cell in thin sections was checked by periodically transferring thick sections of plastic-embedded material (1/2-1 μm) on to glass slides; these were stained with a few drops of hot 1% toluidine blue in 1% sodium borate; they were then washed, dried and photographed as necessary (e.g., Figure 11-7). Live cells were photographed with Zernike or Normarski interference phase contract optics.

Scanning Microscopy

Initially, cells were fixed in 1% OsO_4 as above and then treated with 1% aqueous "Glusulase" (Pickett-Heaps 1974b) for 1 hour before dehydration. Later, cleaner preparations of *Euglena* were obtained (Figure 11-1) by fixing as for transmission microscopy, in glutaraldehyde for 15 minutes, washing, treating for 1 hour with the Glusulase, and *then* post-fixing in the osmium. All cells were collected on Millipore "Solvinert" filters (Pickett-Heaps 1974b), dehydrated slowly in acetone and critical point dried. Specimens were then shadowed with carbon and gold as usual.

RESULTS

Sectioning Problems

Euglena can be regarded as an essentially bilaterally symmetrical organism which has become twisted (Figure 11-1). The position and orientation of the dividing nucleus was variable within wide limits, and so two-dimensional sections rarely contained more than a few features of interest, regardless of how carefully the cells were oriented for sectioning. For example, longitudinal sections of the dividing nucleus never included both sets of the flagellar bases. Consequently, much of our data was derived from assemblages of serial section. Further, the systems of tubules around the reservoir and canal were very complex in organization and we remain uncertain as to their precise orientation and extent.

In *Phacus,* the helical organization of the cell is not as pronounced as in *Euglena* (compare Figures 11-1 and 11-2); the cells are strongly flattened, and precise orientation of individual cells for sectioning was easier.

Quality of Fixation

Euglena and *Phacus* proved easy to fix for scanning microscopy (Figures 11-1 and 11-2). *Euglena* is usually covered with some mucilage removable with Glusulase, an enzyme preparation which worked best when glutaraldehyde was used as the initial fixative rather than OsO_4 (see Materials and Methods).

The quality of fixation achieved for transmission microscopy of *E. gracilis* was

bb - basal bodies
c - chromosomes
cn - canal
cv - contractile vacuole
f - flagellum
i - interzonal spindle

n - nucleus
nc - nucleolus
ne - nuclear envelope
p - paramylon granule
r - reservoir
t - microtubules

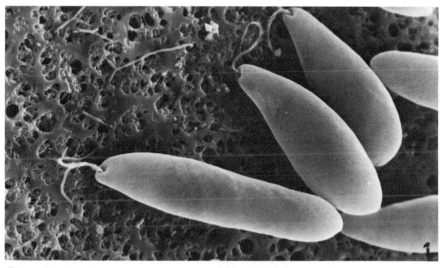

Figure 11-1.
Scanning micrograph of *Euglena,* cleaned with Glusulase. The single flagellum whose flagellar hairs (mastigonemes) have aggregated somewhat, emerge through the cytostome. At this low magnification, the spiral pellicular ridges are only just visible. X 1,500

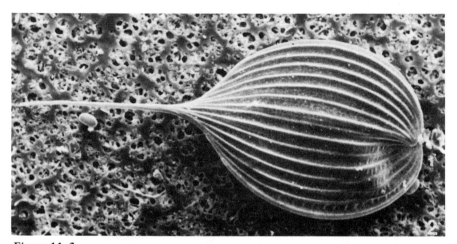

Figure 11-2.
Scanning micrograph of *Phacus;* the long caudus, prominent pellicular ridges and single flagellum (*f*) are obvious. X 730 approx.

variable. Many cells were dense or else displayed considerable internal damage. The results illustrated here, the best obtained, are still suboptimal; mitochondria and chloroplasts often show osmotic damage, as does the contractile vacuolar apparatus. Even so, much could be discerned of the cell's mitotic structure. We suspect osmium fixation alone might ameliorate these osmotic artefacts, but we were unwilling to risk possible destructive effects of such fixation upon microtubular systems. Leedale (e.g., 1967) in particular has achieved excellent fixation of *Euglena* with osmium alone. Our *Phacus* gave consistently better results, displaying less osmotic damage, although the cytoplasm was more dense.

Interphase

Little need be said here concerning the structure of the interphase *Euglena* cell which has been thoroughly described in a number of previous papers (see Introduction). Our *Euglena* almost always had dense material lining the inner surfaces of the reservoir; its origin and significance is unknown, but it usefully indicated the extent of the reservoir during mitosis (e.g., see Figures 11-5A-C, 11-6, 11-8, 11-11B, 11-12, 11-18, 11-21A and B).

P. longicaudus is typical of the genus: the interphase cell is flattened, broadly ovate, and slightly twisted along its longitudinal axis. The posterior tapers abruptly into a long spike or caudus (Figures 11-2-11-4). In face view, the cell is weakly asymmetrical; the cytostome is located in between two slightly overlapping, flattened anterior lobes, one always a little larger than the other (Figures 11-2-11-4). As with other euglenoids, the cell is bounded by a pellicle, which in this genus is relatively rigid, and distinguished by its component series of longitudinal striae (pellicular ridges) which lack the strong spiral orientation typical of the pellicle of *Euglena* (compare Figures 11-1 and 11-2). These ridges, which are situated within the plasmalemma, also exhibit some cross-banding in both light and electron microscopy. Before cytokinesis, the cell appears markedly thicker and rounds up as the pellicle seems to soften (see Weik and Pickett-Heaps 1977). During cytokinesis, the daughter cells are quite fusiform, unlike the normal interphase from which they soon reacquire after division.

Internally, the *Phacus* cell contains numerous discoid chloroplasts lacking pyrenoids. The paramylon reserves in actively growing cells usually form two large and characteristic inclusions (Figure 11-3), as well as many smaller plates distributed throughout the cytoplasm. One of the larger inclusions is disc-shaped and posterior to the nucleus, the other spool-shaped and located between the nucleus and reservoir. During cytokinesis, each daughter cell normally acquires one or the other of these larger paramylon bodies. The eyespot is on the side of the reservoir as usual, invariably that side of the cell possessing the smaller of the two anterior lobes (Figure 11-3); almost directly opposite on the other side of the reservoir, but located at a different level and slightly skewed, are the contractile vacuole and the basal bodies. Only one of the two flagella emerges through the cytostome. The nucleus, located slightly towards the caudal end of the interphase cell, is ovoid and tends to be broadly kidney-shaped and positioned immediately posterior to or

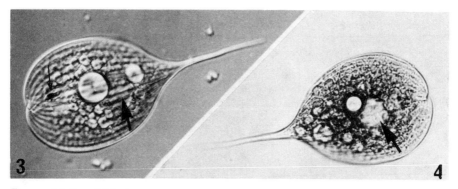

Figures 11-3 and 11-4.
Live *Phacus* Figure 11-3. Typical interphase cell. The small arrow indicates the eyespot on the side of the reservoir, and the large arrow, the nearly invisible nucleus between the two very conspicuous paramylon granules. X 430 approx. Figure 11-4. Early division—the nucleus (arrow) is moving toward the reservoir, around one of the paramylon granules. X 430 approx.

slightly to the side of the spool-shaped granule of paramylon (Figure 11-3) often lying against its curved surface.

Condensed chromatin was uniformly distributed throughout the nucleoplasm except near the central group of several endosomes (nucleoli). The nucleoplasm contained many small granular particles. Our *Phacus* also contained endonuclear bacteria similar to those described by Leedale (1969) in the nucleus of *E. spirogyra.* These bacteria divided by fission during interphase and nuclear division. The presence of the bacterium did not appear to affect the nucleus nor the course of mitosis.

Mitosis

As Leedale points out, mitosis is unusual in *Euglena* and so the classical terms "prophase. . .telophase" have to be used with discretion (Leedale 1967). In particular, Leedale states that by *metaphase* in *E. gracilis,* the doublet chromatids from late prophase have become organized into a circlet of single chromatids; anaphase is unusually prolonged (60-70% of the mitotic cycle) and movement of chromatids to the poles is staggered, some arriving much later than others. Such facts need be remembered during the following section.

Prophase

As has been well documented (Sommer and Blum 1965; Leedale 1968), prophase in *Euglena* is marked by retraction of the emergent flagellum, duplication of the flagellar bases, and movement of the nucleus to the anterior (= flagellar) end of the cell, immediately adjacent to the reservoir. During mitosis, new pellicular ridges are intercalated between the older ones (Sommer and Blum 1964; Leedale 1968); these grow from the cytostome around the cell during mitosis. Thus dividing cells

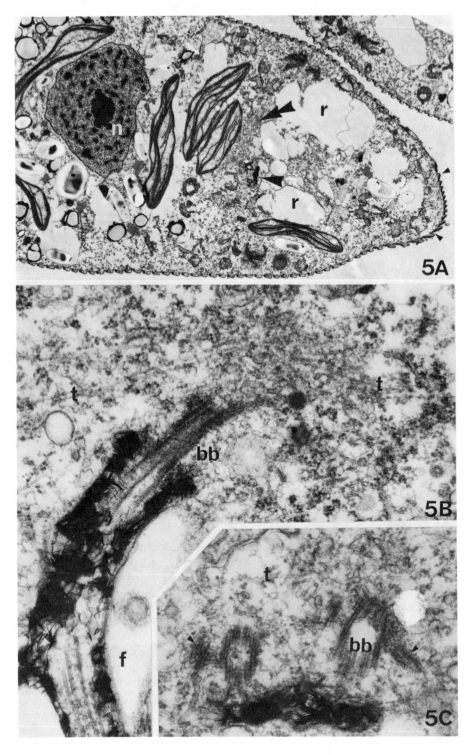

had two patterns of pellicular ridges on their surface, the extent of each pattern depending upon the stage reached in division (Weik and Pickett-Heaps 1977).

The electron microscope clearly shows that in *Euglena,* the flagellar bases had replicated and separated before the nucleus reached them. For example, the cell in Figure 11-5A had two short flagella in its canal and four basal bodies (Figure 11-5B and C); replication of the pellicle via intercalation of new ridges is visible just outside the cytostome, but the nucleus was still central (Figure 11-5A). Numerous microtubules extended into the cytoplasm from, and lay between, each pair of basal bodies (Figure 5B and C).

In *Phacus* too, incipient mitosis is signaled by this anterior nuclear migration. Initially, the nucleus characteristically slid around the spool-shaped paramylon granule from its more posterior interphase position, and so both became slightly displaced, the nucleus toward the larger half of the cell containing the contractile vacuole and the granule toward the side containing the eyespot (Figure 11-4). During this passage, the nucleus became somewhat distorted, but later it resumed its spherical profile with a pronounced anterior depression where it abutted the lower margin of the reservoir.

Once the migrating nucleus reached its anteriormost position, the two pairs of basal bodies have become increasingly separated across the floor of the reservoir (Figure 11-6). The anterior migration placed the nucleus closely adjacent to the pairs of basal bodies, but an obvious direct contact between them was not observed. When exactly basal body replication occurred was not determined but it clearly preceded the arrival of the prophase nucleus at the reservoir. Scattered groups of microtubules soon appeared in the nucleus once it was near the flagellar bases.

Prometaphase

By prometaphase, the chromatin underwent a distinctive change in *Euglena gracilis.* Condensed during interphase as usual for the genus, it soon appeared more dispersed and less stainable under the electron microscope (Figure 11-8). By metaphase, individual chromosomes were difficult to discern in thin sections, but they were clearer in $1/2$ μm sections stained with toluidine blue for light microscopy (Figure 11-7). By now the new pellicular ridges extended posteriorly about half way along the cell. The nucleus was closely appressed to the reservoir (Figure 11-8) and the nuclear envelope was frequently distorted near the flagellar bases. A direct

---

Figure 11-5.

E. gracilis, very early prophase Figure 11-5A. The flagellar bases have replicated; one pair shown in Figure 11-5B, is visible (single arrow), while the other, that in Figure 11-5C, is out of the plane of this section, and is located at the double arrow. The reservoir (*r*) is enlarged, but the nucleus (*n*) is still central. Imminence of cell division is also indicated by the pellicular replication apparent at the cytostome (small arrows). X 4,200 Figures 11-5B and 11-5C. These show the two pairs of flagellar bases of the cell in Figure 11-5A; in both, large numbers of microtubules (*t*) extend from near the basal bodies (*bb*), many lying between them. X 31,000 and 39,000

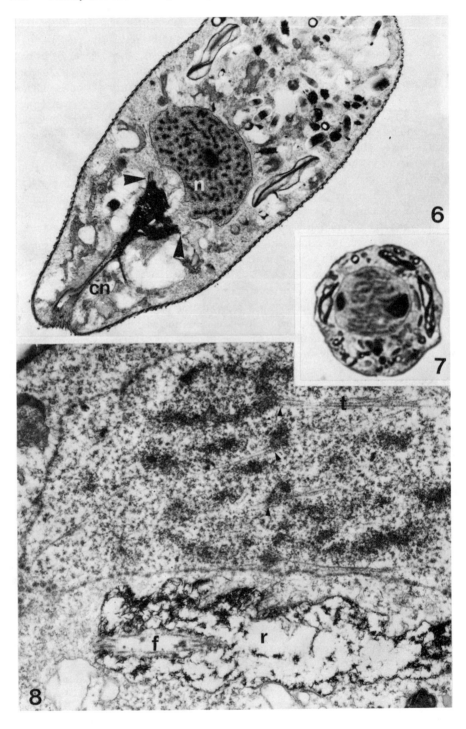

attachment between basal bodies and nuclear envelope was carefully sought but none was ever visible. The space between these usually contained much ill-defined, finely fibrous material, and sometimes coarser fibers ran from near the basal bodies toward the nuclear envelope. The numerous extranuclear microtubules earlier associated with the flagellar bases (Figure 5B and C) had largely disappeared. In both genera, the microtubules now present in the nucleus had become more organized in bundles aligned with the axis of the future spindles. Occasionally, some of these tubules in *E. gracilis* appeared to insert into ill-defined kinetochores (Figure 11-8).

During prophase, the separate endosomes of *Phacus* aggregated to form an essentially single body (visible following its division, in Figure 11-15). In both genera, elongation of the endosome commenced at this stage as the intranuclear tubules proliferated, and while the nucleus was still circular in section. Some tubules were associated with the endosome while the remainder were distributed throughout the nucleoplasm. Like all other euglenoid algae studied so far, the nuclear envelope remained completely intact throughout mitosis.

Metaphase

As emphasized earlier, metaphase appears unusual in *Euglena,* and is a difficult stage to define ultrastructurally. By now, the endosome (nucleolus) is always conspicuously drawn out (Figures 11-7 and 11-11A), so that it extends across the nucleus; as shown by Leedale (1968, 1970), microtubules were frequently appressed to it. The chromatin has now become centrally located, apparently forming a rather broad, loosely organized mass of chromosomes (chromatids), the apparent euglenoid equivalent of the metaphase plate (Figure 11-7; cf., Hall 1937). The nucleus was indented by its arching over the separated flagellar bases (Figure 11-11B), and the polar regions were devoid of chromatin. Microtubules were numerous and organized as before into bundles aligned along the spindle axis (Figure 11-10). Careful examination revealed that several microtubules in some bundles inserted into a distinct, layered kinetochore (as in Figure 11-9). The chromosomes remain ill-defined, however, and true paired (i.e., metaphase) kinetochores were only rarely

---

Figure 11-6.
E. gracilis, prophase 1/2 μm-thick section photographed at 1 MeV. The nuceus (*n*) has now moved to the separated flagellar bases, whose basal bodies (one set visible) are located at the arrows. The canal (*cn*) is still single and narrow. X 3,600

Figure 11-7.
E. gracilis, light micrograph of thick section. The edges of the dense, drawn-out nucleolus is visible at the spindle poles. This is probably representative of metaphase, with the chromosomes central in the nucleus, and aligned along the spindle axis. X 1,800 approx.

Figure 11-8.
E. gracilis. Early metaphase (?). The nucleus is now appressed to the elongated reservoir (*r*). Spindle tubules (*t*) are becoming more oriented and aggregated into bundles; some kinetochores (arrows) visible. X 11,000

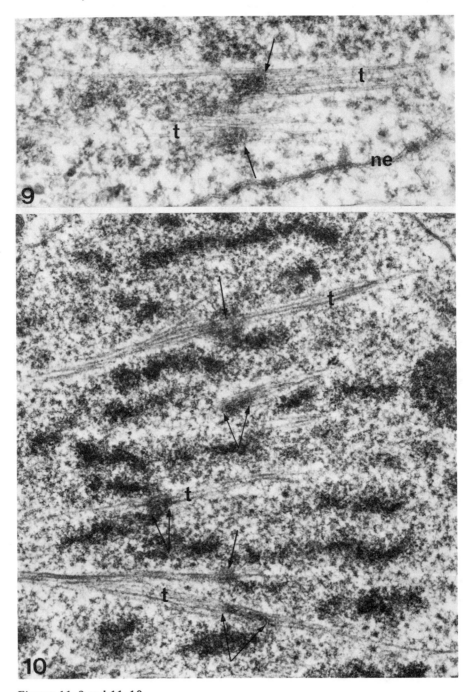

Figures 11-9 and 11-10.
Kinetochores in *E. gracilis*. 11-9. Two single kinetochores, early anaphase (same cell as that in Figures 11A, 11B). × 29,000. 11-10. Presumed metaphase spindle.

detected. (One such nucleus is shown in Figure 11-10). The kinetochores were never found organized in a classical plate configuration, although about six cells examined were judged to be at this "metaphase" stage. The spindle microtubules did not appear to end or insert in any specific polar structure. Because chromosomes were so ill-defined, some cells appearing to be at this stage of metaphase had probably in fact initiated anaphase (e.g., the cell in Figure 11-11A and B). The nucleolus consisted of two quite distinct phases, one dense and the other comprising a homogeneous, less dense granular mass. These two phases were particularly clear in thick sections examined with the high voltage microscope, and they were most noticeable by metaphase.

In *Phacus,* this metaphase stage has also proven difficult to define. Kinetochores were sought, but structural equivalents to those described above for *Euglena*, were not encountered. However, numerous tubules did seem to insert into ill-defined regions of the chromosomes, which remained quite dispersed throughout the nucleus.

Anaphase and Telophase

During anaphase, the nucleus elongated across the cell while remaining in close association with the pairs of flagellar bases. The basal body complexes were never at the spindle poles but to one side of them (Figure 11-12). The nearly localized deformation suffered by the nuclear envelope was often pronounced, but direct contact between basal bodies and the membrane was never apparent. Instead, the multitude of fine fibres around the basal bodies often extended close to the nuclear surface.

In *Phacus,* these fine fibers were clearly organized into two quite distinct rhizoplasts which extend, one from each basal body, to the nucleus. This association is quite well developed before anaphase, during which the fibers seem to lengthen.

During mid-anaphase, the chromosomes appeared now to fill a nucleus traversed by longitudinal bundles of tubules; some of these tubules remained embedded in the single kinetochores in *Euglena* (Figure 11-9). However, by late anaphase when the nucleus had become dumb-bell shaped (e.g., Figure 11-14A), kinetochores were no longer visible, and the polar regions of the nucleus now contained a few scattered microtubules. The tubules in both genera had now become concentrated in the interzone (Figures 11-14A, 11-15, 11-16), particularly near the nuclear envelope (Figure 11-13), so much so that the nuclear envelope here appeared to be lined internally with a peripheral layer of close-space tubules. Concurrently, the greatly elongated endosome became separated into two distinct bodies, each ending up in one daughter nucleus. However, some remnants of it usually remained in the

---

Three pairs of kinetochores (paired arrows) are visible; the adjacent serial section showed two more pairs at the single arrows. The spindle microtubules (*t*) now show a pronounced tendency to aggregate into groups; as in Figure 11-9, some tubules in such groups insert into the kinetochores, while others run past them. X 15,000

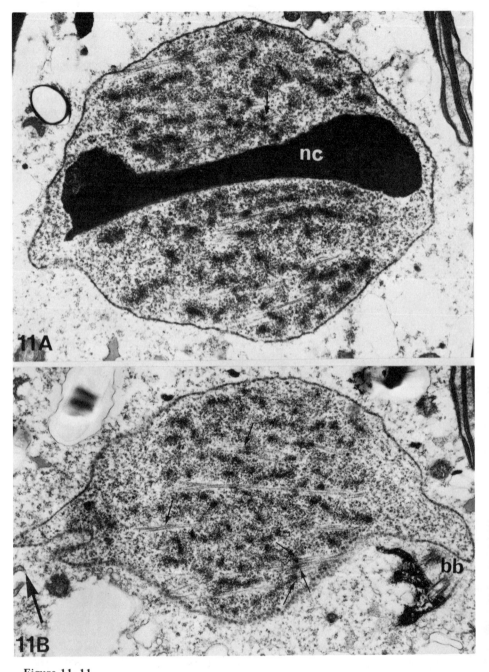

Figure 11-11.
Early anaphase; *E. gracilis*, cell sectioned from one end. Figure 11-11A. Section through the middle of the nucleus and its elongated nucleolus (*nc*). Two groups of single (as in Figure 11-9) kinetochores could be seen amongst the chromosomes

interzone. As the spindle elongated, the dividing nucleus receded slightly from the reservoir (Figure 11-12). The interzonal spindle was apparently quite flattened; when cells were sectioned from one end, the interzone always appeared quite narrow Figures 11-14A, 11-15), while cells sectioned from one side always had a broader interzonal spindle (Figures 11-12, 11-16). The canal was still single as this stage, although it possessed two complements of the sets of pellicular tubules (Figure 11-14B; see Weik and Pickett-Heaps 1977).

The basal bodies were not coplanar with the axis of the spindle, and as the cell proceeded through anaphase cleavage, the angle between them grew. Thus, if the twisted cell at late telphase is viewed from one end (Figure 11-14A), the angle between the axis of the spindle superimposed on the axis of the flagellar bases approaches 90°; this is clear from Figure 11-18, and is also illustrated later for *Phacus.*

The inner face of each nucleus where still attached to the interzonal spindle, developed a pronounced depression (Figures 11-14A, 11-15). Eventually, the tapered, distal ends of the midbody pinched free of each nucleus, leaving the entire interzonal spindle isolated in the cytoplasm (Figure 11-19A and B). Just prior to its separation it contained only microtubules and residual nucleoplasm (Figure 11-19B). This elongated sack of tubules enclosed by the old nuclear envelope, apparently floated around in the cleaving cell for some while; it has been detected in cells at the final stages of cleavage and also in cleaving cells of *Trachelomonas* (unpublished observations).

The organization of tubules within this interzonal spindle was of interest. Although the highly variable orientation and small size of the abandoned interzonal spindle created serious difficulties in attempting to secure good transverse sections of it, a fortunate series of such sections from *Phacus* revealed the presence of 60 to 70 peripherally arranged microtubules with fewer in the center. Three of such sections are shown here (Figure 11-20A-C). The number of tubules became greatly reduced toward the ends of the midbody. They seemed denser than normal and perhaps were sometimes paired up; incomplete "c" tubules were also encountered. We remain uncertain whether some euglenoid version of a midbody was perhaps present. The fate of the detached interzonal spindle was not determined.

Following the separation from the interzonal spindle, the daughter nuclei came back quite close together. In *Phacus,* the nuclei rounded up and moved nearer the central (longitudinal) axis of the cell where they end up diagonally opposite one another on either side of the interzonal remnant (compare Figures 11-15-11-17). However, this condition is temporary, since during cleavage, the nuclei had become widely separated in opposite halves of the cell (Figure 11-19A). In both species the

---

(one is arrowed) in serial sections of this nucleus; note that the poles are still free of chromatin. X 9,500. Figure 11B. Same nucleus as that above; small arrows indicate more kinetochores. The poles of the dividing nucleus curve over the widely separated basal bodies—one set is indicated (*bb*), and the position of the other is shown by the large arrow. X 9,300

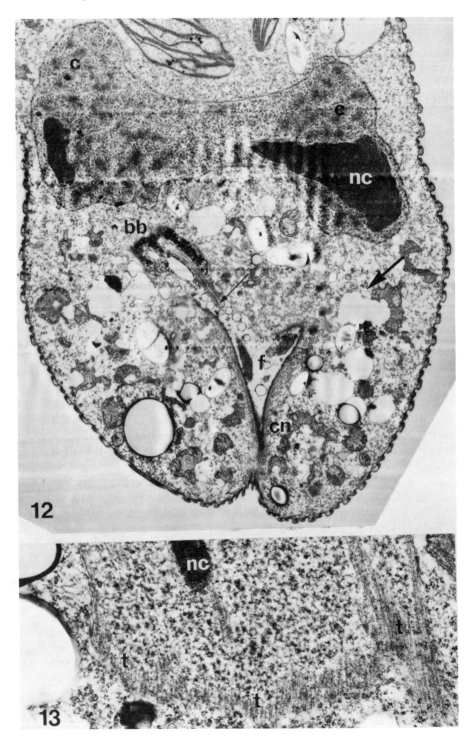

nuclei were found very close to the lateral margins of the pellicle. Furthermore, each nucleus invariably had a pronounced anterior projection directed toward (*Phacus*) or arching over (*Euglena*) its associated pair of basal bodies (partly visible in (Figure 11-18). Our observations suggest that the separation of daughter nuclei into daughter cells during cleavage was effected by the maintenance of this close association. As before, a direct contact between the nuclear projection and the basal bodies was not observed. Nevertheless, the mass of fine filaments, presumably rhizoplasts, between these organelles was still very much in evidence, and they were usually organized into more definite fibers passing close to the nucleus (Figures 11-21A and B). In *Phacus,* a single rhizoplast could be traced from each basal body to the protruding nuclear envelope (i.e., making two rhizoplasts per flagellar apparatus, as in Figure 11-22A-C). The rhizoplasts appeared not to contact the nucleus but to run tangentially past its surface. In at least one instance, the two rhizoplasts indented the daugher nucleus but did not appear to penetrate its envelope, this being clear in thick sections viewed in stereo, from the high voltage microscope.

DISCUSSION

Our results indicate that, while mitosis in *Euglena* and *Phacus* has several unusual features, the structure of the spindle and probably the mechanisms involved in chromatid separation are like that of many other eucaryotic organisms.

In *Astasia,* Sommer and Blum (1965) suggest that following its pre-mitotic anterior movement, the proximity of the nucleus to the flagellar apparatus triggers duplication of basal bodies. It is now clear that the flagellar bases have replicated and become quite widely separated, by the time that the nucleus reaches the reservoir (Figure 11-5A-C). Separation of the bases may be accomplished by the extensive arrays of microtubules that surround each pair of basal bodies (Figure 11B and C). *Euglena* at this stage looks as if it contains two tiny asters, but these microtubules soon mostly disappear as the nucleus enters prophase. Whether these extra-nuclear microtubules end up in the nucleus to form the spindle is uncertain; we have no direct information concerning this possibility.

Elongation and division of the endosome (nucleolus) have long been regarded as unusual features of euglenoid mitosis. In *Phacus,* the many small endosomes

---

Figures 11-12 and 11-13.
Late anaphase in *E. gracilis.* Figure 11-12. The chromosomes (*c*) have separated to the poles, and kinetochores are no longer apparent. The nucleolus (*nc*) is still drawn out. One set of basal bodies (*bb*) is seen beside the nucleus; the other set is situated at the large arrow, out of the plane of this section. There are the two longer (later, to form the emergent) flagella (*f*) inside the single canal (*cn*), but at least one of the two short flagella (small arrow) is also formed at this stage. The edge of the nucleus shows the typical slight distortion near the basal bodies. X 7,400. Figure 11-13. This shows the increasing number of microtubules (*t*) that characteristically line the nuclear envelope (cut tangentially here) around the elongating, flattened interzone of the anaphase/telophase spindle. X 23,000

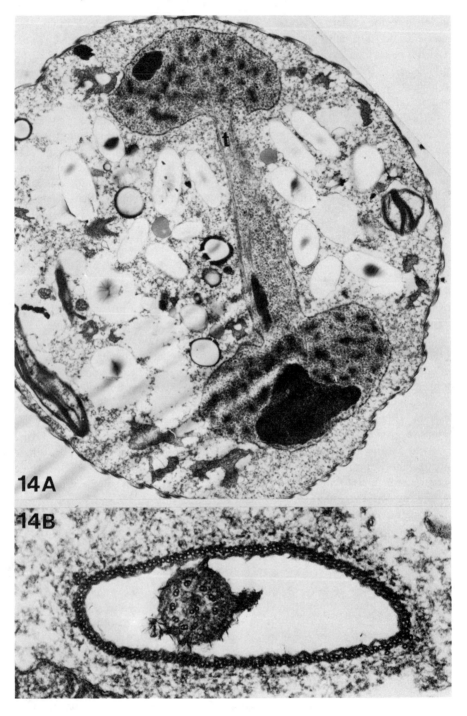

coalesce before mitosis and thereafter the single resultant body behaves like the endosome of *Euglena* (Figures 11-11A, 11-12). Persistence of the nucleolar material through mitosis, and its segregation into daughter cells is not all that unusual in plant cells (e.g., Pickett-Heaps 1970). Recently, Braselton et al. (1975) have described mitosis in *Sorosphaera* (Plasmodiophoromycete), whose metaphase spindle is described as "cruciform" because, while the chromosomes extend across the nucleus, the central nucleolus becomes rod-shaped, perpendicular to and tra-versing this metaphase plate; later it is partitioned into daughter nuclei. Thus, its behavior closely resembles what happens to the euglenoid endosome during mitosis. We feel that the elongation of the endosome in *Euglena* and *Phacus* at metaphase can best be regarded as a manifestation of the forces at work in the living spindle. We confirm Leedale's (1970) observation that numerous parallel tubules line this elongating endosome, and we interpret these tubules as constitu-ting part of the continuous system of the spindle. These tubules may in fact per-form the directed, specific task of partitioning the nucleolus in the daughter cells, but such a phenomenon would appear to be unusual.

The existence of kinetochores in *E. gracilis* (Figures 11-8-11-11) should remove some of the aura of mystery and uniqueness that surrounds mitosis in the genus (Leedale 1967, 1968). The connection of chromatin (individual chromatids being unexpectedly difficult to discern in our samples) to discrete groups of microtubules suggests that the mechanism of mitosis in *Euglena* is essentially the same as in many other eucaryotic organisms. Furthermore, we gained the strong impression that each of such bundles of tubules consisted of kinetochore tubules interspersed with other longer tubules traversing the spindle (i.e., the continuous set of microtubules). Thus each chromatid (or pair of chromatids) could be envisaged as having its own miniature sub-spindle (see particularly Figures 11-9 and 11-10). Such grouping of kinetochore and continuous tubules into discrete bundles again is not unusual; to give but one example, it is often apparent in the spindle of *Oedogonium* (Coss and Pickett-Heaps 1974) and hints at the existence of a lateral interaction between the two types of spindle tubules, an interaction that perhaps is important in achieving mitotic movement. However, in some organisms (e.g., certain fungi), there appears little if any such interaction possible from the disposition of these two sets of tubules, so such an association should not be assumed to be obligatory for achieving mitosis. We cannot definitely state that *Phacus* is equivalent to *Euglena* in this respect. *Phacus* did not reveal distinct kinetochores and so unequivocal attachment

---

Figure 11-14.
Telophase; *E. gracilis* sectioned from one end. Figure 11-14A. This shows the typi-cal telophase nucleus; the nuclear envelope is contracting around the daughter nuclei, isolating the interzonal spindle, containing many tubules (*t*). X 6,600. Figure 11-14B. Cross-section of the canal from this same cell; it has not yet divided (one of the first events in cytokinesis); the microtubules surrounding it, are grouped in doublets and they extend out the cytostome, around the cell under the pellicular ridges. X 63,000

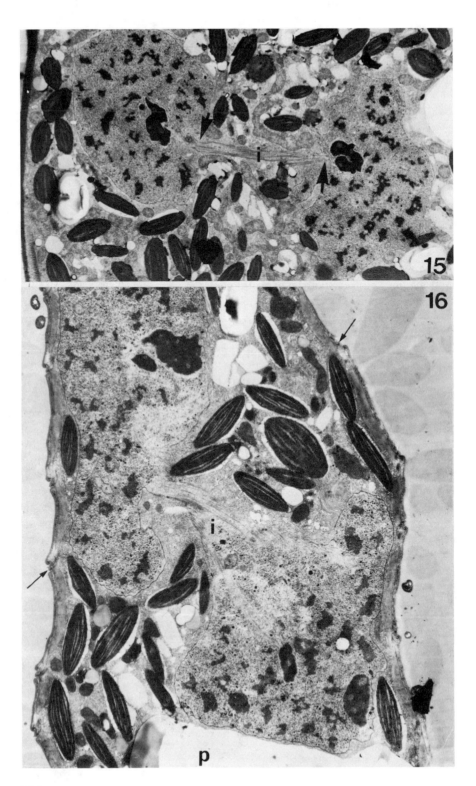

15

16

i

i

p

326

of microtubules to chromosomes is difficult to demonstrate, although our micrographs suggest that such attachments do exist. The general organization of the spindle appears similar in the two species.

Several puzzling features concerning euglenoid mitosis remain, for which we have little information. We cannot explain or further add to Leedale's observations that in *E. gracilis* at "metaphase," the chromatids are single and already separated. However, this may be related to Leedale's observation that chromosomal movement is not synchronous. Hall (1937) suggests that at this apparent "metaphase" stage, the chromosomes are partly split, with their two single ends directed at opposite poles. Several fungi and other organisms have been shown never to form a classical metaphase plate of chromosomes (e.g., Aist and Williams 1972); instead, the paired chromatids become attached to the poles by microtubules in the usual fashion while remaining scattered throughout the nucleus. Thus when anaphase commences, the initial separation of the chromatids appears complex and rather haphazard and later, chromatids arrive at each pole at different times. In *Euglena*, the position of the kinetochores, single or double, can be interpreted to follow a similar pattern of events: by mid-anaphase, they still tend to be irregularly dispersed within each separate group of chromatin.

By late anaphase, the kinetochores and their associated tubules disappear. Microtubules now concentrate in the interzone. At this stage, Leedale (1967, 1968) reports that the elongated endosome undergoes its final partitioning, rapidly flowing into each daughter nucleus. Again, we suspect that this is a manifestation of the interzonal spindle's activity. The tight contraction of the nuclear envelope separates the interzonal spindle from the daughter nuclei (see particularly Figure 11-15). Similar contraction during anaphase probably explains why the interzonal nuclear envelope becomes lined by the layer of microtubules (Figure 11-13). It is unusual that the pinched off central spindle should remain intact so long after mitosis (Figure 11-19A and B), and we have confirmed it many times. This interzonal spindle is distinctly flattened (compare Figures 11-12 and 11-14; 11-15 and 11-16); its structure is of interest since it undergoes appreciable elongation. We suspected that the tubules, often more densely stained than usual, are paired which hints at the sort of interzonal spindle structure already described for *Diatoma* (Pickett-Heaps et al. 1975). Unfortunately, this interzonal spindle is difficult to cut accurately in transverse section because of its highly variable orientation within the cell. We have been unable to track tubules in it, to show what we expect, namely

---

Figures 11-15 and 11-16.
Telophase, *P. longicaudus;* the two sections are taken roughly at right angles to each other through the spindle. Figure 11-15. The nuclear envelope (arrows) is constricting tightly to abcise the interzonal spindle (*i*); the nucleolus, multiple at interphase, aggregates into a single body which is then divided into the daughter nuclei, as here. × 3,100. Figure 11-16. In this view, the interzonal spindle is much flatter than in Figure 11-15; the cell is cut from one end. One daughter nucleus is appressed to the very large paramylon granule (*p*). The pellicle shows new strips intercalated between the old (two with arrows). × 4,600

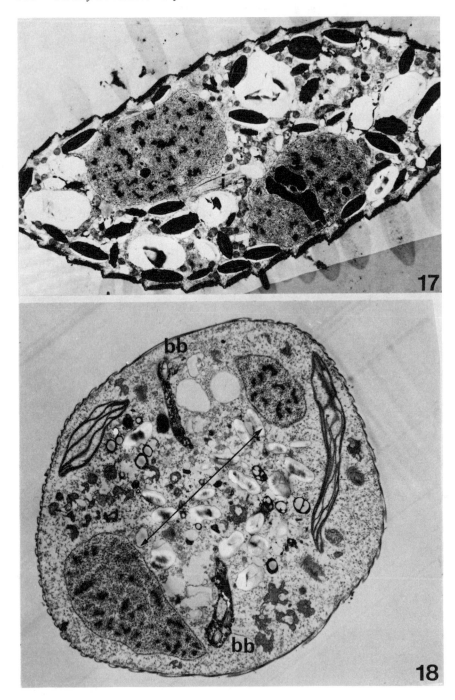

that a given tubule from one daughter nucleus tends to interact specifically with tubules from the other (as in *Diatoma* and *Melosira*). The presence of "c" tubules here in *Phacus* indicates that many of them are breaking down (Figures 11-20A–C). We examined several spindles in which this excision of the interzonal spindle was occurring at the moment of fixation (e.g., Figure 11-15); we could not correlate any feature (such as microfilaments) with the apparently contracting region of the nuclear envelope.

The nature of the connection between the basal body complex and nucleus remains mysterious. Earlier reports suggested that the basal bodies act as some form of division center, an interpretation based on the widespread and now suspect dogma that the basal bodies (equivalent to centrioles) play some necessary role in forming the spindle. Sommer and Blum (1965) suggest that in *Astasia* "these fibers (*the spindle microtubules*) briefly come into direct contact with the kinetosome" (our italics), a phenomenon we have never observed in *Phacus* or *Euglena.* In these two euglenoids (as in certain other organisms), the basal bodies remain on the outside of a completely intact nuclear envelope, and they are never seen at the broad poles of the spindle but always laterally, beside the nucleus (e.g., Figure 11-12). It is difficult to imagine how they could exert an influence on the organization and function of the tubules within the nuclear envelope. Furthermore, since we invariably found the flagellar bases beside the poles, our observations contrast with several early drawings of the euglenoid spindle showing a fiber directly connecting the pole and its adjacent polar flagellar base (e.g., Hall 1937; Johnson 1956). Nevertheless, the continuing, close association of nucleus with basal bodies is obviously of profound importance to successful cell division. The nucleus always suffers distortion near the basal bodies, and later during cytokinesis in *E. gracilis,* there is invariably a projection of the nucleus arching over the basal bodies (Figure 11-21A). In *Phacus,* this projection is in the form of a cone which is accurately directed at the flagellar apparatus (Figure 11-22A–C); we believe that this association (attachment) of each daughter nucleus to its basal body complex is what ensures partitioning of daughter nuclei into daughter cells by effectively attaching each nucleus to one flagellar apparatus so that the cleavage furrow initiated between flagellar bases on the floor of the reservoir, then necessarily has to pass between the daughter nuclei (Weik and Pickett-Heaps 1977). The increasing twist of the spindle axis *vs.* the axis of the flagellar bases during late mitosis (cf., Figures 11-15–11-18) is almost certainly consequent upon the helical nature of the organism, and particular-

---

Figures 11-17 and 11-18.
Twisting of the spindle during late telophase; Figure 11-17; *P. longicaudus,* Figure 11-18; *E. gracilis;* both sectioned from one end. Figure 11-17. The daughter nuclei have both been twisted sideways, around the interzonal spindle (*i*) which lies between them. × 3,300. Figure 11-18 The orientation of the spindle axis between the daughter nuclei, is indicated by the double-headed arrow. In contrast, the base of the reservoir (only partly visible in the section) lies between the two sets of basal bodies (*bb*). Each nucleus arches over its flagellar base. × 4,200.

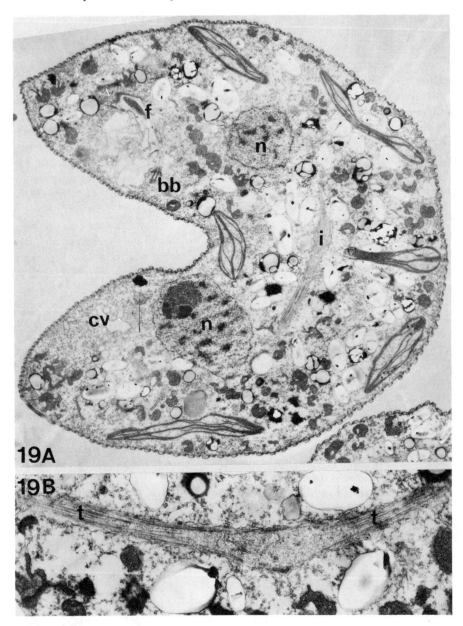

Figure 11-19.
Cleavage, *E. gracilis*. Figure 11-19A. This shows the interzonal spindle (*i*) still present, free in the cytoplasm. Each daughter nucleus (*n*) arches over, and is very closely associated with its flagellar base (the basal bodies, *bb*, and flagellum, *f*, of one and part of the contractile vacuole, *cv*, of the other, are visible in this particular section). × 4,500. Figure 11-19B. The interzonal spindle of the cell shown above (serial section). × 12,000

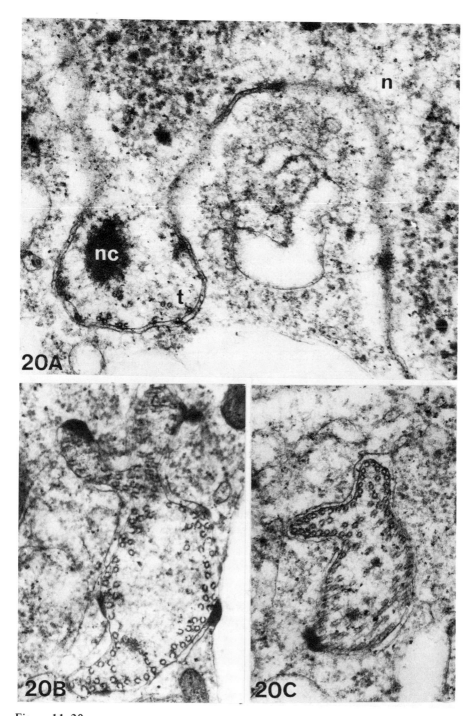

Figure 11–20.
Transverse serial sections through the interzonal spindle of telophase cell of *P. longicaudus*. Figure 11–20A. The spindle is still attached to the daughter nucleus

ly its pellicle, many of whose tubules run into the canal (Figure 11-14B) and up to the reservoir. We are puzzled by Leedale's (1967) comment that daughter nuclei can move "easily" from one daughter cell to another during cleavage. We have not observed living cells as closely as Leedale has; binucleate cells can indeed be found in our population of fixed and embedded material, presumably resulting from such a movement putting both nuclei into one-half of a cleaving cell, but our sectioned material has always shown one nucleus closely associated with one pair of flagellar bases via this protrusion of the nucleus (as in Figure 11-21A and B).

The cytoplasm between the nucleus and basal body complex is filled with fine fibers and often a more discrete fiber, the rhizoplast, is visible which in *Euglena* once demonstrated a faint banding pattern reminiscent of the rhizoplast of other algae. In *Phacus,* there were in numerous dividing cells two distinct fibers demonstrably present, one running from each basal body towards the mitotic or daughter nucleus, but as always, these fibers approached the nuclear envelope tangentially, lying very close to it (Figure 22A-C). The older literature often mentions fibers connecting the nucleus with basal bodies (e.g., Hall 1937), and some authors (e.g., Johnson 1956) claim that these fibers even enter the nucleus and end near the endosome. Electron microscopy confirms their existence in *Phacus,* but their nature is more obscure. Premitotic movement of the nucleus to the reservoir is also mysterious, and we have no reason to invoke an involvement of such fibers.

SUMMARY

Mitosis commences with replication of the basal bodies and the initiation of new pellicular strips intercalated between the old at the cytostome. The enlarged nucleus soon migrates to the floor of the reservoir, where the separating basal body complexes are situated, surrounded by numerous microtubules. Soon the nucleus itself comes to contain increasing numbers of tubules, and these aggregate into bundles aligned along the spindle axis. The endosome, ensheathed by microtubules, begins to elongate at this stage. The spindle remains entirely closed during mitosis. In *Euglena,* distinct kinetochores are detectable, with several tubules from each

---

at this end, although twisted tantential to it (as in Figure 17). Microtubules (t) are few; nearby is part of the nucleolus (nc). X 39,000. Figure 20B and 20C. Tubules are more numerous here, toward the center of the interzonal remnant; most lie close to the nuclear envelope (often in pairs ?), and many appear to be breaking up into "c" tubules. X 42,000

Figure 11-21.
Serial sections, cleaving cell of *E. gracilis.* Figure 11-21A. This shows part of one daughter nucleus (n) arching toward the basal bodies; a prominent fibrous band (rhizoplast ?) runs close to the nuclear envelope (small arrow). The large arrow indicates a set of rootlet tubules (also shown in Figure 11-21B). X 17,000. Figure 11-21B. Nearby serial section, revealing the two sets (small arrows) of fibrous bands running from the basal bodies to the nucleus (not in the plane of this section). X 17,000

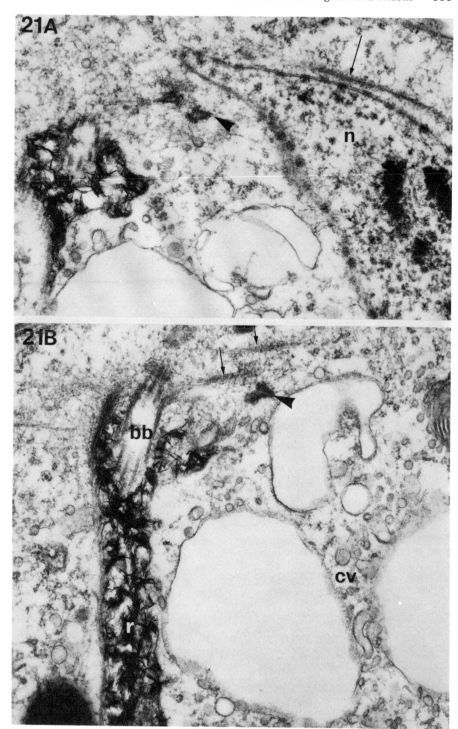

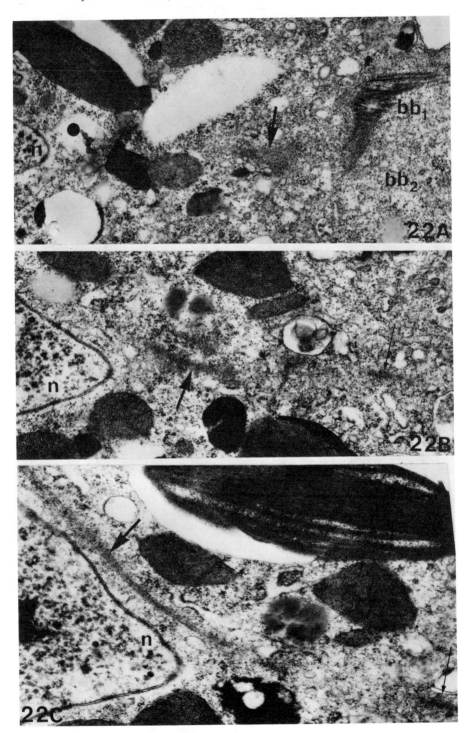

group inserted here. Metaphase appears normal in many respects. Anaphase spindle elongation is followed by disappearance of the kinetochores and their tubules; meanwhile, increasing numbers of tubules appear in the interzone, mostly lining the contracting nuclear envelope. By telophase, the latter contracts tightly around each daughter nucleus, abscising the interzonal spindle as a long sack of tubules which persists in the cleaving cell for some time. Throughout mitosis and cleavage, the nucleus (or each daughter nucleus) remains closely associated with the separated pairs of basal bodies, which are invariably laterally situated beside the spindle. The nature of the association of the nucleus and basal bodies remains obscure. Physical contact, or a direct interconnection between the two, was never observed. Instead, the basal bodies were surrounded by diffuse complexes of fine fibers, which in both species were partly aggregated into at least one, and probably two (definitely two in *Phacus*), rhizoplasts. These emanate from the basal bodies and pass very close tangentially to the nuclear envelope; the nucleus is deformed somewhat at this site, and by telophase, the daughter nuclei always have a projection, surmounted by the rhizoplast, arching over the basal bodies. This association is presumed to ensure partitioning of the daughter nuclei into daughter cells during cleavage.

ACKNOWLEDGEMENT

The authors gratefully acknowledge the grant support to Jeremy D. Pickett-Heaps, which enabled this work to be carried out, from the National Science Foundation (#GB-32034).

LITERATURE CITED

Aist, J. R. and P. H. Williams. 1972. Ultrastructure and time course of mitosis in the fungus *Fusarium oxysporum*. *J. Cell Biol.* **55**:368–389.
Braselton, J. P., C. E. Miller, and D. G. Pechak. 1975. The ultrastructure of cruciform nuclear division in *Sorosphaera veronicae* (Plasmodiophoromycete). *Amer. J. Bot.* **62**:349–358.
Buetow, D. E. 1968a. *The Biology of* Euglena. Vol. I Academic Press, New York.
Buetow, D. E. 1968b. *The Biology of* Euglena. Vol. II. Academic Press, New York.
Buetow, D. E. 1968c. Morphology and ultrastructure of *Euglena*. In *The Biology of* Euglena, Vol. I, ed. D. E. Buetow. Academic Press, New York, pp. 110–184.
Coss, R. A., and J. D. Pickett-Heaps. 1974. The effects of isopropyl-N-phenyl carbamate on the green alga *Oedogonium cardiacum*. I. Cell division. *J. Cell Biol.* **63**: 84–98.

Figure 11-22.

Three of a series of serial sections through a cleaving cell of *P. longicaudus*. Figure 11-22A. One basal body (bb_1) is obvious, near the reservoir. The other is out of the plane of this section; its position is indicated by bb_2, and part of its rhizoplast can be seen (large arrow). X 27,000 approx. Figures 11-22A–C. The rhizoplast from bb_2 (large arrows) can be traced running tangentially to the surface of the nuclear projection (*n*). The other rhizoplast, from bb_1, is indicated by the small arrows and it too approaches the nucleus in this manner. X 30,000 approx.

Hall, R. P. 1937. A note on behaviour of the chromosomes in *Euglena. Trans. Am. Microscop. Soc.* **56**:288–290.

Johnson, L. P. 1956. Observations on *Euglena fracta* sp. nov., with special reference to the locomotor apparatus. *Trans. Am. Microscop. Soc.* **75**:271–281.

Kubai, D. F. 1975. The evolution of the mitotic spindle. *Int. Rev. Cytol.* (In press.)

Kubai, D. F. and H. Ris. 1969. Division of the dinoflagellate *Cyrodinium cohnii* (Schiller). A new type of nuclear reproduction. *J. Cell Biol.* **40**:508–528.

Leedale, G. F. 1967. *Euglenoid Flagellates.* Prentice-Hall, Inc., New Jersey.

Leedale, G. F. 1968. The nucleus in *Euglena.* In *The Biology of* Euglena, Vol. I, ed. D. E. Buetow. Academic Press, New York, pp. 185–243.

Leedale, G. F. 1969. Observations on endonuclear bacteria in euglenoid flagellates. *Österr. Bot. Z.* **116**:279–294.

Leedale, G. F. 1970. Phylogenetic aspects of nuclear cytology in the algae. *Ann. N. Y. Acad. Sci.* **175**:429–453.

Miller, K. R. and L. A. Staehelin. 1973. Fine structure of the chloroplast membranes of *Euglena gracilis* as revealed by freeze-cleaving and deep-etching techniques. *Protoplasma* **77**:55–78.

Pickett-Heaps, J. D. 1970. The behaviour of the nucleolus during mitosis in plants. *Cytobios.* **6**:69–78.

Pickett-Heaps, J. D. 1974a. The evolution of mitosis and the eucaryotic condition. *BioSystems* **6**:37–48.

Pickett-Heaps, J. D. 1974b. Scanning electron microscopy of some cultured desmids. *Trans. Am. Microscop. Soc.* **93**:1–23.

Pickett-Heaps, J. D., K. L. McDonald, and D. H. Tippit. 1975. Cell division in the pennate diatom *Diatoma vulgare. Protoplasma* **86**:205–242.

Sommer, J. R. and J. J. Blum. 1964. Pellicular changes during division in *Astasia longa. Exp. Cell Res.* **35**:423–425.

Sommer, J. R. and J. J. Blum. 1965. Cell division in *Astasia longa. Exp. Cell Res.* **39**:504–527.

Soyer, M. O., and O. K. Haapala. 1974. Division and function of dinoflagellate chromosomes. *J. Microscopie* **19**:137–146.

Spurr, A. R. 1969. A low-viscosity epoxy resin embedding medium for electron microscopy. *J. Ultrastruct. Res.* **26**:31–43.

Weik, K. L. and J. D. Pickett-Heaps. 1977. Cell division in *Euglena* and *Phacus.* II. (In preparation)

Meiotic and Mitotic Divisions in the Basidiomycotina

Kenneth Wells   University of California, Davis

INTRODUCTION

The basidium and basidiospore were first accurately described and illustrated by Léveillé (1837). The studies of Rosenvinge (1886), Wager (1893, 1894), Dangeard and Sapin-Trouffly (1893), Rosen (1893), Dangeard (1895), Poirault and Racibor-ski (1895), Maire (1900a,b), Harper (1902), Holden and Harper (1902), and others established that the basidium is the structure in which karyogamy and meiosis take place. These early studies were made of fixed material embedded in paraffin and usually stained with gentian violet, safranin, haematoxylin, carmine, or Fleming's triple stain. Although these techniques were not especially conducive to an accurate description of the changes in chromosome morphology during prophase I, they established that the probasidium* initially contains two nuclei that subsequently fuse to form a single diploid nucleus. Except in those taxa forming resistant, thick walled probasidia, such as the Uredinales, the diploid nucleus soon undergoes meiosis. In most species the resulting haploid nuclei migrate into four basidiospore initials. Often each haploid nucleus undergoes a single, postmeiotic division before the basidiospore matures (Olive 1953, 1965).

---

*The basidial terminology adopted here follows Martin (1957) and Wells (1964b). The hetero-basidium is understood as a basidium that develops internal hypobasidial segments or forms well defined epibasidia, or both; whereas homobasidia are considered as basidia lacking distinct epibasidia and hypobasidial segments.

The interest in chromosome morphology in other organisms stimulated similar studies in the Basidiomycotina (Wakayama 1930, 1932; Colson 1935; Olive 1942, 1949; Sanwal 1953); however, these studies of prophase I in the Basidiomycotina continued to be made of stained paraffin sections or similar embedding material (Duncan and Galbraith 1973) until the introduction of propiono-iron carmine (Lu 1962), aceto-iron haematoxylin (Lowry 1963), and propiono-iron haematoxylin (Henderson and Lu 1968) smear techniques.

The lack of widely accepted concepts of the mechanics of mitosis and meiosis in the Basidiomycotina, and in other fungal taxa, is due to (1) the failure of fungal nuclei to stain with most conventional nuclear stains (Robinow and Bakerspigel 1965), (2) the relatively small sizes of both haploid and diploid nuclei, (3) the small sizes of hyphae, budding cells, and basidia, making it difficult to spread the chromosomes with conventional smear techniques (Shatla and Sinclair 1966), and (4) the lack of clearly defined stages of division as occur, for example, in the onion root tip or in the lily anther.

The introduction of glutaraldehyde and OsO_4 in the preparation of thin sections for electron microscopy resulted in renewed interest in meiosis and mitosis in the Basidiomycotina and overcame some of the inherent difficulties of cytological studies in the basidiomycetes. Girbardt's (1968) studies of mitosis during which electron micrographs were prepared of the same nucleus seen with phase contrast optics permitted the identification of the sequence of mitosis during clamp formation in dikaryotic hyphae. The studies of heterobasidiomycetous yeasts by McCully and Robinow (1972a,b) were based on observation of aceto-orcein stained cells by light microscopy, of living cells with phase contrast optics, and of thin sections with electron microscopy. The correlation of these observations also resulted in a better understanding of the sequence of mitosis.

Because basidial development in the species of *Coprinus* is approximately synchronized within an individual basidiocarp (Buller 1924), the studies of meiosis in the several species of this genus have been most productive. In such species, light and electron microscopic studies of separate portions of the same basidiocarp fixed at intervals permits an understanding of the sequence of meiosis and also allows the observer to examine numerous nuclei at approximately the same stage of meiosis (Lu 1967, 1969, 1970; Lerbs and Thielke 1969; Lerbs 1971; Thielke 1974). Lu (1967) and Raju and Lu (1973b) noted that 67–76% of the basidia in a basidiocarp are at the same stage of ontogeny.

There is certainly a need for many additional studies of both meiosis and mitosis in the Basidiomycotina, especially those that correlate the results of available techniques of observation, before a firm understanding of the nuclear behavior during division is possible. However, a discussion at this time of recent studies of nuclear divisions in the basidiomycetes will profitably correlate the recent studies of Lu (1964, 1967) and Lu and Raju (1970) of chromosome morphology during prophase I of meiosis, and recent works utilizing light and electron microscopy with the results obtained by earlier workers. Because the earlier studies of nuclear divisions in the Basidiomycotina have been reviewed in detail by Olive (1953, 1965) and by

Robinow and Bakerspiegel (1965), only selected studies from the earlier literature will be mentioned here.

INTERPHASE NUCLEUS

The haploid nucleus is delimited by a pair of membranes interrupted by annulated pores and, as in other eukaryotic organisms, the outer membrane is often continuous with the endoplasmic reticulum (Wells 1964a and b; 1965; Lu 1967; Motta 1969; Raudaskoski 1970; Girbardt 1970; McLaughlin 1971; Setliff et al. 1972; Setliff et al. 1974). In somatic nuclei of *Polyporus versicolor*, (≡ *Polystictus versicolor*) Girbardt (1970) estimated that nuclear pores account for 8-10% of the nuclear surface. In yeast-like cells and in basidia, the interphase nuclei are frequently subglobose to oval, but in narrow hyphae they are often irregular in form (Saksena 1961b; Flentje et al. 1963; Duncan and Macdonald 1965; Brushaber et al. 1967; Finley 1970; McCully and Robinow 1972a,b; Thielke 1973) and usually vary in diameter from 1.5-4 μm (Wakayama 1930, 1932; Colson 1935; Flentje et al. 1963; Duncan and Macdonald 1965; Shatla and Sinclair 1966; McLaughlin 1971; Setliff et al. 1974). The somatic nuclei in the apical hyphal segments of *Poria monticola* are 3-3.5 μm in diameter and 10-13 μm in length, with narrow protrusions extending up to 10 μm into the adjacent cytoplasm; however, the nuclei in the distal hyphal segments are only approximately 3 μm in length (Brushaber and Jenkins 1971). Girbardt (1970) reported that the somatic nuclei of the apical hyphal segments of *Polyporus versicolor* average 2.2 × 12 μm and those of *Flammulina velutipes* and *Fomes fomentarius* were of similar dimensions. The nuclei of *Pellicularia filamentosa* (=*Rhizoctonia solani*) and *Coprinus micaceus* were, according to Girbardt (1970), about 1.5 × 3-3.5 μm. Most interphase nuclei contain a single nucleolus (Figures 12-1 and 12-12) that is composed of granular components, fibrillar components, and an amorphous matrix (Lu 1967; Motta 1969; Girbardt 1970; Dunkle et al. 1970; Setliff et al. 1974) and is usually located eccentrically (Wakayama 1930; Dunkle et al. 1970; Thielke 1972b, 1973; Setliff et al. 1974). Within the nucleoplasm of the interphase nucleus, the chromatin is often concentrated adjacent to the spindle pole body (Girbardt 1970, 1971; Dunkle et al. 1970; McLaughlin 1971; Coffey et al. 1972; Thielke 1972b, 1973; Setliff et al. 1974). This eccentric concentration of the chromatin has been suggested (Harper 1905; Aist and Williams 1972) as representing a direct, permanent attachment of the chromosomes to the spindle pole body during interphase. During division the spindle pole body and kinetochores theoretically remain attached via the chromosomal microtubles (Harper 1905; Aist and Williams 1972).

BASIDIOMYCETE SPINDLE POLE BODY

In one of the best early studies of nuclear divisions in the basidium, Wager (1893) noted deeply staining granules at the poles of the spindle during division I and II in *Stropharia stercoraria* and *Amanita muscaria*. In a later study of the basidia of

Mycena galericulata, Wager (1894) described archoplasmic bodies, or centrospheres, associated with the prefusion and diploid nuclei. According to Wager (1894), following the displacement of the nucleus to the apex of the basidium, the archoplasmic body disappeared. Wager (1894) postulated that this body was a precursor to the centrosomes that appeared as densely staining bodies at the poles of the division I and division II spindles.

Subsequent workers have noted similar organelles at the spindle poles in light and electron microscopic studies of meiosis and mitosis in the Basidiomycotina and have applied a variety of terms to such structures: e.g., centrosome was used by Wager (1893, 1894), Wakayama (1930, 1932), Lu (1967), Motta (1969), Lu and Raju (1970), McLaughlin (1971), and Raju and Lu (1970); centriole by Lu (1964), Lu and Brodie (1964), Wilson et al. (1967), and Lerbs (1971); centriole-like body by Motta (1969) and Lerbs and Thielke (1969); centriolar plaque by Brushaber and Jenkins (1971); kinetochore equivalent (=KCE) by Girbardt 1968, 1971); microtubular organizing center (=MTOC) by Pickett-Heaps (1969) and Rogers (1973); centriole equivalent by Thielke (1974); spindle pole body (SPB) by Aist and Williams (1972), Raju and Lu (1973a), Setliff et al. (1974), and Gull and Newsam (1975a); and nucleus-associated organelle by Girbardt and Hädrich (1975).

While none of the terms that have been used is, in my opinion, completely satisfactory, *spindle pole body* (SPB) has been adopted in several recent studies (Raju and Lu 1973a, Setliff et al. 1974, Gull and Newsam 1975a) and was recommended by a group of participants at the First International Mycological Congress, Exeter, England, in 1971 (Aist and Williams 1972). Fulton (1971) reviewed the terminology of centriole-like structures and recommended that the definition of a centriole, and similar bodies, should be based on structure rather than function. Although the term spindle pole body does imply some function and, as Girbardt and Hädrich (1975) have pointed out, does emphasize only those phases of the nuclear cycle during which a spindle apparatus is present, it is adopted here. Being an approximate English translation of "Spindelpolkörperchen," a descriptive designation used by Harper (1895), it does have the added advantage of priority over most other terms. I do agree with Girbardt and Hädrich (1975) that a revision of the terminology may become necessary when the isolation and chemical characterization of the SPB have been achieved; however, until then spindle pole body seems the most appropriate term because of its wide usage, its early introduction, and because such structures have been reported at meiotic and mitotic spindle poles in all recent studies in the Basidiomycotina.

Since it is obvious that the spindle pole body found in the Ascomycotina (Wells 1970; Zickler 1970) differs structurally and in its behavior during nuclear division from the spindle pole body reported in the investigated species of the Basidiomycotina (Girbardt 1968, 1971; McLaughlin 1971; Raju and Lu 1973a; Gull and Newsam 1975a; Girbardt and Hädrich 1975), I believe that it would be helpful to use the terms *basidiomycete spindle pole body* and *ascomycete spindle pole body* in order to underline the ultrastructural and behavior differences between these two organelles.

Girbardt (1968, 1971) and Girbardt and Hädrich (1975) have described the ultrastructure of the basidiomycete SPB on the haploid nuclei of *Polyporus versicolor* and other basidiomycetes. The dumbell-shaped form of the SPB (Figures 12-1, 12-10, 12-12, 12-14) consists of two globular elements composed of granular or fibrillar material connected by an electron dense, plate- or rod-like middle piece (Girbardt 1968, 1971; McLaughlin 1971; Raju and Lu 1973a; Gull and Newsam 1975a; Girbardt and Hädrich 1975). Rather than the biglobular form reported by most authors, Motta (1969) and Setliff et al. (1974) described and illustrated a plaque-like SPB at interphase.

In the Uredinales the biglobular form of the SPB has not been clearly illustrated. In *Puccinia graminis* f. sp. *tritici* (Dunkle et al. 1970) and *Uromyces phaseoli* (Heath and Heath 1975), the SPB appears more as a disk-shaped structure; but in *Puccina helianthi* and *Melampsori lini*, Coffey et al. (1972) illustrated the SPB as a rod-shaped layer structure connecting two terminal, amorphous regions closely associated with the nuclear envelope.

The biglobular form of the SPB (Figures 12-1, 12-10, 12-12, 12-14) is about 0.6-0.9 μm in length (Lerbs and Thielke 1969; Girbardt 1968; McLaughlin 1971). In a detailed analysis of serial sections of somatic nuclei of *Polyporus versicolor* at varied stages of the nuclear cycle, Girbardt and Hädrich (1975) reported that the SPB attained a maximum length, width, and thickness of 0.6 × 0.15 × 0.05 μm, respectively, during interphase and that the globular elements reached a maximum diameter of 0.5 μm. The biglobular SPB (Figures 12-1, 12-14) is always closely associated with the nuclear envelope, is quite frequently located in a fold of the nuclear envelope (Girbardt 1968, 1971; Lerbs and Thielke 1969; Lerbs 1971; McLaughlin 1971; Rogers 1973; Thielke 1973, 1974; Setliff et al. 1974; Gull and Newsam 1975a, Girbardt and Hädrich 1975), and is Feulgen negative (Girbardt 1968, 1971). The perinuclear space of the nuclear envelope at the point of attachment of the biglobular SPB (Figure 12-14) is filled with an electron dense material (Girbardt 1968; McLaughlin; 1971; Coffey et al. 1972; Gull and Newsam 1975a). The membranes of the nuclear envelope (Figure 12-14) adjacent to the SPB are evenly spaced, more electron dense, and lack nuclear pores (Coffey et al. 1972; Gull and Newsan 1975a). Chromatin-like material is often found adjacent to the inner membrane of the nuclear envelope adjacent to the SPB (Figure 12-14) in interphase nuclei (Girbardt 1968, 1971; McLaughlin 1971; Coffey et al. 1972; Gull and Newsam 1975a).

Most workers agree that the monoglobular SPB is derived from the biglobular SPB at the time of spindle formation. The globular elements of the biglobular SPB (Figure 12-11) separate to form the two monoglobular SPBs of the spindle (Girbardt 1968; Lerbs and Thielke 1969; McLaughlin 1971; Girbardt and Hädrich 1975).

The monoglobular form of the basidiomycete SPB (Figures 12-3, 12-5, 12-12, 12-13, 12-16, 12-17, 12-19) is generally spherical, fibrillar to granular, and appoximately 0.15-0.5 μm in diameter (Girbardt 1968; 1971; Motta 1969; Lerbs and Thielke 1969; Lerbs 1971; McLaughlin 1971; Thielke 1974; Setliff et al. 1974; Gull and Newsam 1975a). Most workers report that the monoglobular SPB is largest during anaphase I and telophase I.

The reported structure of the monoglobular form is varied. Raju and Lu (1973a)

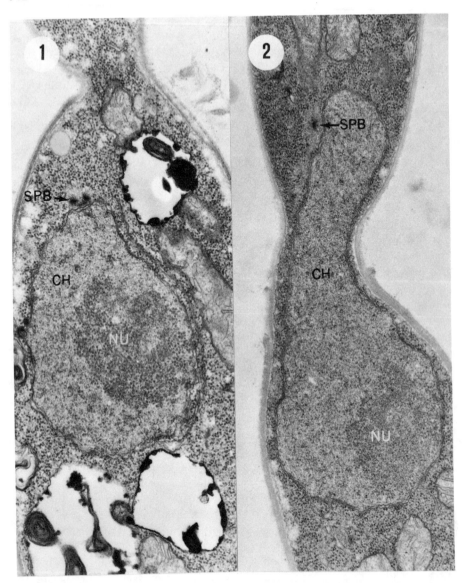

Figures 12-1 and 12-2.

Leucosporidium scottii, micrographs of mitosis (From McCully and Robinow 1972b). Figure 12-1. Premitotic stage. Longitudinal section of mother cell (below) and portion of bud (above). The nucleus is pear-shaped with the SPB at the tip. Granular and amorphous components are visible in the nucleolus. X 25,750. Figure 12-2. Premitotic stage. Longitudinal section of mother cell (below) and bud (above). Chromatin containing portion of the nucleus with the attached SPB has migrated into the bud while the nucleolus remains in the rounded base of the mother cell. X 26,000. CH, chromatin region of the nucleus; NU, nucleolus; SPB, spindle pole body.

state that it consists of a relatively dense core surrounded by a dense band that is, in turn, surrounded by a less dense outer zone. Lu (1967) and Lerbs and Thielke (1969) report that the monoglobular SPB is composed of a dense core, whereas McLaughlin (1971) describes it as uniformly dense.

Both the monoglobular (Figures 12-3, 12-5) and biglobular SPBs (Figures 12-1, 12-2) are usually surrounded by a ribosome-free zone (Girbardt, 1968, 1971; McLaughlin 1971; McCully and Robinow 1972a,b; Girbardt and Hädrich 1975). During the nuclear cycle in the apical hyphal segments of *Polyporus versicolor*, Girbardt (1968, 1971) and Girbardt and Hädrich (1975) have illustrated a distinct organization of the cytoplasm bordering the monoglobular SPB, which is formed during mitosis and disappears during the subsequent interphase. The SPB is separated from the cytoplasm by a perforated cap of endoplasmic reticulum (ER) that is continuous with the outer membrane of the nuclear envelope. In the cytoplasm adjacent to the ER cap is a layer of mutlivesicular bodies surrounded by several mitochondria in a radiating pattern. During many stages, microtubules radiate from the periglobular, ribosome-free zone surrounding the SPB into the surrounding cytoplasm. The perforated ER cap and multivesicular bodies were illustrated by McCully and Robinow (1972a,b) in several heterobasidiomycetous yeasts. The perforated ER cap was also observed by Setliff et al. (1974) during mitosis in *Poria latemarginata*. McLaughlin (1971) noted multivesicular bodies in the vicinity of the SPBs during meiosis in *Boletus rubinellus*, and Lerbs (1971) and Thielke (1974) illustrated the perforated ER cap outside the SPB of meioticially dividing nuclei of *Coprinus radiatus*. Thus, it is conceivable that a similar organization of organelles and membranes is associated with the monoglobular SPB during some phases of meiosis.

As has been pointed out (Girbardt 1971; McLaughlin 1971; Gull and Newsan 1975a), single thin sections at right or oblique angles to the long axis of the biglobular SPB give the impression of the monoglobular stage; therefore, serial sections or large numbers of random sections are necessary before the form of the SPB can be determined. Girbardt (1971) noted that numerous fixation trials are essential for each species studied in order to obtain an optimum image of the SPB. It seems probably, therefore, that some of the different images obtained by various workers may well be due to problems of fixation.

The biglobular SPB has been reported on prefusion nuclei (McLaughlin 1971; Gull and Newsam 1975a) and on diploid nuclei prior to the fusion of the nucleoli (Gull and Newsam 1975a). Setliff et al. (1974) reported that the SPB of the prefusion nucleus was spherical. Although Raju and Lu (1973a) reported that they were unable to detect the SPB from prefusion to late pachytene, Lu (personal communication) subsequently found it at all stages of prophase I. Rogers (1973), Setliff et al. (1974), and Gull and Newsam (1975a) noted the SPB on nuclei with synaptonemal complexes. Because it is difficult to distinguish the biglobular SPB in all thin section and to identify the stage of prophase I in electron micrographs, the opinions of the form of the SPB in prophase I are varied. In my opinion, the two biglobular SPBs of the prefusion nuclei fuse after karyogamy (Wager 1894), pos-

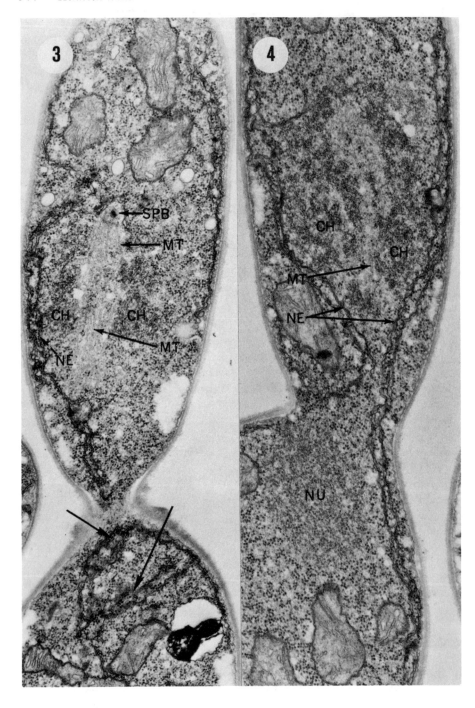

sibly during synapsis, to form a single biglobular SPB; however, the evidence is admittedly very sparse. The available evidence suggests that the two globular elements of the biglobular SPB separate during meiosis at prometaphase I and pass to the opposite poles of the nucleus (Lerbs and Thielke 1969; Raju and Lu 1973a; Rogers 1973; Thielke 1974). Most workers agree that the monoglobular SPB is present at metaphase I and II, anaphase I and II, and telophase I and II (McLaughlin 1971; Lerbs and Thielke 1969; Raju and Lu 1973a; Thielke 1974).

Although it is not firmly established, the biglobular form seems to be present at interphase I or prophase II (McLaughlin 1971; Raju and Lu 1973a). The images of biglobular SPBs on interphase I and prophase II nuclei that have been published do not demonstrate the typical, dumbbell form of the biglobular form found on interphase nuclei; therefore, these images may represent a transitional duplicating stage. In any case, presumably, the two globular elements again separate in prophase II to form the poles of the spindles of metaphase II and anaphase II. Gull and Newsam (1975a) note the biglobular SPB on postmeiotic nuclei.

Certainly, the basidiomycete spindle pole bodies seem to function as the poles of the spindle apparatus during meiotic and mitotic division. Girbardt (1961, 1968, 1971), Wilson and Aist (1967), Wilson et al. (1967), Lerbs and Thielke (1969), Thielke (1972b), Rogers (1973), and Setliff et al. (1974) have suggested that the SPB is associated with nuclear movement during divisions in hyphae, cellular thalli, and basidia. During active movement of the nuclei in dikaryotic hyphae of *Polyporus versicolor*, Girbardt (1968) found that cytoplasmic microtubules were constantly associated with the SPB. During movement, especially in dikaryotic hyphae, the nucleus usually becomes attenuate and appears as a sac-like structure with the SPB at the narrow, anterior end and the nucleolus in the rounded, posterior portion (Girbardt 1968; Wilson and Aist 1967).

MITOSIS

Recent comparative light and electron microscopic studies (Girbardt 1961, 1968, 1971; Motta 1969; McCully and Robinow 1972a,b; Thielk 1972b, 1973; Setliff et al. 1974; Girbardt and Hädrich 1975) have provided a good deal of information on somatic nuclear divisions in both mycelial and cellular basidiomycete thalli. The difficulties of interpreting images seen in the light microscope of fixed and stained nuclei and arranging the images in the proper sequence has led to diver-

---

Figures 12-3 and 12-4.
Leucosporidium scottii, micrographs of mitosis (From McCully and Robinow 1972b). Figure 12-3. Longitudinal section of mother cell (below) and daughter cell (above). Only one SPB is visible in the section. The nucleus is near metaphase. McCully and Robinow (1972b) interpreted the membranes in the mother cell (arrows) as remnants of the nuclear envelope. X 30,300. Figure 12-4. Longitudinal section of mother cell (below) and bud (above). The chromatin in the bud is dividing and has separated from the disintegrating nucleolus in the mother cell. X 31,500. CH, chromosome; MT, microtubule; NE, nuclear envelope; SPB, spindle pole body.

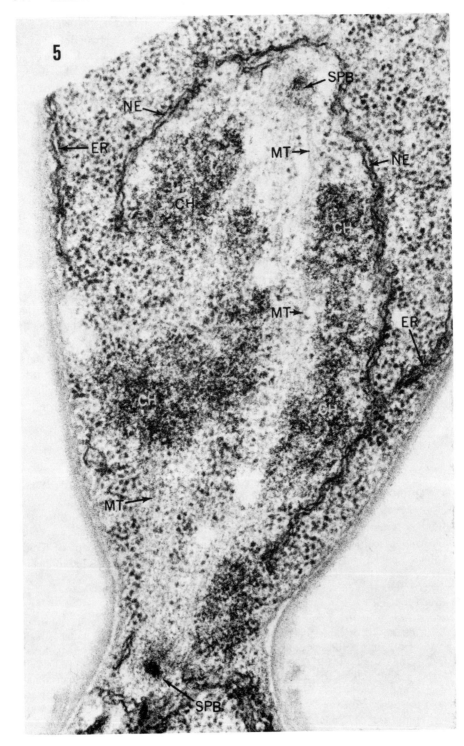

gent opinions of somatic divisions in the Basidiomycotina. For example, Saksena (1961b) studying nuclear divisions in the hyphae and basidiospores of *Pellicularia praticola* concluded that normal mitosis did not occur in these structures but that the chromatin divided by elongation, constriction, and separation into two equal portions. In contrast, Lu (1964) examined both monokaryotic and dikaryotic hyphae of *Cyathus stercoreus* and reported that the somatic nuclei divided by a process similar to classical mitosis. A clear concept of the correct sequence of the nuclear stages during divisions of the dikaryon was possible in Lu's (1964) study and in others of clamp bearing hyphae (Girbardt 1961, 1968, 1971; Thielke 1972b, 1973; Girbardt and Hädrich 1975) because it is possible to correlate the images of nuclear division with the stage of clamp and septum formation. These processes are synchronized in such thalli.

Prophase, as shown by Girbardt (1961, 1968), involves an unusual polarization and separation of the interphase nucleus into a chromatin containing part and a part containing the nucleolus. Girbardt's early (1961) study of *Polyporus versicolor* was based on observations of living dikaryotic hyphae with phase contrast optics. In his later study Girbardt (1968) employed a technique in which the same nuclei observed with phase contrast optics were subsequently fixed, thin sectioned, and examined with the elctron microsope. In these studies Girbardt (1961, 1968) observed that prior to chromosome division, the nucleus becomes bipolar with the chromatin concentrated in that portion to which the SPB is attached and the nucleolus in the other portion. The two portions then separate with the part containing the nucleolus soon disappearing. This predivision polarization of the nucleus was apparently first reported by Wager (1893, 1894) during prophase I of meiosis. It has subsequently been detected by Girbardt (1961, 1968), Thielke (1972b, 1973) and Brushaber and Jenkins (1971) during mitosis in both monokaryotic and dikaryotic hyphae in a number of Homobasidiomycetes and by McCully and Robinow (1972a,b) in several heterobasidiomycetous yeasts. It has also been shown to occur during the postmeiotic mitotic divisions in the basidia of a number of Homobasidiomycetes by Duncan (1970) and by Galbraith and Duncan (1972). Girbardt (1968) reported that the volume of the chromatin containing portion [= karyokinetic nucleus (Girbardt 1968) or dividing nucleus (Thielke 1972b, 1973)] of the prophase nucleus was approximately one-tenth of that of the interphase nucleus in *Polyporus versicolor*.

The nucleus in a dividing, yeast-like heterobasidiomycetous cell becomes oriented with the SPB in the daughter cell and the portion of the nucleus containing the nucleolus in the mother cell (Figures 12-1, 12-2) (McCully and Robinow 1972a,

---

Figure 12-5.
Leucosporidium scottii, micrograph of mitosis (From McCully and Robinow 1972b). Longitudinal section of mother cell (below) and bud (above) showing a meta-anaphase spindle. Note the presence of ribosome-like particles within the spindle region and the interrupted nuclear envelope. X 76,900. CH, chromosome; ER, endoplasmic reticulum; MT, microtubule; NE, nuclear envelope; SPB, spindle pole body.

b). In a dikaryotic, clamped hypha the clamp proliferates laterally and towards the distal portion of the hypha. One nucleus enters the clamp with the SPB oriented towards the apex of the clamp, whereas the other member of the dikaryon remains in the hypha with the SPB oriented towards the apex of the hypha (Girbardt 1961, 1968; Brushaber and Jenkins 1971; Thielke 1972b, 1973). This orientation of the dikaryon in subterminal hyphae segments is sometimes maintained during interphase (Lu 1964b) with the SPBs facing each other and the nucleoli at the opposite ends of the nuclei. In the terminal hyphal segment, the SPBs of the dikaryon are often on that end of the nuclei closest to the hyphal tip (Girbardt 1968) and one nucleus rotates prior to mitosis.

Following the reduction in volume of the nucleus and the loss of the nucleolus, the chromosomes become contracted and stained more intensely (Figures 12-3, 12-4). The globular elements of the biglobular SPB present on the interphase and early prophase nuclei separate and pass to the opposite poles of the spindle (Lu 1964; Girbardt 1968, 1971; Brushaber and Jenkin 1971; McCully and Robinow 1972a,b; Thielke 1973). Concurrent with the formation of the spindle, the SPBs lose their close association with the outer membrane of the nuclear envelope, and the nuclear envelope then breaks down in the vicinity of the SPBs (Figures 12-3–12-5) (Girbardt 1968, 1971; McCully and Robinow 1972a,b). Thielke (1973) noted that the nucleolus, following separation from the nucleus, may remain visible until telophase but often disappears earlier; whereas Galbraith and Duncan (1972) reported that the nucleolus was often visible as a detached entity at anaphase.

Reports on the continuity of the nuclear envelope are varied. During the late anaphase-telophase stage observed of somatic divisions in the hymenium of *Poria latemarginata*, Setliff et al. (1974) reported that the envelope remained intact except in the vicinity of the SPBs. Thielke (1972b, 1973), on the basis of phase contrast observations of mitotic divisions in several basidiomycetes, reported that mitosis was intranuclear with the SPBs passing into the nucleoplasm prior to the formation of the spindle. Motta (1969) reported that the nuclear envelope disappeared at metaphase and was reformed at telophase during mitosis in the hyphae of *Armillaria mellea*, but Girbardt (1968) reported that the new nuclear envelope formed prior to chromosome separation remained intact until late anaphase.

At *metaphase* a spindle is formed (Motta 1969; McCully and Robinow 1972a, b), and the chromosomes form a densely stained cluster in the media region of the spindle in appropriately stained preparations (Lu 1964, 1974b; Shatla and Sinclair 1966; Finley 1970; Valkoun and Bartos 1974). The spindle (Figure 12-5) seems to be composed of SPBs at the poles (Girbardt 1961, 1968, 1971; Lu 1964; Finley 1970; Raudaskoski 1970; McCully and Robinow 1972a,b), which are connected by continuous microtubules (Girbardt 1968; Motta 1969; McCully and Robinow 1972a,b). Microtubules arising from the SPB and radiating into the cytoplasm were reported by Girbardt (1968), Setliff et al. (1974), and Girbardt and Hädrich (1975) during division.

Anaphase, as in meiosis, involves the asynchronous disjunction and movement of the chromosomes to the poles (Lu 1964, 1974b; Motta 1969; Finley 1970). The central bundle of continuous microtubules has been seen by a number of investigators (Girbardt 1961, 1968; Motta 1969; McCully and Robinow 1972a,b; Setliff et al. 1974). Motta (1969) also reported chromosomal fibers at this stage. It seems probable that the so-called "two track," "double bar," "double bridge," or "double stranded" stages (Duncan and Macdonald 1965; Brushaber et al. 1967; Brushaber and Jenkins 1971; Day 1972) represents an optical section of anaphase. In photomicrographs the continuous microtubules form a clear zone between the migrating anaphase chromosomes when the line of vision is perpendicular to the plane of focus and the plane of focus is in the median region of the dividing nucleus. The anaphase chromosomes above and below the plane of focus in such "two track" stages can be seen by varying the focus, as noted by Setliff et al. (1974) during anaphase I of meiosis.

Anaphase movement of the chromosomes involves both a movement of the chromosomes to the polar SPBs presumably due to the shortening of the chromosome microtubules; and a separation of the SPBs, which is accompanied by an elongation of the continuous microtubules. The separation of the SPBs is evident in the electron micrographs of McCully and Robinow (1972a,b) and in the photomicrographs of Lu (1974) and was reported by Girbardt (1961, 1968) and Thielke (1972b, 1973).

During early anaphase in *Rhodosporidium* sp., McCully and Robinow (1972a) reported that the nuclear envelope was intact, but at late anaphase it ruptured in the intranuclear region and was open at the poles in the vicinity of the SPBs. In *Leucosporidium scottii* (Figure 12-5) (McCully and Robinow 1972b) the original nuclear envelope was absent on one side of the spindle during anaphase. Girbardt (1968) reported that the nuclear envelope remained intact until late anaphase. Thielke (1972b, 1973) proposed that the nuclear envelope was continuous throughout division, except possibly during prophase when the SPBs passed from the cytoplasm into the nucleoplasm.

At *telophase* the chromosomes become diffused in the nucleoplasm and a nucleolus is reformed (Lu 1964). Girbardt (1968) was of the opinion that the nuclear envelope "tears" in the intranuclear region and "immediately closes," but Motta (1969) proposed that the nuclear envelope was resynthesized at telophase. McCully and Robinow (1972b) were of the opinion that a considerable portion of the original nuclear envelope was utilized in the formation of the nuclear envelopes of the daughter nuclei.

Mitosis in *Polyporus adustus, Stropharia rugosoannulata*, and the other species examined by Thielke (1973) is completed in 3-7 minutes. In *Pellicularia praticola*, according to Saksena (1961b), mitosis is completed in 8-10 minutes. In a study of *Pellicularia filamentosa* and *P. practicola*, Flentje et al. (1963) reported that mitosis lasted for 3-5 minutes.

It is possible to obtain a rather precise impression of the time sequence of mitosis in a basidiomycete from Girbardt's (1961) meticulous analysis of ten "normal" somatic divisions in *Polyporus versicolor* by interpreting his "Teilungsphasen" in terms of conventional terminology of the mitotic phases. Thus, division phase III, comparable to metaphase (Aist and Williams, 1972), was determined by Girbardt as lasting 1.9 minutes; division phase IV, approximating anaphase, lasted 1.3 minutes; and phase V, comparable to telophase, required 2.1 minutes. Prophase would seem to encompass division phase II and possibly, most of phase I, as defined by Girbardt (1961). The duration of phases I and II in Girbardt's study was 2.5 minutes. Thus, mitosis in *P. versicolor* is completed in approximately 8 minutes (Girbardt 1961, 1968).

The available evidence (Girbardt 1968, 1971; Motta 1969; McCully and Robinow 1972a,b; Girbardt and Hädrich 1975) suggests that in the species of basidiomycetes studied, the nuclear envelope is not as persistent during mitosis as in the Ascomycotina. Thielke (1972b, 1973) attributes the reports of nuclear envelope discontinuities to the reaction of the delicate membranes to the fixative; however, it seems to me that the many reports of discontinuities cannot be disregarded. It is significant, in my opinion, that McCully and Robinow (1972a,b) noted that the phase contrast between the chromatin-containing portion of the nucleus and the cytoplasm was almost completely lost after the premitotic separation of the nucleolar containing portion of the nucleus. Other workers (Flentje et al. 1963) have noted that the nuclei become "invisible" during division. As suggested by McCully and Robinow (1972b), it is quite conceivable that this loss of contrast is a result of the breakdown of the nuclear envelope. While it is possible that the behavior of the nuclear envelope might vary in different taxa, it does not appear logical that there are fundamental differences in closely related species. It seems probably that a more uniform concept of nuclear envelope behavior will emerge with additional comparative studies utilizing light, phase contrast, and electron optics.

Although the question of the continuity of the nuclear envelope during mitosis cannot be settled at this time, electron microscopic studies have established the consistent presence of a spindle apparatus. The spindle consists of continuous microtubules and spherical spindle pole bodies. It seems likely, also, that microtubules radiate from the SPBs into the cytoplasm and that chromosomal microtubules are present; however, distinct kinetochores during mitosis have not been recorded. Although clear metaphase plates cannot be seen and anaphase movement is clearly asynchronous, there is little doubt that somatic division in the Basidiomycotina is mitotic, as opposed to amitotic or endomitotic. I agree with Lu (1964) that it is most unlikely that mitosis and amitosis take place in related species and certainly not in the same species.

MEIOTIC S-PHASE

Noting that meiosis is normally initiated at night in the strains of *Coprinus cinereus* (=*Coprinus lagopus*) and other species of *Coprinus* examined by him, Lu (1972, 1974a,c) began a series of studies on the effects of light and temperature on meiosis. He found that meiosis occurs normally in the basidiocarps of *Coprinus cinereus* at 25°C in continuous light or in a diurnal light/dark cycle but does not take place in basidiocarps at 35°C in continuous light. Since hymenial ontogeny is approximately 67-76% synchronized in *C. cinereus* (Lu 1967; Raju and Lu 1973b), it is possible to monitor the basidia cytologically. Lu (1972, 1974a,c) determined that a 2-hour sensitive period occurs during the 6th and 7th hours prior to karyogamy. The inhibiting effect of 35°C and continuous light can be reversed by exposure of the basidiocarps to a period of darkness or to 25°C within 20 hours of inhibition (Lu 1974a,c). At 25°C the minimum period of darkness to effect reversal is 2 hours (Lu 1974c), and karyogamy takes place within 6-7 hours following reversal (Lu 1972a,c). A period of inhibition of more than 20 hours exposure to 35°C and continuous light induces a reversion to mitosis in the basidia (Lu 1974a,c).

In further studies of *Coprinus cinereus* in which replication was studied by measuring the rate of incorporation of ^{32}P in DNA, Lu and Jeng (1975) determined that the premeiotic S-phase occurs prior to karyogamy. The kinetics of ^{32}P incorporation indicate that the meiotic S-phase takes place 8 hours prior to karyogamy at 25°C with diurnal lighting; however, if meiosis is blocked initially by exposing the basidiocarps to 35°C in continuous light, the S-phase takes place upon exposure to 25°C and diurnal lighting only 6 hours prior to karyogamy. Mitotic DNA replication is apparently not affected by 35°C and continuous light (Lu and Jeng 1975).

These studies (Lu 1972, 1972a,c; Lu and Jeng 1975) suggest that the process sensitive at 35°C and continuous light is premeiotic replication, which occurs 6-8 hours prior to karyogamy; and that it is this sensitivity of the S-phase to light that is probably responsible for meiosis taking place at night in *C. cinereus* and similar species of *Coprinus*.

KARYOGAMY

Nuclear fusion occurs in the median portion of most homobasidia (Wager 1893; Wakayama 1930; Saksena 1961a; Wells 1965). In heterobasidia karyogamy usually takes place in the median region of the probasidium (Holden and Harper 1902; Olive 1953; Wells 1964b; Furtado 1968); however, in the species of *Helicogloea* fusion occurs in a unique sac-like extension of the probasidium (Baker 1936).

During karyogamy the nuclei approach each other and fuse to form a dumbbell-shaped to oval nucleus with two nucleoli. The nucleoli soon fuse and the subglobose, diploid nucleus is formed (Wager 1893; Holden and Harper 1902; Wakayama 1930; Baker 1936; Wells 1965). Most diploid nuclei are 3-5 μm in diameter during prophase I (Wakayama 1930, 1932; McLaughlin 1971), but the diploid nuclei of the species of *Coleosporium* (Uredinales) attain a diameter of 8.5-13 μm (Holden and Harper 1902; Olive 1949; 1965; Sanwal 1953).

The diploid nucleus in most homobasidiomycetous species remains in the middle part of the basidium until late prophase I, or approximately diplotene or diakinesis (Saksena 1961a; Thielke 1974). The nucleus then migrates to the apex of the basidium and undergoes the two divisions of meiosis (Wells 1965; Setliff et al. 1974; Thielke 1974).

MEIOSIS

A synopsis of the published approximations of the time course of the stages of meiosis is given in table 12-1.

Table 12-1

Time course of meiosis and basidial development in *Coprinus*

Stage	Time[1]	Species studied	Reference
Dikaryotic basidium[2]	24 hr or longer	*C. atramentarius, C. comatus, C. cinereus, C. micaceus*	Lu and Raju 1970; Raju and Lu 1970
Zygotene[3]	4 hr	*C. cinereus*	Raju and Lu 1970
	6-8 hr	*C. radiatus*	Lerbs 1971
Pachytene	4-5 hr	*C. atramentarius, C. comatus, C. cinereus, C. micaceus*	Lu and Raju 1970; Raju and Lu 1970
Diplotene-diakinesis	4.5 hr	*C. cinereus*	Raju and Lu 1970
Metaphase I–Anaphase I	40 min	*C. cinereus*	Raju and Lu 1973a
Metaphase I–Interphase I	1 hr	*C. cinereus*	Raju and Lu 1970
Division II[4]	1 hr	*C. cinereus*	Raju and Lu 1970
Sterigmata formation	2 hr	*C. cinereus*	Raju and Lu 1970
Basidiospore maturation	6-8 hr	*C. cinereus*	Raju and Lu 1970

1. Time measurements are based on populations of basidia rather than on individual basidia.
2. This refers to that portion of basidial development from the differentiation of the basidium until karyogamy.
3. Zygotene here is understood as the developmental phase from karyogamy to the completion of synapsis.
4. Lerbs (1971) reports that both division I and II are completed in 30–45 min.

Leptotene

Wakayama (1930, 1932) considered that the presynaptic chromosomes of the diploid nucleus represented the leptotene stage, whereas Raju and Lu (1973b) proposed that the chromosomes in prefusion nuclei were comparable to leptotene in plants and animals. Since synapsis seems to occur immediately following karyogamy, most other workers, however, have not attempted to define a leptotene stage.

Zygotene (Figure 12-7)

In a study of several species of *Coprinus* (i.e., *C. atramentarius, C. comatus, C. micaceus,* and *C cinereus*), Lu and Raju (1970) reported that the chromosomes are elongated in the prefusion nuclei and remain unchanged in length during synapsis (Figures 12-6 - 12-8). Wakayama (1930, 1932) in his study of a number of homobasidiomycete species, Lu (1964) in a study of *Cyathus stercoreus*, Lu and Brodie (1962) working with *Cyathus olla*, Saksena (1961a) working with *Pellicularia praticola*, and Wilson et al. (1967) in their study of *Fomes annosus* also did not note presynaptic contraction and pachytene elongation of the chromosomes as has been reported in several species of the *Coleosporium* of the Uredinales (Olive 1949, 1953, 1965; Sanwal 1953) and in most species of the Ascomycotina (Harper 1905; McClintock 1945; Singleton 1953; Olive 1953, 1965). This would suggest that chromosome behavior during prophase I in the Uredinales is more similar to the Ascomycotina than to the other orders normally included in the Basidiomycotina. Furtado (1968) did, however, report that the presynaptic chromosomes in *Myxarium nucleatum* of the Tremellales were contracted and elongated following pairing, as did Huffman (1968) in his study of *Collybia maculata*.

Wakayama (1930, 1932) reported that the homologues in several species of the Homobasidiomycetes aggregated on one side of the nucleus during early synapsis resembling, somewhat, a leptotene "bouquet" arrangement noted in a number of plants and animals.

In *Coprinus cinereus* and in several other species of *Coprinus*, Lu and Raju (1970) and Raju and Lu (1970) reported that nucleolar fusion preceded synapsis. Lu (1970) noted that nucleolar fusion preceded the formation of the complete synaptonemal complexes, but the lateral components of the homologues of the nucleolar chromosomes were present during nucleolar fusion. Olive (1942, 1949, 1953) and Sanwal (1953) suggested that fusion of the nucleoli was induced by the synapsis of the nucleolar chromosomes, which, in view of the available evidence, seems reasonable.

Pachytene (Figures 12-8, 12-16)

During pachytene the chromosomes contract and are probably best counted at this stage if they can be properly spread (Wakayama 1930; Saksena 1961a; Lu 1964; Lu and Raju 1970). In *Cyathus stercoreus* Lu (1964) was able to distinguish the morphology of individual chromosomes during this stage. In several species of *Coprinus*, Lu and Raju (1970) demonstrated with photomicrographs the kinetochores of a number of bivalents.

As noted above, several studies of species of the Uredinales (Olive 1949, 1953, 1965; Sanwal 1953) have reported that the chromosomes elongated during pachytene and were more tenuous at this stage then during synapsis.

Synaptonemal complexes (Figure 12-16) were first reported in the Basidiomycotina by Lu (1966, 1967) in *Coprinus cinereus* and have subsequently been seen in several other species by other workers [Volz et al. (1968), Sundberg (1971), and Radu et al. (1974) in *Schizophyllum commune*; McLaughlin (1971) in *Boletus rubinellus*; Wells (1971) in *Myxarium subhyalinum* (= *Sebacina sublilacina*); Thielke (1972a, 1974) in *Stropharia rugosoannulata*, *Agaricus bisporus*, and *Coprinus radiatus*; Setliff et al. (1974) in *Poria latemarginata*; Gull and Newsam (1975b) in seven species of the Agaricales].

In *Coprinus cinereus*, Lu (1967, 1970) found that the lateral components appeared prior to the formation of the central component. He suggested that the

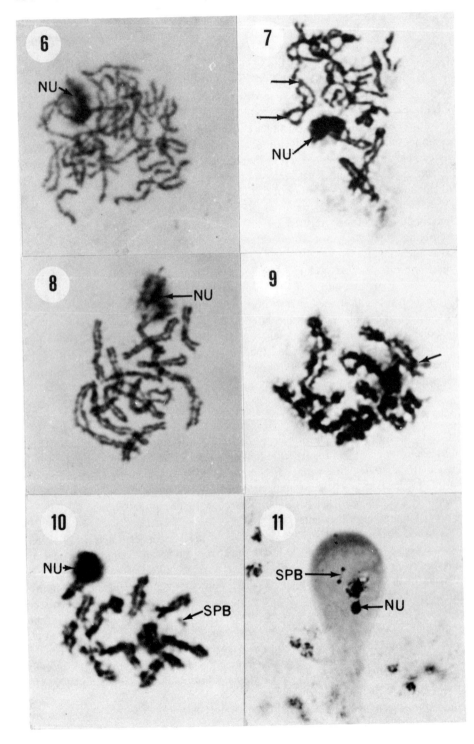

formation of the synaptonemal complex involved initially the pairing of the homologous chromosomes followed by the formation of the central component.

In the studies published to date the basidiomycete synaptonemal complex is ultrastructurally similar to those reported in most other organisms (Westergaard and Wettstein 1972; Gull and Newsam 1975b). The lateral components are ribbon-like. Although Gull and Newsam (1975b) reported that the lateral components are not banded, Sundberg (1971) and Radu et al. (1974) reported and illustrated banded lateral components in *Schizophyllum commune* similar to those found in the Ascomycotina (Westergaard and Wettstein 1972). Gull and Newsam (1975b) reported that the lateral components are joined to the central component by fibrils and also noted, in several of the species studied, electron dense thickenings (i.e., nodes) along the central component. As in other organisms (Westergaard and Wettstein 1972), all three components of the synaptonemal complexes often seem to terminate at the inner membrane of the nuclear envelope where the chromatin regions associated with the outer regions of the lateral components are often broader and more electron dense (Setliff et al. 1974; Gull and Newsam 1975b). Lu (1967), on the other hand, maintained that only the lateral components terminated on the nuclear envelope.

Gull and Newsam (1975b) also found that the central component of a complex was often associated with the nucleolus. Because of the synchronized meiosis in the lamellae of *Coprinus cinereus*, Gull and Newsam (1975b) were able to determine that this association with the nucleolus occurred at pachytene; however, in the other species studied the stage of prophase I could not be accurately determined. In some cases the central component traversed the nucleolus, but in the species of *Russula* studied the central component was adjacent to the periphery of the nucleolus (Gull and Newsam 1975b).

Lu (1969, 1970) has demonstrated in *Coprinus cinereus* a close association between crossing over and the presence of synaptonemal complexes. Lu (1969) exposed basidiocarps of *C. cinereus* to 35°C for 3 hours or to 5°C for 13 hours and determined that the rate of crossing over between two factors of the same linkage group increased 95% during the heat treatment and 220% during the cold treat-

---

Figures 12-6–12-11.
Photomicrographs of meiosis in species of *Coprinus*. Figure 12-6. *Coprinus micaceus* (From Lu and Raju 1970). Postkaryogamy nucleus prior to synapsis. X 3,150. Figure 12-7. *Coprinus micaceus* (From Lu and Raju 1970). Beginning of synapsis, zygotene. One pair seems to have begun pairing at the ends of the homologues (arrows). X 2,700. Figure 12-8. *Coprinus micaceus* (From Lu and Raju 1970). Synapsis nearly complete at early pachytene. X 3,000. Figure 12-9. *Coprinus micaceus* (From Lu and Raju 1970). Diplotene with lampbrush appearance. Note the chiasma (arrow). X 2,800. Figure 12-10. *Coprinus micaceus* (From Lu and Raju 1970). Diakinesis. Note the biglobular SPB. X 3,150. Figure 12-11. *Coprinus comatus* (From Raju and Lu 1973a). Diakinesis showing dividing biglobular SPB and nucleolus. X 1,900. All nuclei were stained with propiono-iron-haematoxylin (Henderson and Lu 1968). CH, chromosomes; NU, nucleolus; SPB, spindle pole body.

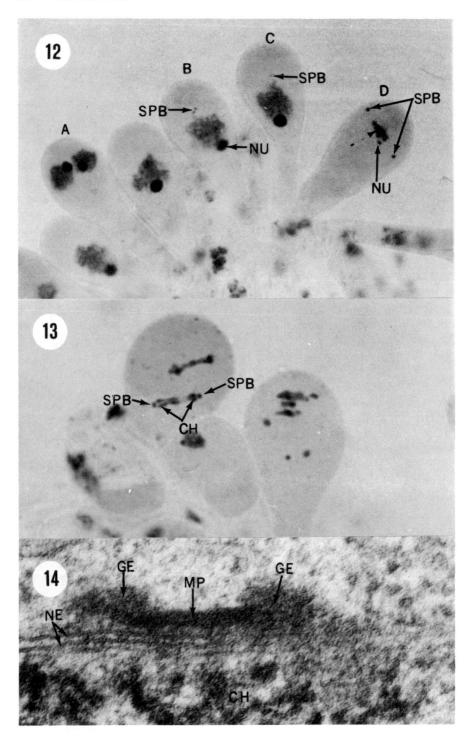

ment. By sampling the basidiocarps at progressive stages of ontogeny before and after the temperature treatments, Lu (1969, 1970) determined that the sensitive stage began with the appearance of the synaptonemal complex at zygotene and ended with its disappearance at diplotene.

In a later study on the effects of 5 krad of gamma irradiation on recombination in *Coprinus cinereus*, Raju and Lu (1973b) reported a maximum recombination of 63% when the basidiocarps were treated with the majority of basidia at prekaryogamy. They also found that cold treatments (5°C) at pachytene coupled with radiation treatments at prekaryogamy were additive, whereas combined radiation treatments at prekaryogamy and heat treatments (35°C) at pachytene did not significantly alter the recombination rate of the heat treatment alone. They proposed that radiation and high temperature act on the same process, while radiation and low temperature influence different steps of recombination.

Diplotene (Figure 12-9)

At diplotene the bivalents become shorter and the homologous chromosomes begin to pull apart (Saksena 1961a; Lu and Raju 1970; Lu 1964). In several species of *Coprinus*, Lu and Raju (1970) reported that the homologous chromosomes have a fluffy appearance and they interpreted this as a lampbrush stage of diplotene.

Diakinesis (Figures 12-10, 12-11)

During diakinesis the chromosomes contract still further (Colson 1935; Wakayama 1930, 1932; Saksena 1961a), the homologous centromeres become further separated, and the chiasmata are terminalized (Lu 1964; Lu and Brodie 1962; Lu and Raju 1970). The fluffy nature of the homologues reported in some species of *Coprinus* disappears at diakinesis (Lu and Raju 1970).

Prometaphase (Figure 12-11)

Prior to or during the formation of the spindle and the disappearance of the nucleolus in most homobasidiomycete species studied, the diploid nucleus is dis-

---

Figures 12-12-12-14.
Figure 12-12. *Coprinus comatus*, photomicrograph of diplotene, prekaryogamy, and metaphase I (From Raju and Lu 1973a). Basidium A is at prekaryogamy, B and C are at diplotene, and D is at metaphase I. A biglobular SPB is visible in basidium B and monoglobular SPBs are seen in basidium D. X 1,800. Figure 12-13. *Coprinus cinereus*, photomicrograph of anaphase II (From Raju and Lu 1973a). The monoglobular SPBs are visible in the lower nucleus of the middle basidium. Notice that the chromosomes appear in "double rows." X 2,400. CH, chromosomes; NU, nucleolus; SPB spindle pole body. Figure 12-14. *Pholiota terrestris*, micrograph of meiosis (Wells, unpublished). Basidiomycete spindle pole body on nuclear envelope of post-meiotic, haploid nucleus in the basal region of a basidium. X 132,000. CH, chromatin; GE, globular element of spindle pole body; MP, middle piece of spindle pole body; NE, nuclear envelope.

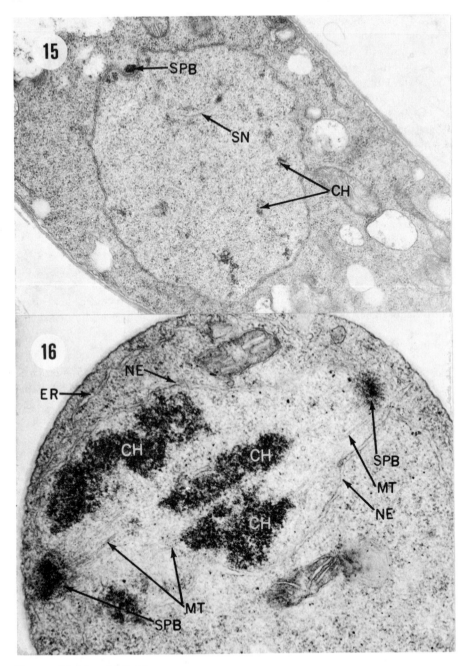

Figures 12-15 and 12-16.
Poria latemarginata, micrographs of meiosis (From Setliff et al. 1974). Figure 12-15. Diploid nucleus at pachytene. Longitudinal section of basidium showing SPB and synaptonemal complexes. X 24,000. Figure 12-16 Meta-anaphase I. Cross-

placed to the apex of the basidium (Wager 1893, 1894; Wakayama 1930, 1932; Lerbs and Thielke 1969; Lerbs 1971; Thielke 1974). The nucleus then becomes bipolar with the nucleolus oriented towards the base of the basidium (Wager 1893, 1894; Baker 1936; Lu 1967; Lerbs 1971) and the SPB adjacent to the apex of the basidium (Wager 1893, 1894; Wakayama 1930, 1932; Lu 1967; Lerbs 1971). The nucleolus, along with, presumably, a portion of the nucleoplasm surrounded by a segment of the nuclear envelope, separates from the chromatin containing portion of the nucleus prior to or at metaphase I (Wager 1893; Colson 1935; Lu 1964; Lu and Brodie 1962; Lerbs 1971; Galbraith and Duncan 1972). The nucleolus in some species disappears altogether prior to metaphase I (Wager 1893; Olive 1953; Lu and Brodie 1962) but can, apparently, linger in the adjacent cytoplasm until anaphase I in others (Wager 1893; Baker 1936; Galbraith and Duncan 1972).

When the chromosomes are at diakinesis, the globular ends of the biglobular spindle pole body separate (Figure 12-11) and pass to the opposite poles of the diploid nucleus (Lu 1967; Lerbs and Thielke 1969; McLaughlin 1971; Raju and Lu 1973a; Thielke 1974). During this time the spindle is formed, and the nuclear envelope is interrupted (Wager 1893, 1894; Wakayama 1930, 1932; Lu 1967; Lerbs and Thielke 1969; Raju and Lu 1973a). In contrast Thielke (1974) maintains that the SPBs pass through the nuclear envelope and that the spindle is intranuclear. Initially the SPBs are close to the chromatic mass but pull away to form the metaphase I spindle (Figure 12-12) (Thielke 1974).

In my opinion, the phase of division during which the nucleus becomes polarized, the nucleolus and a portion of the nucleoplasm separate from the chromatin containing portion of the nucleus, and the spindle form is appropriately termed prometaphase.

Metaphase I (Figures 12-12, 12-16)

In most homobasidia metaphase I takes place near the apex of the basidium (Wager 1893, 1894; Wakayama 1930, 1932; Lerbs and Thielke 1969; Lerbs 1971; Galbraith and Duncan 1972; Setliff et al. 1974; Thielke 1974). In cylindrical heterobasidia, such as those of the Uredinales and Auriculariales, metaphase I occurs in the median region of the cylindrical metabasidium.

The spindle is generally formed perpendicular to the longitudinal axis of the stichic basidia of the species of the Tulasnellales, Tremellales, Agaricales, and the gasteromycete taxa. In the species of the Dacrymycetales, Auriculariales, Uredinales, and Septobasidiales the spindle is usually parallel to the longitudinal axis of the chiastic basidia. In some taxa of the Aphyllophorales (Petersen 1973) and Brachybasidiales (McNabb and Talbot 1973) both stichic and chiastic basidia have been reported. In some species the spindle axis varies from transverse through oblique to

section of basidium and spindle showing both monoglobular SPBs. X 40,400. CH, chromosome; ER, endoplasmic reticulum; MT, microtubules; NE, nuclear envelope SN, synaptonemal complex; SPB, spindle pole body.

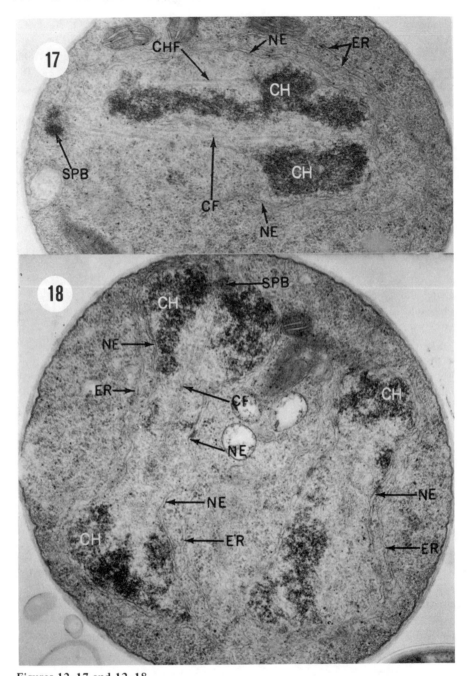

Figures 12-17 and 12-18.
Poria latemarginata, micrographs of meiosis (From Setliff et al. 1974). Figure 12-17. Anaphase I. Cross-section through basidium with only a portion of the dividing nucleus shown. X 32,900. Figure 12-18. Late anaphase II. Cross-section of

longitudinal (Wakayama 1932; Saksena 1961a). Thus, Juel's (1898) suggestions that the orientation of the spindle is of taxonomic significance has not been widely accepted (Wakayama 1932; Rogers 1934; Olive 1953).

Photomicrographs (Figure 12-12) and drawings of metaphase I illustrate this stage as a dense cluster of chromosomes between two SPBs (Wager 1893; Holden and Harper 1902; Wakayama 1930, 1932; Colson 1935; Baker 1936; Lu 1967; McClaren 1967; Lerbs 1971; Raju and Lu 1973a; Setliff et al. 1974). A similar image is seen in the electron microscope (Lu 1967; Thielke 1974).

The nuclear envelope seems to be disorganized at metaphase I (Wager 1893, 1894; Wakayama 1930, 1932; Lu 1967; Lerbs and Thielke 1969; Lerbs 1971; McLaughlin 1971; Wells 1971; Raju and Lu 1973a). Setliff et al. (1974) report, however, that the nuclear envelope remains intact except in the vicinity of the SPBs. Thielke (1974), who maintains that division is intranuclear, notes that ribosomes are present within the nuclear envelope during division. In some of the electron micrographs published by Setliff et al. (1974) there are indications of ribosomes within the spindle at meta-anaphase I and anaphase I (Figures 12-16, 12-17). The presence of ribosomes, or ribosome-like particles, within the dividing nucleus suggest, in my opinion, that the nuclear envelope does not function as an effective differentiating barrier between the nucleoplasm and cytoplasm during metaphase I, anaphase I, and early telophase I.

The spindle at metaphase I (Figure 12-16), and metaphase II, consists of two spherical, granular SPBs, continuous fibers extending between the polar SPBs, and chromosomal fibers connecting the chromosomes to the SPBs (Wager 1893, 1894; Juel 1898; Holden and Harper 1902; Wakayama 1930, 1932; Colson 1935; Lu 1967; Lerbs and Thielke 1969; Lerbs 1971; McLaughlin 1971; Wells 1971; Raju and Lu 1973a; Rogers 1973; Setliff et al. 1974; Thielke 1974). In *Poria latemarginata* the spindle is approximately 4–5.5 μm from pole to pole (Setliff et al. 1974).

Microtubules extending from the SPBs into the adjacent cytoplasm were not noted by Setliff et al. (1974); however, Juel (1898), Holden and Harper (1902), and Olive (1949, 1953, 1965) in their studies of species of the Uredinales reported conspicuous astral rays during both divisions of meiosis. Wakayama (1930, 1932) also noted astral rays in several species of the Homobasidiomycetes, and Lerbs and Thielke (1969), Lerbs (1971), and Rogers (1973) stated that astral rays were sometimes present during meiosis in the species of *Coprinus* studied by them. Wells (1971) found cytoplasmic microtubules in both divisions of meiosis in the heterobasidia of *Myxarium subhyalinum* (Tremellales).

Flared ends of the chromosomal fibers at the point of attachment to the chromosomes were described by Setliff et al. (1974) as "kinetochore-like attachments" (Figure 12-18). These attachments appear somewhat similar to those described by McCully and Robinow (1972b) in their study of mitosis in *Rhodosporidium* sp.

---

basidium including both spindles. Only one SPB is visible. X 33,200. CF, continuous fiber; CH, chromosome; CHF, chromosomal fiber; ER, endoplasmic reticulum; NE, nuclear envelope; SPB, spindle pole body.

Distinct kinetochores have not been demonstrated with the electron microscope in the Basidiomycotina; however, Lu and Raju (1970) clearly illustrated kineto-chores in their photomicrographs of the pachytene chromosomes of *Coprinus micaceus*.

Lu (1967) noted that the continuous fibers in *Coprinus cinereus* consisted of "more than two dozen microtubules." He suggested that a chromosomal fiber possibly consisted of only two microtubules. In *Boletus rubinellus* McLaughlin (1971) found that 200–250 microtubules were associated with a single SPB. These 200–250 microtubules composed both the continuous and chromosomal fibers.

Anaphase I (Figure 12–17)

During anaphase I the chromosomes separate and move asynchronously to the opposite poles of the spindle forming an elongate to spindle-shaped chromatic mass (Holden and Harper 1902; Wakayama 1930, 1932; Colson 1935; Baker 1936; Olive 1942, 1949, 1953, 1965; Saksena 1961a; Duncan and Macdonald 1965; Lu 1967; Wilson et al. 1967; Huffman 1968; Setliff et al. 1974; Thielke 1974). In electron micrographs the continuous fibers form a dense bundle in the center of the spindle with the migrating chromosomes around the periphery of the continuous fibers at varying distances from the spindle pole bodies (Setliff et al. 1974; Thielke 1974). If the anaphase I, or anaphase II (Figure 12–13), nucleus is viewed at right angles to the plane of division with a light microscope and if the plane of focus intersects the continuous fibers, the "parallel rows" of chromosomes are seen (Setliff et al. 1974). Focusing up and down demonstrates that the migrating chromosomes are oriented around the periphery of the continuous fibers (Setliff et al. 1974).

As in mitosis, anaphase I movement of the chromosomes is accompanied by a shortening of the chromosomal fibers and an elongation of the continuous fibers (Holden and Harper 1902; Wakayama 1930, 1932; Setliff et al. 1974).

Some recent studies with the electron microscope report that the nuclear enve-lope is disorganized during anaphase I (McLaughlin 1971; Raju and Lu 1973a), whereas others (Setliff et al. 1974) maintain that it remains intact except in the vicinity of the SPBs. Thielke (1974), as noted above, has suggested that the spindle is intranuclear.

Telophase I (Figure 12–19)

At telophase I the nuclear envelope re-forms around the condensed chromosomes, and the monoglobular SPB is separated from the chromatin by the nuclear envelope (Lu 1967; Lerbs 1971; Setliff et al. 1974). In a study of *Agaricus bisporus* and *Stropharia rugosoannulata*, Thielke (1972a) presented evidence that during telo-phase I and II the envelopes of the daughter nuclei were formed within the rem-nants of the envelope of the mother nucleus. In the early stages of reformation the chromatin, closely surrounded by the nuclear envelope, is reniform with the monoglobular SPB in an indentation of the nuclear envelope adjacent to the basidial wall (Lu 1967; McLaughlin 1971; Raju and Lu 1973a; Setliff et al. 1974).

The nucleus enlarges, becomes more spherical, and the chromsomes become diffused to form the interphase I nucleus (Wager 1893; Lu 1967; Lerbs 1971; Setliff et al. 1974; Thielke 1974) and the nucleolus is reformed (Wager 1893; Lerbs 1971; Galbraith and Duncan 1972; Thielke 1974).

Interphase I

Although Wakayama (1930, 1932) reported that distinct interphase I nuclei were not formed in a number of homobasidiomycetes that he studied, most authors (Wager 1893; Colson 1935; Duncan and Macdonald 1965; Lerbs 1971; Galbraith and Duncan 1972; Setliff et al. 1974) have described a short interphase between the two divisions of meiosis. Lerbs (1971) reported that the interphase I nuclei migrated to the median region of the basidium and remained there for approximately 15 minutes. The nuclei then migrate back to the apex and undergo division II. Displacement at this time has not been recorded by other authors.

Division II (Figures 12-13, 12-18, 12-19)

Except for the short prophase II and the smaller spindles, division II seems to be basically equivalent to division I (Wager 1893; Wakayama 1930, 1932; Baker 1936; Lerbs 1971; Lerbs and Thielke 1969; Setliff et al. 1974). Wager (1893) did note that the nuclear envelope was more persistent during division II. A similar separation of the nucleolus from the chromatin containing portion of the nucleus prior to metaphase II has been noted several times (Wager 1893; Lu 1964; Galbraith and Duncan 1972). The phases of division of the two nuclei are usually closely synchronized (Wager 1893; Holden and Harper 1902; Wakayama 1930, 1932; Wilson et al. 1967; Lerbs 1971; Setliff et al. 1974) The planes of division are usually parallel; however, variations are not uncommon (Wager 1893; Wakayama 1930, 1932; Saksena 1961a; Lerbs 1971; Setliff et al. 1974).

Following division II in homobasidia (Figures 12-19, 12-20), the four haploid nuclei migrate to the central or basal region of the basidium (Figure 12-20) and remain there during the formation of the sterigmata and basidiospores initials (Wager 1893; Wakayama 1930, 1932; Colson 1935; Wells 1965; Lerbs 1971; Galbraith and Duncan 1972; Setliff et al. 1974).

Setliff et al. (1972, 1974) noted that the nuclei during this phase are surrounded by perinuclear ER (Figure 12-20). It is appropriate to recall here Wager's (1893) observation that the postmeiotic nuclei were in close contact with each other and "appear as if fused together in a homogeneous mass." It is very possible that Wager (1893) had seen a close association of the haploid nuclei within the perinuclear ER in the basal portions of the basidia of *Stropharia stercoraria* and *Amanita muscaria* similar to that which was first clearly illustrated and described by Setliff et al. (1972, 1974) in their electron microscopic studies of *Poria latemarginata*. Others (Wells 1965; Thielke 1968, 1969) have published electron micrographs with suggestions of the presence of perinuclear ER in other species. The perinuclear ER surrounding the postmeiotic nuclei in the Basidiomycotina is not the same as the

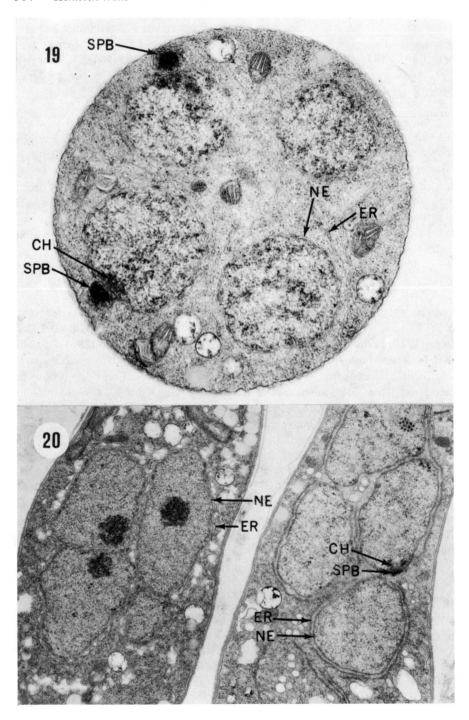

"perinuclear ER" described by Pickett-Heaps (1972) that surrounds dividing nuclei in some species of the algal order Chlorococcales.

POSTMEIOTIC MITOTIC DIVISIONS

In his study of *Hygrocybe conica*, Maire (1902) noted that the nucleolus was lost from the postmeiotic nuclei during their migration into the spore initial. Duncan (1970) made similar observations in a number of species of the Boletaceae and documented these observations with photomicrographs, as did Galbraith and Duncan (1972) in their light microscopic studies of the basidia of *Oudemansiella mucida* and *Nolanea cetrata*.

A mitotic division is regularly completed in the basidiospore or, less often, in the sterigma or in the apex of the basidium in those species of the Homobasidiomycetes that were studied by Duncan and Galbraith (1972). In the Boletaceae examined by Duncan (1970), one set of chormosomes resulting from the mitotic division in the spore moves back into the basidium and degenerates. The mature spores are uninucleate. In *Myxarium nucleatum* (Tremellales) Furtado (1968) presented evidence that a postmeiotic mitotic division occurs in the epibasidium. His photomicrographs suggest that only one set of daughter chromosomes migrates into the basidiospore initial. Duncan (1970) and Duncan and Galbraith (1972) regard the migration of the postmeiotic nucleus through the sterigma as the initiation of mitosis, which is them completed in the spore. In those species of Homobasidiomycetes examined by Duncan and Galbraith (1972) in which the postmeiotic mitosis occurs in the basidium or sterigma, one daughter set of chromosomes, either at anaphase or telophase, mitrates into the spore and one remains in the basidium and degenerates. Of those species in which the mitotic division is completed in the basidiospore, in some species one resulting set of chromosomes migrates into the basidium and degenerates, as in the Boletaceae (Duncan 1970); whereas in other species both sets of chromosomes differentiate into interphase nuclei and remain in the basidiospore, which is binucleate at maturity (Duncan and Galbraith 1972). A few species seem to exhibit different types of postmeiotic nuclear behavior (Duncan and Galbraith 1972).

These studies suggest that the postmeiotic, haploid nucleus does not migrate intact through the sterigma. Rather, the chromatin of the postmeiotic nucleus or one of the daughter nuclei is in a contracted stage of mitosis during migration through the narrow sterigmatic passage, which in *Schizophyllum commune* (Wells 1965) is less than 1 μm in width.

Figures 12-19 and 12-20.
Poria latemarginata, micrographs of meiosis (From Setliff et al. 1974). Figure 12-19. Late telophase II. Cross-section of basidium showing four haploid nuclei with perinuclear ER. Two monoglobular SPBs are visible. $\times$ 26,500. Figure 12-20. Postmeiotic nuclei. Longitudinal section of two basidia. Note the perinuclear endoplasmic reticulum. $\times$ 12,000. CH, chromatin; ER, endoplasmic reticulum; N, nucleus; NU, nucleolus; SPB, spindle pole body.

CHROMOSOME NUMBERS

The haploid number of chromosomes usually reported from the earlier studies of paraffin sections of the basidium varies from two to eight (Wager 1893; Maire 1902; Wakayama 1930, 1932; Olive 1953). More recently, however, Furtado (1968) tentatively recorded the haploid number of chromosomes in *Myxarium nucleatum* as 10 to 12, Lu and Raju (1970) reported 12 to 16 as the haploid number in five species of *Coprinus*, and Lu (1962, 1964) and Lu and Brodie (1964) clearly demonstrated that the haploid number in *Cyathus olla* and *C. stercoreus* is 12. Other more recent studies (Olive 1965) have reported higher numbers. In *Puccinia recondita*, Valkoun and Bartoš (1974) concluded that 6 chromosomes are present in the haploid nucleus. It seems likely that the use of the better smear techniques of Lu (1962), Lowry (1963), Henderson and Lu (1968), Furtado (1968), and McCully and Robinow (1972a,b) will result in higher counts than the early studies that were based on examinations of sectioned material.

DISCUSSION

The evidence provided by recent light and electron microscopic studies of nuclear divisions in the Basidiomycotina has established that the essential features of classical mitosis and meiosis occur in this taxon. These studies have essentially verified the conclusions reached by Olive (1953) in his detailed review of fungus cytology, except perhaps for his proposal that all divisions are intranuclear, and have complimented the early light microscopic studies of meiosis by Wager (1893, 1894), Wakayama (1930, 1932), Colson (1935), Baker (1936), and Lu (1964). As suggested by Lu (1964), it seems unlikely that moth mitosis and amitosis occur in the same thallus or in closely related species. It appears, also, that the mechanics of mitosis and meiosis in the Basidiomycotina are basically identical (Lu 1974b).

A sufficient amount of detail is known to demonstrate that nuclear divisions in the Ascomycotina and in the Basidiomycotina differ in several important features. The studies of Girbardt (1968, 1971), Lerbs and Thielke (1969), Lerbs (1971), McLaughlin (1971), Thielke (1974), Girbardt and Hädrich (1975), and Gull and Newsam (1975a) have established that the spindle pole body on the interphase nucleus in the species of the Basidiomycotina that have been studied, with the possible exception of the Uredinales, is a dumbbell-shaped structure, which remains closely associated with the nuclear envelope. The SPB in these species is composed of two granular to fibrillar globular elements connected by an electron dense middle piece. In contrast, the spindle pole body in the Ascomycotina seems to be a plaque-like structure closely associated with the nuclear envelope, or only partly attached to the envelope with the remaining portion extending into the adjacent cytoplasm, during both interphase and divisions (Wells 1970; Zickler 1970). During division the biglobular basidiomycete SPB forms two monoglobular SPBs that do not remain linked to the nuclear envelope.

A chemical distinction between the basidiomycete SPB and the ascomycete SPB is suggested by the studies of Girbardt and Zickler. Girbardt (1968, 1971) reported that the SPB of *Polyporus versicolor* is Feulgen negative. Zickler (1973) found that the SPBs of a number of ascomycetes are Feulgen positive, as had McDonald and Weijer (1966) in their study of *Neurospora crassa*. Zickler (1973) using enzyme digestion techniques with electron microscopy demonstrated that the SPB of several ascomycetes is sensitive to DNAase.

Most studies (Wells 1970; Zickler 1970) of meiosis in the Ascomycotina have reported an intranuclear spindle, whereas the majority of studies of both mitosis and meiosis in the Basidiomycotina have reported that the nuclear envelope is interrupted during some phases of division.

The loss of the nucleolus during prophase and the concomitant reduction in nuclear volume noted by Wager (1893, 1894), Girbardt (1961, 1968), McCully and Robinow (1972a,b), and Thielke (1972b, 1974) have not been documented in the Ascomycotina.

Even though the actual numbers are unknown, most studies of nuclear divisions in the Ascomycotina (Harper 1905; McClintock 1945; Singleton 1953; Olive 1953; Wells 1970; Zickler 1970) indicate that astral rays are a more conspicuous feature of the ascomycete spinde, although it is intranuclear, than they are of the basidio-mycete spindle (Wakayama 1930, 1932; Lerbs and Thielke 1969; Lerbs 1971; Wells 1971; Rogers 1973). While more precise observations are needed, the available evidence suggests, except for the Uredinales, that the relative number of cytoplas-mic microtubules arising from the ascomycete spindle is greater than those arising from the basidiomycete spindle.

Even though nuclear divisions in the Ascomycotina and Basidiomycotina appear somewhat similar at the light microscopic level, recent electron microscopic studies indicate that the spindle pole bodies in the two taxa differ structurally, differ in behavior during division, and, possibly, differ chemically; therefore, it seems appropriate to employ different terms to designate the spindle pole bodies of these two subdivisions. The dissimilar spindle pole bodies and the other distinguishing characters noted above support the belief that nuclear divisions in the two taxa, with the exception of the Uredinales, are characteristic of the individual taxa.

The recent studies by Heath and Heath (1975) of mitosis in *Uromyces phaseoli* of the Uredinales suggest that nuclear divisions in this species differ from those in other Basidiomycotina. The SPB, according to Heath and Heath (1975) is disk-like rather than dumbbell-shaped or spherical as in the other orders of the Basidiomyco-tina. It is worthy of note here that Holden and Harper (1902) found that the SPBs during mitosis in the rust *Coleosporium sonchiarvensis* sometimes appear "saucer-shaped with the concave side towards the spindle." In addition, a number of studies in the Uredinales (Holden and Harper 1902; Olive 1942, 1949, 1953; Sanwal 1953) have indicated that astral rays are a more conspicuous feature of the meiotic spindle than in other basidiomycetes. Further evidence that nuclear divisions in the Uredinales differ from those of other Basidiomycotina is provided by several studies of chromosome morphology during prophase I. In their studies of several species of *Coleosporium*, Olive (1949) and Sanwal (1953) found that the homologues contract prior to synapsis and elongate following pairing, as in the Ascomycotina; whereas most studies of chromosome morphology during meiosis in the Homobasidiomycetes (Wakayama 1930, 1932; Lu 1964; Lu and Brodie 1962; Lu and Raju 1970) have indicated that synapsis takes place between elongated chromosomes and contraction does not occur until pachytene. Since there are reports of presynaptic contraction and pachytene elongation in the Tremellales (Furtado 1968) and Agaricales (Huffman 1968), additional studies of chromosome morphology during prophase I are needed in all major taxa of the Basidiomycotina.

If these differences between the Uredinales and the remaining orders of the Basidiomycotina become more firmly established, then it would seem that the characteristics of nuclear divisions in the Uredinales are intermediate between those of the Ascomycotina and those of the remaining orders of the Basidiomycotina. Even though the evidence is very fragmentary, the characteristics of nuclear divisions in these taxa, along with the differences in the fine structure of the septal pores (Bracker 1967), support the hypothesis proposed by Linder (1940) and others that the Uredinales are the most primitive taxon in the Basidiomycotina.

The available studies of meiosis and mitosis in the Basidiomycotina do not answer completely the question of the behavior of the nuclear envelope during nuclear divisions in this group. The presence of ribosomes, or ribosome-like particles, within the spindle during division (Setliff et al. 1974; Thielke 1974) and the presence of cytoplasmic microtubules during division (Lerbs and Thielke 1969; Lerbs 1971; Wells 1971; Rogers 1973) suggest that the nuclear envelope is discontinuous during meoisis. Most studies of mitosis (Girbardt 1968, 1971; Motta 1969; McCully and Robinow 1972a,b; Setliff et al. 1974; Girbardt and Hädrich 1975) report discontinuities in the nuclear envelope during mitosis, especially in the polar regions adjacent to the SPBs.

There are a number of unusual features of nuclear divisions in the Basidiomycotina that appear to be adaptations to permit nuclear divisions to occur in narrow thalli, especially narrow hyphae. The predivisional separation of the nucleus into a chromatin containing portion and a nucleolar containing portion has been well documented by Girbardt (1961, 1968), McCully and Robinow (1972a,b), and Thielke (1972b, 1973) during mitosis and has been reported in several species during meiosis (Wager 1893; Thielke 1974). In *Polyporus versicolor*, Girbardt (1968) estimated that this process reduces the volume of the haploid nucleus to a tenth of the volume of the interphase nucleus. As noted by Thielke (1973) this volume reduction serves not only to separate the chromatin and nucleolus but would seem to facilitate division in the restricted space of the somatic thallus.

The narrow spindle formed during mitosis and meiosis is seemingly associated with the lack of a broad and distinct equatorial plate. In photomicrographs or drawings of metaphase during both mitosis (Lu 1966; Shatla and Sinclair 1966; Finley 1970; McCully and Robinow 1972a,b) and meiosis (Wager 1893; Wakayama 1930, 1932; Colson 1935; Baker 1936; Lu 1967; Raju and Lu 1973a; Setliff et al. 1974) the equitorial plate appears more as a dense cluster of chromosomes than as a plate-like distribution of chromosomes. In my opinion, this arrangement of the chromosomes at metaphase in the Basidiomycotina, and in other taxa of the Eumycota in which it occurs, is to be interpreted as an adaptation to the narrow confines of the somatic thallus and basidium and not necessarily as a primitive feature of nuclear divisions in these taxa.

Similarly the asynchronous disjunction of the chromatids during mitosis and meiosis is, in my opinion, a further accommodation to the spatial restrictions imposed by a slender thallus and the relatively minute, perfect reproductive structures. In part, asynchronous disjunction during anaphase would also seem to be associated with the narrow equatorial plate. Thus, instead of two distinct bands of chromosomes, the anaphase chromosomes surround the central core of continuous fibers at varying distances from the SPBs. In photomicrographs and drawings the anaphase nucleus is elongated or spindle-shaped. It is quite possible, in my opinion, that a number of the reports of chromosome bridges and lagging chromosomes are based on observations of normal, basidiomycete anaphase figures. As noted earlier, it also seems likely that the reports of "two track" stages and "parallel rows" of chromosomes are median views of anaphase chromosomes along the continuous fibers.

ACKNOWLEDGMENTS

I am indebted to Dr. B. C. Lu for his review of the manuscript and for the loan of the negatives used to reproduce Figures 12-6-12-13, to Dr. C. F. Robinow for the use of the negatives used to prepare Figures 12-1-12-5, and to Dr. H. C. Hoch for the loan of the negatives used in preparing Figures 12-15-12-20. I am grateful to the editors of *Chromosoma* and *Journal of Cell Science* for permission to reproduce Figures 12-1-12-13. Figures 12-15-12-20 are reproduced by permission of the National Research Council of Canada from the *Canadian Journal of Botany*.

Funds from National Science Foundation Research Grant GB 40071 were used during this study. The skillful assistance of Mr. George Gaboury is gratefully acknowledge.

LITERATURE CITED

Aist, J. R. and P. H. Williams. 1972. Ultrastructure and time course of mitosis in the fungus *Fusarium oxysporum. J. Cell Biol.* **55**:368–389.

Baker, G. E. 1936. A study of the genus *Helicogloea. Ann. Mo. Bot. Gard.* **23**:69–128.

Bracker, C. E. 1967. Ultrastucture of fungi. *Ann. Rev. Phytopath.* **5**:343–374.

Brushaber, J. A. and S. F. Jenkins, Jr. 1971. Mitosis and clamp formation in the fungus *Poria monticola. Am. J. Bot.* **58**:273–280.

Brushaber, J. A., C. L. Wilson, and J. R. Aist. 1967. Asexual nuclear behavior of some plant-pathogenic fungi. *Phytopathology* **57**:43–46.

Buller, A. H. R. 1924. *Researches on Fungi.* Vol. III. Longmans, Green, and Co., London. 611 pp.

Coffey, M. D., B. A. Palevitz, and P. J. Allen. 1972. The fine structure of two rust fungi, *Puccinia helianthi* and *Melampsora lini. Can. J. Bot.* **50**:231–240.

Colson, B. 1935. The cytology of the mushroom *Psalliota campestris* Quel. *Ann. Bot.* (Lond.) **49**:1–18.

Dangeard, P. A. 1895. Mémoire sur la Reproduction sexuelle des Basidiomycètes. *Botaniste* **4**:119–181.

Dangeard, P. A. and P. Sapin-Trouffly. 1893. Une pseudo-fecondation chez les Uredinees. *Compt. Rend. Hebd. Séances Acad. Sci.* **16**:267–269.

Day, A. W. 1972. Genetic implications of current models of somatic nuclear division in fungi. *Can. J. Bot.* **50**:1337–1347.

Duncan, E. G. 1970. Post-meiotic events in boleti. *Trans. Br. Mycol. Soc.* **54**: 367–370.

Duncan, E. G. and M. H. Galbraith. 1972. Post-meiotic events in the Homobasidiomycetidae. *Trans. Br. Mycol. Soc.* **58**:387–392.

Duncan, E. G. and M. H. Galbraith. 1973. Improved procedures in fungal cytology utilizing Giemsa. *Stain Technol.* **48**:107–110.

Duncan, E. G. and J. A. Macdonald. 1965. Nuclear phenomena in *Marasmius androsaceus* (L. ex Fr.) Fr. and *M. rotula* (Scop. ex Fr.) Fr. *Trans. R. Soc. Edinb.* **66**:129–141.

Dunkle, L. D., W. P. Wergin, and P. L. Allen. 1970. Nucleoli in differentiated germ tubes of wheat rust uredospores. *Can. J. Bot.* **48**:1693–1695.

Finley, D. E. 1970. Somatic mitosis in *Ceratobasidium flavescens* and *Pellicularia koleroga*. *Mycologia* **62**:474–485.

Flentje, N. T., H. M. Stretton, and E. J. Hawn. 1963. Nuclear distribution and behaviour throughout the life cycles of *Thanatephorus*, *Waitea*, and *Ceratobasidium* species. *Aust. J. Biol. Sci.* **16**:450–467.

Fulton, C. 1971. Centrioles. In *Results and Problems in Cell Differentiation*. Vol. 2. *Origin and Continuity of Cell Organelles*, eds. W. Beerman, J. Reinert, and H. Ursprung. Springer-Verlag, New York, pp. 170–221.

Furtado, J. S. 1968. Basidial cytology of *Exidia nucleata*. *Mycologia* **60**:9–15.

Galbraith, M. H. and E. G. Duncan. 1972. Nucleolar behavior during division in the Homobasidiomycetidae. *Trans. Bot. Soc. Edinb.* **41**:469–473.

Girbardt, M. 1961. Licht- und Elektronenoptische Untersuchungen an *Polystictus versicolor* (L.). VII. Lebendbeobachtung und Zeitdauer der Teilung des vegetativen Kernes. *Exp. Cell Res.* **23**:181–194.

Girbardt, M. 1968. Ultrastructure and dynamics of the moving nucleus. In *Aspects of Cell Motility*. 22nd Symp. Soc. Exp. Biol., Oxford, ed. P. L. Miller. Cambridge Univ. Press, London, pp. 249–259.

Girbardt, M. 1970. Die Ultrastruktur des Pilzkernes. I. Der Interphasekern. *Z. Allg. Mikrobiol.* **10**:451–468.

Girbardt, M. 1971. Ultrastructure of the fungal nucleus. II. The kinetochore equivalent (KCE). *J. Cell Sci.* **2**:453–473.

Girbardt, M. and H. Hädrich. 1975. Ultrastruktur des Pilzkernes. III. Genese des Kern-assoziierten Organells (NAO = "KCE"). *Z. Allg. Mikrobiol.* **15**:157–173.

Gull, K. and R. J. Newsam. 1975a. Meiosis in basidiomycetous fungi. I. Fine structure of spindle pole body organization. *Protoplasma* **83**:247–257.

Gull, K. and R. J. Newsam. 1975b. Meiosis in basidiomycetous fungi. II. Fine structure of the synaptonemal complex. *Protoplasma* **83**:259–268.

Harper, R. A. 1895. Beitrag zur Kenntniss der Kerntheilung und Sporenbildung im Ascus. *Ber. Dtsch. Bot. Ges.* **13**:67–78.

Harper, R. A. 1902. Binucleate cells in certain Hymenomycetes. *Bot. Gaz.* **33**:1–25.

Harper, R. A. 1905. *Sexual Reproduction and the Organization of the Nucleus in Certain Mildews*. Carnegie Inst. Washington Publ. 37. 104 pp.

Heath, I. B. and M. C. Heath. 1975. Ultrastructural observations on mitosis and microtubules in the rust *Uromyces phaseoli*. In *Program 26th Ann. AIBS Meeting Biol. Soc.*, Corvallis, Oregon, p. 166. (Title only.)

Henderson, S. A. and B. C. Lu. 1968. The use of haematoxylin for squash preparations of chromosomes. *Stain Technol.* 43:233–236.

Holden, R. J. and R. A. Harper. 1902. Nuclear divisions and nuclear fusion in *Coleosporium sonchi-arvensis*, Lev. *Trans. Wis. Acad. Sci., Arts, Lett.* 14:63–82.

Huffman, D. M. 1968. Meiotic behavior in the mushroom *Collybia maculata* var. *scorzonerea*. *Mycologia* 60:451–456.

Juel, H. O. 1898. Die Kerntheilungen in den Basidien und die Phylogenie der Basidiomyceten. *Jahrb. Wiss. Bot.* 32:361–388.

Lerbs, V. 1971. Licht- und elektronenmikroskopische Untersuchungen an meiotischen Basidien von *Coprinus radiatus* (Bolt). Fr. *Arch. Mikrobiol.* 77:308–330.

Lerbs, V. and C. Thielke. 1969. Die Entstehung der Spindel während der Meiose von *Coprinus radiatus*. *Arch. Mikrobiol.* 68:95–98.

Léveillé, J. H. 1837. Recherches sur l'Hymenium des Champignons. *Ann. Sci. Nat. Bot.*, Ser. 2, 8:321–338.

Linder, D. H. 1940. Evolution of the Basidiomycetes and its relation to the terminology of the basidium. *Mycologia* 32:419–447.

Lowry, R. J. 1963. Aceto-iron-hematoxylin for mushroom chromosomes. *Stain Technol.* 38:199–200.

Lu, B. C. 1962. A new fixative and improved propionocarmine squash technique for staining fungus nuclei. *Can. J. Bot.* 40:843–847.

Lu, B. C. 1964. Chromosome cycles of the basidiomycete *Cyathus stercoreus* (Schw.) de Toni. *Chromosoma* (Berl.) 15:170–184.

Lu, B. C. 1966. Fine structure of meiotic chromosomes of the basidiomycete *Coprinus lagopus*. *Exp. Cell Res.* 43:224–227.

Lu, B. C. 1967. Meiosis in *Coprinus lagopus:* a comparative study with light and electron microscopy. *J. Cell Sci.* 2:529–536.

Lu, B. C. 1969. Genetic recombination in *Coprinus* I. Its precise timing as revealed by temperature treatment experiments. *Can. J. Genet. Cytol.* 11:834–847.

Lu, B. C. 1970. Genetic recombination in *Coprinus*. II. Its relations to the synaptinemal complexes. *J. Cell Sci.* 6:669–678.

Lu, B. C. 1972. Dark dependence of meiosis at elevated temperatures in the basidiomycete *Corprinus lagopus*. *J. Bacteriol.* 111:833–834.

Lu, B. C. 1974a. Genetic recombination in *Coprinus*. IV. A kinetic study of the temperature effect on recombination frequency. *Genetics* 78:661–677.

Lu, B. C. 1974b. Meiosis in *Coprinus*. V. The role of light on basidiocarp initiation, mitosis, and hymenium differentiation in *Coprinus lagopus*. *Can. J. Bot.* 52:299–305.

Lu, B. C. 1974c. Meiosis in *Coprinus*. VI. The control of the initiation of meiosis. *Can. J. Genet. Cytol.* 16:155–164.

Lu, B. C. and H. J. Brodie. 1962. Chromosomes of the fungus *Cyathus*. *Nature* (Lond.) 194:606.

Lu, B. C. and H. J. Brodie. 1964. Preliminary observations of meiosis in the fungus *Cyathus*. *Can. J. Bot.* 42:307–310.

Lu, B. C. and D. Y. Jeng. 1975. Meiosis in *Coprinus*. VII. The prekaryogamy S-phase and the postkaryogamy DNA replication in *C. lagopus. J. Cell Sci.* **17**:461–470.

Lu, B. C. and N. B. Raju. 1970. Meiosis in *Coprinus*. II. Chromosome pairing and the lampbrush diplotene stage of meiotic prophase. *Chromosoma* (Berl.) **29**: 305–316.

Maire, R. 1900a. Sur la cytologie des Hyménomycètes. *Compt. Rend. Hebd. Séances Acad. Sci.* **131**:121–124.

Maire, R. 1900b. Sur la cytologie des Gasteromycètes. *Compt. Rend. Hebd. Séances Acad. Sci.* **131**:1246–1248.

Maire, R. 1902. Recherches cytologiques et taxonomiques sur les Basidiomycètes. *Annexe du Bull. Soc. Mycol. Fr.* **18**:1–209.

Martin, G. W. 1957. The tulasnelloid fungi and their bearing on basidial terminology. *Brittonia* **9**:25–30.

McClaren, M. 1967. Meiosis in *Coprinus atramentarius. Can. J. Bot.* **45**:215–219.

McClintock, B. 1945. *Neurospora.* I. Preliminary observations on the chromosomes of *Neurospora crassa. Am. J. Bot.* **32**:671–678.

McCully, E. K. and C. F. Robinow. 1972a. Mitosis in heterobasidiomycetous yeasts. II. *Rhodosporidium* sp. (*Rhodotorula glutinis*) and *Aessosporon salmonicolor*). *J. Cell Sci.* **11**:1–31.

McCully, E. K. and C. F. Robinow. 1972b. Mitosis in heterobasidiomycetous yeasts. I. *Leucosporidium scottii* (*Candida scottii*). *J. Cell Sci.* **10**:857–881.

McDonald, B. R. and J. Weijer. 1966. The DNA content of centrioles of *Neurospora crassa* during divisions I and IV of ascosporogenesis. *Can. J. Genet. Cytol.* **8**: 42–50.

McLaughlin, D. J. 1971. Centrosomes and microtubules during meiosis in the mushroom *Boletus rubinellus. J. Cell. Biol.* **50**:737–745.

McNabb, R. F. R. and P. H. B. Talbot. 1973. Holobasidiomycetidae: Exobasidiales, Brachybasidiales, Dacrymycetales, and Tulasnellales. In *The Fungi, an Advanced Treatise,* vol. IVB, eds. G. C. Ainsworth, F. K. Sparrow, and A. S. Sussman. Academic Press, New York, pp. 317–325.

Motta, J. J. 1969. Somatic nuclear division in *Armillaria mellea. Mycologia* **61**: 873–886.

Olive, L. S. 1942. Nuclear phenomena involved at meiosis in *Coleosporium helianthi. J. Elisha Mitchell Sci. Soc.* **58**:43–51.

Olive, L. S. 1949. Karyogamy and meiosis in the rust *Coleosporium vernoniae. Am. J. Bot.* **36**:41–54.

Olive, L. S. 1953. The structure and behavior of fungus nuclei. *Bot. Rev.* **19**:439–586.

Olive, L. S. 1965. Nuclear behavior during meiosis. In *The Fungi, an Advanced Treatise.* Vol. I, eds. G. C. Ainsworth and A. S. Sussman. Academic Press, New York, pp. 143–161.

Petersen, R. H. 1973. Aphyllophorales II: The clavarioid and cantharelloid Basidiomycetes. In *The Fungi, and Advanced Treatise,* Vol. IVB, eds. G. C. Ainsworth, F. K. Sparrow, and A. S. Sussman. Academic Press, New York, pp. 351–368.

Pickett-Heaps, J. D. 1969. The evolution of the mitotic apparatus: an attempt at comparative ultrastructural cytology in dividing plant cells. *Cytobios* **3**:257–280.

Pickett-Heaps, J. D. 1972. Variation in mitosis and cytokinesis in plant cells: its significance in the phylogeny and evolution of ultrastructural systems. *Cytobios* **5**:59–77.

Poirault, G. and M. Raciborski. 1895. Sur les noyeaux des Urédinées. *J. Bot.* (Morot) **9**:318-324, 325-332.

Radu, M., R. Steinlauf, and Y. Koltin. 1974. Meiosis in *Schizophyllum commune.* Chromosomal behavior and the synaptinemal complex. *Arch Microbiol.* **98**: 301-310.

Raju, N. B. and B. C. Lu. 1970. Meiosis in *Corpinus.* III. Timing of meiotic events in *C. lagopus* (sensu Buller). *Can. J. Bot.* **48**:2183-2186.

Raju, N. B. and B. C. Lu. 1973a. Meiosis in *Coprinus.* IV. Morphology and behaviour of spindle pole bodies. *J. Cell Sci.* **12**:131-141.

Raju, N. B. and B. C. Lu. 1973b. Genetic recombination in *Coprinus.* III. Influence of gamma-irradiation and temperature treatments on meiotic recombination. *Mutat. Res.* **17**:37-48.

Raudaskoski, M. 1970. Occurrence of microtubules and microfilaments, and origin of septa in dikaryotic hyphae of *Schizophyllum commune. Protoplasma* **70**: 415-422.

Robinow, C. F. and A. Bakerspigel. 1965. Somatic nuclei and forms of mitosis in fungi. In *The Fungi, an Advanced Treatise*, Vol. I, eds. G. C. Ainsworth and A. S. Sussman. Academic Press, New York, pp. 119-142.

Rogers, D. P. 1934. The basidium. *Stud. Nat. Hist. Iowa Univ.* **16**:160-182.

Rogers, M. A. 1973. Ultrastructure of meiosis in *Coprinus stercorarius. Am. J. Bot.* **60**(suppl.):21-22.

Rosen, F. 1893. Beiträge zur Kenntniss der Pflanzenzellen. II. Studien über die Kerne und die Membranbildung bei Myxomyceten und Pilzen. *Beitr. Biol. Pflanzen* **7**:237-266.

Rosenvinge, L. K. 1886. Sur les noyaux des Hyménomycètes. *Ann. Sci. Nat. Bot.,* Sér. 7, **3**:75-93.

Saksena, H. K. 1961a. Nuclear phenomena in the basidium of *Ceratobasidium praticolum* (Kotila) Olive. *Can. J. Bot.* **39**:717-725.

Saksena, H. K. 1961b. Nuclear structure and division in the mycelium and basidiospores of *Ceratobasidium praticolum. Can. J. Bot.* **39**:749-756.

Sanwal, B. D. 1953. The development of the basidium in *Coleosporium sidae. Bull. Torrey Bot. Club* **80**:205-216.

Setliff, E. C., H. C. Hoch, and R. F. Patton. 1974. Studies on nuclear division in basidia of *Poria latemarginata. Can. J. Bot.* **52**:2323-2333.

Setliff, E. C., W. L. MacDonald, and R. F. Patton. 1972. Fine structure of percurrent badidial proliferations in *Poria latemarginata. Can. J. Bot.* **50**:1697-1699.

Shatla, M. N. and J. B. Sinclair. 1966. *Rhizoctonia solani*: mitotic division in vegetative hyphae. *Am. J. Bot.* **53**:119-123.

Singleton, J. R. 1953. Chromosome morphology and the chromosome cycle in the ascus of *Neurospora crassa. Am. J. Bot.* **40**:124-144.

Sundberg, W. J. 1971. A study of basidial ontogeny and meiosis in *Schizophyllum commune* utilizing light and electron microscopy. Ph.D. Thesis, University of California, Davis. 155 pp.

Thielke, C. 1968. Membransysteme in meiotischen Basidien. *Ber. Dtsch. Bot. Ges.* **81**:183-186.

Thielke, C. 1969. Die Substruktur der Zellen im Fruchtkörper von *Psalliota bispora. Mush. Sci.* **7**:23-30.

Thielke, C. 1972a. New investigations on the fine structure of mushrooms. *Mush. Sci.* **8**:285-293.

Thielke, C. 1972b. Reduktion des Kernvolumens vor der Mitose bei Basidiomyceten. *Naturwissenschaften* **59**:471.

Thielke, C. 1973. Intranucleäre Mitosen in homokaryotischen und dikaryotischen Mycelien der Basidiomyceten. *Arch. Mikrobiol.* **94**:341–350.

Thielke, C. 1974. Intranucleäre Spindeln und Reduktion des Kernvolumens bei der Meiose von *Coprinus radiatus* (Bolt). Fr. *Arch. Microbiol.* **98**:225–237.

Valkoun, J. and P. Bartoš. 1974. Somatic chromosome number in *Puccinia recondita. Trans. Br. Mycol. Soc.* **63**:187–189.

Volz, P. A., C. Heintz, R. Jersild, and D. J. Niederpruem. 1968. Synaptinemal complexes in *Schizophyllum commune. J. Bacteriol.* **95**:1476–1477.

Wager, H. 1893. On nuclear divisions in the Hymenomycetes. *Ann. Bot.* (Lond.) **7**: 489–514.

Wager, H. 1894. On the presence of centrospheres in fungi. *Ann. Bot.* (Lond.) **8**: 321–334.

Wakayama, K. 1930. Contributions to the cytology of fungi. I. Chromosome number in Agaricaceae. *Cytologia* (Tokyo) **1**:369–388.

Wakayama, K. 1932. Contributions to the cytology of fungi. IV. Chromosome number in Autobasidiomycetes. *Cytologia* (Tokyo) **3**:260–284.

Wells, K. 1964a. The basidia of *Exidia nucleata.* I. Ultrastructure. *Mycologia* **56**: 327–341.

Wells, K. 1964b. The basidia of *Exidia nucleata.* II. Development. *Am. J. Bot.* **51**:360–370.

Wells, K. 1965. Ultrastructural features of developing and mature basidia and basidiospores of *Schizophyllum commune. Mycologia* **57**:236–261.

Wells, K. 1970. Light and electron microscopic studies of *Ascobolus stercorarius.* I. Nuclear divisions in the ascus. *Mycologia* **62**:761–790.

Wells, K. 1971. Basidial development in *Sebacina sublilacina* (Tremellales). In *Abstracts 1st. Intern. Mycol. Congress,* Exeter, England, p. 101.

Westergaard, M. and D. von Wettstein. 1972. The synaptinemal complex. *Annu. Rev. Genet.* **6**:71–110.

Wilson, C. L. and J. R. Aist. 1967. Motility of fungal nuclei. *Phytopathology* **57**: 769–771.

Wilson, C. L., J. C. Miller, and B. R. Griffin. 1967. Nuclear behavior in the basidium of *Fomes annosus. Am. J. Bot.* **54**:1186–1188.

Zickler, D. 1970. Division spindle and centrosomal plaques during mitosis and meiosis in some Ascomycetes. *Chromosoma* (Berl.) **30**:287–304.

Zickler, D. 1973. Evidence for the presence of DNA in the centrosomal plaques of *Ascobolus. Histochemie* **34**:227–238.

Index

About the Editors

THOMAS L. ROST received the Ph.D. degree in plant morphology from Iowa State University in 1971. Following a National Cancer Institute postdoctoral fellowship at Brookhaven National Laboratory, he became a member of the faculty of the University of California at Davis where he is currently an assistant professor. Dr. Rost is a co-author of the introductory text, *Botany: A Brief Introduction to Plant Biology*. He is a member of Sigma Xi, Gamma Sigma Delta and numerous professional societies. His research interests include, regulation of the cell cycle in plant meristems, the effects of stress on cell cycle, and seed anatomy and physiology.

ERNEST M. GIFFORD, JR., professor of botany, has been a member of the faculty of the University of California at Davis since 1949, after receiving the Ph.D. in botany from U. C. Berkeley. He is currently chairman of the botany department and Editor-in-Chief of the *American Journal of Botany*. He was a Merck Senior postdoctoral fellow at Harvard University in 1956 and 1957 and has received Guggenheim, Fulbright, and NATO Senior Science Fellowships. Dr. Gifford is co-author of a textbook, *Comparative Morphology of Vascular Plants*, and is currently an advisor to the Editor of *Encyclopaedia Britannica*. His research interests are in plant meristems, the flowering process, and spermatogenesis.